Value-Distribution Theory
(in two parts)

Part A

PURE AND APPLIED MATHEMATICS

A Series of Monographs and Textbooks

1. K. YANO. Integral Formulas in Riemannian Geometry (1970)
2. S. KOBAYASHI. Hyperbolic Manifolds and Holomorphic Mappings (1970)
3. V. S. VLADIMIROV. Equations of Mathematical Physics (A. Jeffrey, editor; A. Littlewood, translator) (1970)
4. B. N. PSHENICHNYI. Necessary Conditions for an Extremum (L. Neustadt, translation editor; K. Makowski, translator) (1971)
5. L. NARICI, E. BECKENSTEIN, and G. BACHMAN. Functional Analysis and Valuation Theory (1971)
6. D. S. PASSMAN. Infinite Group Rings (1971)
7. L. DORNHOFF. Group Representation Theory (in two parts). Part A: Ordinary Representation Theory. Part B: Modular Representation Theory (1971, 1972)
8. W. BOOTHBY and G. L. WEISS (eds.). Symmetric Spaces: Short Courses Presented at Washington University (1972)
9. Y. MATSUSHIMA. Differentiable Manifolds (E. T. Kobayashi, translator) (1972)
10. L. E. WARD, JR. Topology: An Outline for a First Course (1972)
11. A. BABAKHANIAN. Cohomological Methods in Group Theory (1972)
12. R. GILMER. Multiplicative Ideal Theory (1972)
13. J. YEH. Stochastic Processes and the Wiener Integral (1973)
14. J. BARROS-NETO. Introduction to the Theory of Distributions (1973)
15. R. LARSEN. Functional Analysis: An Introduction (1973)
16. K. YANO and S. ISHIHARA. Tangent and Cotangent Bundles: Differential Geometry (1973)
17. C. PROCESI. Rings with Polynomial Identities (1973)
18. R. HERMANN. Geometry, Physics, and Systems (1973)
19. N. R. WALLACH. Harmonic Analysis on Homogeneous Spaces (1973)
20. J. DIEUDONNÉ. Introduction to the Theory of Formal Groups (1973)
21. I. VAISMAN. Cohomology and Differential Forms (1973)
22. B.-Y. CHEN. Geometry of Submanifolds (1973)
23. M. MARCUS. Finite Dimensional Multilinear Algebra (in two parts) (1973)
24. R. LARSEN. Banach Algebras: An Introduction (1973)
25. R. O. KUJALA and A. L. VITTER (eds.). Value Distribution Theory: Part A; Part B. Deficit and Bezout Estimates by Wilhelm Stoll (1973)

In Preparation:

K. B. STOLARSKY. Algebraic Numbers and Diophantine Approximation

Value-Distribution Theory

(in two parts)

Proceedings of the Tulane University Program on Value-Distribution Theory in Complex Analysis and Related Topics in Differential Geometry

Edited by

Robert O. Kujala and Albert L. Vitter III

Department of Mathematics
Tulane University
New Orleans, Louisiana

Part A

MARCEL DEKKER, INC. New York 1974

MARCEL DEKKER, INC.
305 East 45th Street, New York, New York 10017

LIBRARY OF CONGRESS CATALOG CARD NUMBER: 73-89281

ISBN: 0-8247-6124-3

Current printing (last digit):
10 9 8 7 6 5 4 3 2 1

PRINTED IN THE UNITED STATES OF AMERICA

CONTENTS

PREFACE

In the spring of 1973 a special semester-long program on value-distribution theory in complex analysis and related topics in differential geometry was held at Tulane University. The purpose of the program was to bring together mathematicians working in this rapidly developing field to clarify its most important problems and stimulate further research.

Professor Wilhelm Stoll of Notre Dame University was Visiting Professor at Tulane during the entire semester. He gave a course concerning his research on deficit and Bezout estimates, the notes from which form Part B of these proceedings. Professors Michael Cowen of Princeton University and Ivan Cnop of Vrije Universiteit Brussel were visiting at Tulane during the whole 1972-73 academic year. Professor Cowen lectured in the fall of 1972 on the work of Bott and Chern on the value-distribution theory of hermitian vector bundles. The manuscript for this course has appeared as Tulane University Lecture Notes.

The other participants in this program visited Tulane for two-week periods to engage in seminars and informal discussions and to lecture on their own work. Some of the topics and results of this activity are reflected in the papers in these proceedings. Visiting Tulane during the semester were S. S. Chern, Mark Green, and H. H. Wu (University of California, Berkeley), Phillip Griffiths (Harvard University), Bernard Shiffman (Yale University), James King and I. M. Singer (Massachusetts Institute of Technology), Michael Atiyah (Oxford University and Institute of Advanced Study), Peter Kiernan (University of British Columbia), James Carlson (Brandeis University), John Hirschfelder (University of Washington), Reese Harvey and R. O. Wells, Jr. (Rice University).

We wish to thank all these mathematicians for participating in the program and for their contributions to this volume. We especially want to thank Professor Stoll for his fine lectures, for many interesting discussions, and for personally supervising the production of Part B of these proceedings. Special thanks also go to Michael Cowen for his intensive participation in our program throughout the year.

We thank the National Science Foundation for funding our special semester through a portion of a Science Development Grant Renewal awarded to Tulane University.

It is a pleasure to thank our colleagues in the Department of Mathematics here for their interest in the program. Professors Frank Birtel and Frank Quigley were particularly instrumental in the organization and accomplishment of the semester.

We also thank Hester Paternostro, Joanne Fink, and Deborah Casey of the staff of the Department of Mathematics at Tulane for their patience, cooperation, and, above all, their skill to which is due the physical beauty of this volume.

Finally, we express appreciation to the staff of Marcel Dekker Inc. for their patience and understanding during the preparation of this volume.

Robert O. Kujala

Albert L. Vitter III

Tulane University
New Orleans, La.

CONTRIBUTORS TO PART A

James A. Carlson, Brandeis University

Shiing-Shen Chern, University of California, Berkeley

I. Cnop, Vrije Universiteit Brussel

Michael J. Cowen, Princeton University

Mark L. Green, University of California, Berkeley

R. E. Greene, University of California, Los Angeles

Phillip A. Griffiths, Harvard University

Reese Harvey, Rice University

John J. Hirschfelder, University of Washington

Peter Kiernan, University of British Columbia

A. W. Knapp, Cornell University

James R. King, Massachusetts Institute of Technology

Robert O. Kujala, Tulane University

Bernard Shiffman, Johns Hopkins University

Albert L. Vitter, III, Tulane University

R. O. Wells, Jr., Rice University

H. Wu, University of California, Berkeley

CONTENTS OF PART B

Value-Distribution Theory

(in two parts)

Part A

SOME REMARKS ON NEVANLINNA THEORY

Phillip Griffiths
Harvard University

0. Nevanlinna theory is a beautiful subject whose results include some of the most striking and subtle in complex analysis. However, it is my feeling that most of the deeper theorems are essentially of a one-variable nature, and the most important problem in the subject is to examine those questions in several variables which are naturally posed rather than being analogues of the one-variable type results. In this paper I shall give a brief and incomplete survey of that part of Nevanlinna theory dealing with defect relations, the purpose being an attempt to substantiate my claim that the subject has a one-variable character. Then I shall turn to several variable questions, discussing some pitfalls, a few positive indications, and finally some naturally posed problems. As general references to value distribution theory, I suggest [7] for the classical case, [8] for First Main Theorems, [2] for defect relations in the equi-dimensional case, and [10] for the Ahlfors' theory of holomorphic curves.

1. In its most general setting, Nevanlinna theory deals with the global study of holomorphic mappings $f: U \to M$ between complex manifolds. Perhaps the most profitable case is when $U = \mathbb{C}^n$ and M is a projective algebraic variety. Then the growth of the mapping f has an intrinsic meaning, and in Nevanlinna theory one studies f by seeing how the image $f(\mathbb{C}^n)$ meets the subvarieties of M. The classical case is $f: \mathbb{C} \to \mathbb{P}^1 = \mathbb{C} \cup \{\infty\}$, an *entire meromorphic function*. Here Nevanlinna theory deals with the distribution of the roots of the equation

$$f(z) = a \qquad (z \in \mathbb{C},\ a \in \mathbb{P}^1). \tag{1}$$

In brief outline, the highlights are the following (c.f. [7] for details):

Upper Bound. Let $n(a,r)$ be the number of solutions of (1) in the disc $|z| \leqq r$. Suppose first that f is holomorphic and let

$$M(f,r) = \max_{|z| \leqq r} \log |f(z)|$$

measure the growth of f. Then it follows, e.g., from *Jensen's theorem* that

$$n(a,r) \leqq CM(f,2r) + O(1,a). \tag{2}$$

[*Note*: For reasons arising from Jensen's theorem, it is convenient to consider the logarithmically averaged form of $n(a,r)$, called the *counting function*

$$N(a,r) = \int_0^r n(a,\rho)\frac{d\rho}{\rho}.$$

Then (2) follows from the inequality

$$N(a,r) \leqq M(f,r) + O(1,a).] \tag{3}$$

In case f is meromorphic, we define the *order function*

$$T(r) = \int_{a \in \mathbb{P}^1} N(a,r)d\mu(a)$$

where

$$d\mu(a) = (\sqrt{-1}/2\pi)\left[1 + |a|^2\right]^{-2} da \wedge d\bar{a}$$

is the non-Euclidean area element on the Riemann sphere. If f is holomorphic, then one proves the inequalities

$$T(r) \leqq M(f,r) + O(1)$$
$$M(f,r) \leqq CT(2r) + O(1),$$

so that the two ways of measuring the growth of f are essentially equivalent. The *First Main Theorem* (F.M.T.) gives

$$N(a,r) \leqq T(r) + O(1,a), \tag{4}$$

estimating the number of solutions of (1) by the growth of f.

Lower Bound. Historically, this proceeded in three stages:

(a) The *Liouville theorem**, which says that the image $f(\mathbb{C})$ does not omit an open set in $\mathbb{P}^1$. Thus (1) has a solution for almost all a;

(b) The *Picard theorem*, which says that (1) has solutions for all but at most *two* points $a \in \mathbb{P}^1$; and

(c) The *Nevanlinna defect relation*, which loosely stated says that for any *three* distinct points a_1, a_2, a_3 on $\mathbb{P}^1$,

$$N(a_1,r) + N(a_2,r) + N(a_3,r) \geqq T(r) + \varepsilon(r) \tag{5}$$

$$\text{where} \qquad \lim_{r\to\infty} \left[\frac{\varepsilon(r)}{T(r)}\right] = 0.$$

On the whole, the situation is reminiscent of algebra: $T(r)$ plays the role of the degree of a polynomial, the upper bound (4) is similar to the (obvious) bound on the number of roots of a polynomial in terms of its degree, and the more subtle lower bound (5) is the analogue of the fundamental theorem of algebra. The only major difference is that *three* points, instead of *two* as in the polynomial case, are required in (5). As we shall see below, in several variables this similarity between algebraic and general holomorphic mappings is lost.

2. As for generalizations of the classical case to $f: \mathbb{C}^n \to M$, the situation regarding the position of $f(\mathbb{C}^n)$ vis à vis the *divisors* on M is in reasonably good shape. Now divisors on M are locally the zeroes of one holomorphic function, and it is pretty clear that the study of the zeroes of one function in several variables is about the same as in the good old days of one complex variable. From the analytic point of view, both are centered around the distributional equation (cf. [2])

$$\Delta \log |f| = \{f = 0\} \tag{6}$$

for an analytic function f and where Δ is the Laplacian and $\{f = 0\}$ is integration over the zero set of f. Thus my contention in the introduc-

* From our point of view, the Liouville theorem is obtained by integrating (4) over the image $f(\mathbb{C})$ in $\mathbb{P}^1$, and using that $T(r) \to \infty$ as $r \to \infty$ and $\int_{\mathbb{P}^1} d\mu = 1$.

tion should perhaps be amended to say that Nevanlinna theory in *codimension one* is in pretty good shape because the intuition and analytic methods from the one-variable case carry over, the new ingredient being mostly formalism.

3. Let me describe in outline what the situation regarding divisors is and mention one outstanding problem. To do this, it is convenient to use the relationship between divisors, line bundles, and Chern classes (this is the basis for the formalism mentioned above). The reference here is §0 of [5]. Given a *holomorphic line bundle* $L \to M$ and holomorphic section $\sigma \in \Gamma(M, \mathcal{O}(L))$, the zero set $\{\sigma = 0\}$ is a divisor D on M. Conversely, given D there is an associated line bundle $L = [D]$ and holomorphic section σ whose zero set is D. We denote by $|L|$ the *complete linear system* of all divisors D coming from sections σ of $L \to M$. Since σ and σ' determine the same divisor D exactly when $\sigma = \lambda\sigma' (\lambda \in \mathbb{C}^*)$, $|L|$ is the projective space of lines in $\Gamma(M, \mathcal{O}(L))$. For example, if $M = \mathbb{P}^m$ and L is the hyperplane line bundle, then $|L| = (\mathbb{P}^m)^*$ is the dual projective space of hyperplanes in $\mathbb{P}^m$.

Now given a holomorphic line bundle $L \to M$, we choose a metric in the fibres and denote by $|\tau|^2$ the square length of a local section τ. Then

$$dd^c \log \frac{1}{|\tau|^2} = dd^c \log \frac{1}{|\tau'|^2}$$

for any two non-vanishing holomorphic sections τ, τ' of L over an open set on M, and thus

$$dd^c \log \frac{1}{|\tau|^2} = \omega$$

is a global C^∞ (1,1) form on M. Obviously ω is closed, and the deRham class of ω in $H^2(M, \mathbb{R})$ is the *Chern class*, $c_1(L)$, of $L \to M$. For any $D \in |L|$, ω is the Poincaré dual of the homology cycle $\{D\} \in H_{2m-2}(M, \mathbb{Z})$ carried by D. The line bundle $L \to M$ is *positive* in case there is a fibre metric such that ω is a positive (1,1) form on M; this means that locally

$$\omega = \frac{\sqrt{-1}}{2}\left[\sum_{i,j} g_{ij}\, dz_i \wedge d\bar{z}_j\right]$$

where the Hermitian matrix g_{ij} is positive definite; we write $\omega > 0$ in this situation.

The formalism of line bundles and Chern classes is introduced for two reasons: One is to have an analytic way of measuring the *size* of a divisor: Given line bundles L and L', we say that

$$c_1(L) > c_1(L')$$

in case there are fibre metrics whose Chern classes satisfy

$$\omega > \omega';$$

and then for divisors D and D' we say that

$$D > D'$$

in case $c_1([D]) > c_1([D'])$. The second reason is that the equation of currents [5]

$$dd^c \log \frac{1}{|\sigma|^2} = D \tag{7}$$

gives an analytic method of relating the size of a divisor to the defining equation. Note that (7) is the global intrinsic form of (6).

Given $f:\mathbb{C}^n \to M$ and divisor D on M defined by the zeroes of $\sigma \in \Gamma(M, \mathcal{O}(L))$, we set (pardon the flood of notations)

$$\phi = \frac{\sqrt{-1}}{2\pi}\left[\sum_{i=1}^{m} dz_i \wedge d\bar{z}_i\right] \qquad \textit{(Kähler form on } \mathbb{C}^n\textit{)}$$

$$\phi^q = \phi\wedge\ldots\wedge\phi \text{ (q-times)}, \quad \Phi = \phi^n$$

$$B_r = \{z \in \mathbb{C}^n : |z| \leqq r\}$$

$$D_f = f^{-1}(D) \text{ and } \omega_f = f^*\omega$$

$$n(D,\rho) = \int_{D_f \cap B_\rho} \phi^{n-1}$$

$$N(D,r) = \int_0^r n(D,\rho)\rho^{1-2n}\, d\rho \qquad \textit{(counting function)}$$

$$t(L,\rho) = \int_{B_\rho} \omega_f \wedge \phi^{n-1}$$

$$T(L,r) = \int_0^r t(L,\rho)\rho^{1-2n}\, d\rho \qquad \textit{(order function)}$$

The F.M.T., which is proved using (7) exactly as in the one-variable case, gives

$$N(D,r) \leqq T(L,r) + O(1,D), \tag{8}$$

which is an upper bound on the size of $f^{-1}(D)$ in terms of the average or expected value $T(L,r)$. From (8) we easily obtain* the *generalized Liouville theorem*: If $\omega > 0$ and f is non-constant, then the image $f(\mathbb{C}^n)$ meets almost all $D \in |L|$.

Consequently, for divisors we always have a good upper bound and at least a crude lower bound. The more refined Picard theorems and defect relations have been proved in essentially two cases (thus far, the latter holds whenever the former does):

(a) The *equidimensional case* of $f: \mathbb{C}^m \to M_m$ where the Jacobian $J(f) \not\equiv 0$; then the image $f(\mathbb{C}^m)$ meets any D such that [2]

$$c_1([D]) + c_1(K_M) > 0 \quad (K_M = \textit{canonical line bundle of } M); \text{ and}$$
$$D \text{ has simple normal crossings} \tag{9}$$

We remark that in case $M = \mathbb{P}^m$, the canonical line bundle

---

* The proof follows from (8) by the same argument as outlined in the first footnote.

$K_M = H^{-(m+1)}$ where $H \to \mathbb{P}^m$ is the hyperplane line bundle; for $m = 1$, this is the number *two* in the Picard theorem. As we shall see below, the canonical divisor also appears in higher codimensional questions.

(b) The case of a *non-degenerate holomorphic curve* $f: \mathbb{C} \to \mathbb{P}^m$. Here the *Borel theorem* states that the image $f(\mathbb{C})$ can miss at most $m + 1$ hyperplanes in general position. The corresponding defect relation is due to Ahlfors [1], whose proof has great geometric subtlety, while at the same time being based on the equation (6) above. The Ahlfors theorem has been generalized by Stoll to the position of the image of $f: \mathbb{C}^n \to \mathbb{P}^m$ relative to the linear hyperplanes in projective space. Stoll's results are given in his paper in Acta Math., Vol. 90 (1953), 1-115 and Vol. 92 (1954), 55-169. His methods are applied in his paper "Deficit and Bezout Estimates" which appears in Part B of these Proceedings.

4. Turning now to higher codimension problems, we want to study a holomorphic mapping $f:\mathbb{C}^n \to M$ where M is a smooth projective variety. Let ω be a Kähler metric on M, ϕ the usual Kähler form on $\mathbb{C}^n$, and set (c.f. §5 of [5])

$$t_q(r) = \frac{1}{r^{2n-2q}} \int (f^*\omega)^q \wedge \phi^{n-q}$$
$$T_q(r) = \int_0^r t_q(\rho)\frac{d\rho}{\rho} \qquad (10)$$

If $\lambda_1 \leq \cdots \leq \lambda_n$ are the eigenvalues of $f^*\omega$ relative to ϕ, then clearly

$$t_q(r) = \frac{1}{r^{2n-2q}} \int_{B_r} \sigma_q(\lambda_1,\ldots,\lambda_n)\cdot\Phi \qquad (11)$$

where σ_q is the qth elementary symmetric function of $\lambda_1,\ldots,\lambda_n$. The *order functions* $T_q(r)$ are the basic quantities regulating the growth of f. Given positive increasing functions $A(r)$, $B(r)$ we write

$$A \sim B$$

to mean that $A(r) \leqq CB(r) \leqq C'A(r)$ for positive constants C, C'. Then clearly

> $T_1(r) \sim T(L,r)$ where $L \to M$ is a positive line bundle; and the $T_q(r)$ are intrinsically defined in the sense of the equivalence relation $\sim$.

(Actually, the $T_q(r)$ are intrinsic in a somewhat more refined sense, but we won't be concerned with that here.) To see better what the $T_q(r)$ are, consider the case of

$$f:\mathbb{C}^2 \to \mathbb{P}^2$$

and let ω be the standard Kähler metric on $\mathbb{P}^2$. Then

$$\begin{aligned} T_1(r) &= \int_{D\in\mathbb{P}^{2*}} N(D,r)d\mu(D) \\ T_2(r) &= \int_{A\in\mathbb{P}^2} N(A,r)d\mu(A), \end{aligned} \tag{12}$$

where in the first equation D runs over the lines in $\mathbb{P}^2$, and in the second A varies over the points in $\mathbb{P}^2$.

If we now look to see what happens to the lower and upper bounds in the general case, then there is considerable trouble. To begin with, Cornalba and Shiffman [4] gave an f such that, for a suitable $A \in \mathbb{P}^2$,

$$\lim_{r\to\infty} \left[\frac{T_2(r)}{N(A,r)}\right] = 0 .$$

Thus, the lower bound (8) and the analogy with algebraic mappings are gone. Secondly, there is a famous example of Fatou-Bieberbach giving $f:\mathbb{C}^2 \to \mathbb{P}^2$ such that image $f(\mathbb{C}^2)$ omits an open set in $\mathbb{P}^2$, and consequently the Liouville theorem fails in higher codimension.

Roughly speaking, the reason for these troubles might be explained as follows: A map $f:\mathbb{C}^2 \to \mathbb{C}^2 \subset \mathbb{P}^2$ is given by

$$f(z_1,z_2) = (f_1(z_1,z_2), f_2(z_1,z_2))$$

where f_1, f_2 are two arbitrary entire functions on $\mathbb{C}^2$. As we go to infinity, these two functions may not "interact" properly, and the study of the common zeroes $\{f_1 = 0, f_2 = 0\}$ doesn't at first sight seem to be that much more promising than in the case of two real analytic functions on $\mathbb{R}^2$. Further evidence for this point of view is provided by the theorem of Chern [3] in the following form due to Wu:

Given $f: \mathbb{C}^2 \to \mathbb{P}^2$, suppose that

$$\lim_{r\to\infty} \left[\frac{t_1(r)}{T_2(r)}\right] = 0. \tag{13}$$

Then the image $f(\mathbb{C}^2)$ is dense in $\mathbb{P}^2$.

Geometrically, $t_1(r)$ is the integral of $\lambda_1 + \lambda_2$ whereas $T_2(r)$ has to do with $\lambda_1\lambda_2$. Thus, if the mapping functions f_1, f_2 do interact to the extent that, e.g., $\lambda_2 \leqq c\,\lambda_1$ (so that f is *quasi-conformal*), then the Liouville theorem holds.

After all these negative remarks, I should like to mention a little theorem which, it seems to me, bodes well for the study of value distribution theory in higher codimension, although not necessarily in the codimension one format. Namely, for a holomorphic mapping $f: \mathbb{C}^2 \to \mathbb{P}^2$, we may ask what, if any, relation holds between the two order functions $T_1(r)$ and $T_2(r)$? For $f: \mathbb{C}^2 \to \mathbb{C}^2 \subset \mathbb{P}^2$ of the form

$$f(z_1, z_2) = (z_1 + h(z_2), z_2),$$

the Jacobian $J(f) \equiv 1$ and it follows that

$$\begin{aligned} T_2(r) &= 0(\log r) \\ T_1(r) &\sim M(h_2, r) \end{aligned} \tag{14}$$

Consequently, $T_1(r)$ is independent of $T_2(r)$, which may seem a little surprising in view of (12). On the other hand it may be proved that in case f omits a divisor D with $\mu c_1(D) + c_1(K_{\mathbb{P}^2}) \geqq 0$

$$\log T_2(r) \leqq 2 \log T_1(\theta r) + \mu T_1(\theta r) + O(1) \qquad (\theta > 1), \tag{15}$$

and in case the image $f(\mathbb{C}^2)$ misses an ample divisor

$$T_2(r) \leqq C\, T_1(\theta r)^2 \qquad (\theta > 1) \tag{16}$$

To me this result says two things:

(i) Holomorphic mappings do have in higher codimension a certain amount of symmetry, since an estimate (15) certainly doesn't hold in the real-analytic case.*

(ii) The canonical line bundle plays a very special role in the study of holomorphic mappings, not only for the study of divisors in the equi-dimensional case as mentioned in (9), but in higher codimensional questions also, as evidenced by (15).

5. As for how the study of Nevanlinna theory in higher codimension should proceed, I might offer the following suggestions: To begin with, value distribution theory in the classical situation of an entire meromorphic function was always balanced by, and frequently motivated by, a wealth of special functions whose behavior one wished to study. For example, the decisive upper bound (2) was originally found by Hadamard in connection with his study of the zeroes of $\zeta(s)$ vis à vis the distribution of prime numbers. Moreover, for such application it was obviously necessary to have an upper bound for every value a, and not just the average statements which may be proved in several variables (c.f. Stoll [9]).

Now in several variables it seems to me that the naturally posed problems deal more with *classes of holomorphic mappings* rather than with *specific functions*. As an example of this, suppose that A is an affine alge-

---

* In general, the only relation between $T_1(r)$ and $T_2(r)$ arises from the inequalities

$$\int(\lambda_1 + \lambda_2) \leqq \left[\int \lambda_1 + \lambda_2)^2\right]^{\frac{1}{2}} \left(\int 1\right)^{\frac{1}{2}} \qquad \text{(Cauchy-Schwarz)}$$

$$\int(\lambda_1\lambda_2) \leqq \int(\lambda + \lambda_2)^2 \qquad \text{(arithmetic and geometric means),}$$

so that both $T_1(r)$ and $T_2(r)$ are dominated by the intrinsically defined, but non-geometric, quantity $\int_0^r \left(\int_{B_\rho} (\lambda_1 + \lambda_2)^2\, \Phi\right)\frac{d\rho}{\rho}$. The proofs of (15) and (16) are given in the paper by J. Carlson and the author in these proceedings.

braic variety. Then the even rational cohomology ring $H^{ev}(A,\mathbb{Q}) = \sum_{q=0}^{n} H^{2q}(A,\mathbb{Q})$ is realized by homotopy classes of holomorphic mappings $f: A \to \mathrm{Grass}(r,N)$ of A into a Grassmannian [6]. Such maps have an intrinsic notion of growth, and it is reasonable to ask how much growth must f have in order to realize a given class $\zeta \in H^{2q}(A,\mathbb{Q})$? For example, we cannot take f to be algebraic unless non-trivial conditions on the Hodge type of ζ are satisfied. Closely related to this existence question is the problem of uniqueness: If $f: A \to \mathrm{Grass}(r,N)$ is algebraic and can be holomorphically deformed to a constant, then can this be done algebraically? In all these questions, value distribution theory in higher codimension should play an essential role, and conversely such algebra-geometric considerations furnish us with a wealth of naturally given classes of holomorphic mappings to study.

REFERENCES

1. Ahlfors, L., *The theory of meromorphic curves*, Acta Soc. Sci. Fenn., Ser. A., vol. 3(1940).

2. Carlson, J. and P. Griffiths, *A defect relation for equidimensional holomorphic mappings between algebraic varieties*, Ann. of Math., 95(1972), 557-584.

3. Chern, S. S., *Holomorphic curves in the plane*, Differential Geometry, in honor of K. Yano, Kiriokunija, Tokyo, 1972 , 73-94.

4. Cornalba, M., and B. Shiffman, *A counterexample to the transcendental Bezout problem*, Ann. of Math., 96(1972), 400-404.

5. Griffiths, P. and J. King, *Nevanlinna theory and holomorphic mappings between algebraic varieties*, Acta Math., 130(1973), 145-220.

6. Griffiths, P., *Function theory of finite order on algebraic varieties*, Jour. Diff. Geom., 6(1972), 285-306, and 7(1972), 45-66.

7. Nevanlinna, R., *Analytic Functions*, Springer-Verlag, Berlin and New York, 1970.

8. Stoll, W., *Value distribution of holomorphic maps into compact complex manifolds*, Lecture Notes in Mathematics, vol. 135, Springer-Verlag, Berlin and New York, 1970.

9. ________, *A Bezout estimate for complete intersections*, Ann. of Math., 96(1972), 361-401.

10. Wu, H., *The equidistribution theory of holomorphic curves*, Ann. of Math. Studies, no. 64, Princeton Univ. Press, 1970.

NONSTANDARD ANALYSIS IN A NUTSHELL

John J. Hirschfelder

University of Washington

INTRODUCTION

A central feature of value distribution theory in its present form is the presence of very delicate convergence problems, arising when singular currents are pulled back under holomorphic maps. These problems arise because currents do not "naturally" pull back under maps; only forms pull back, while currents push down. The observation of Abraham Robinson ([20], p. 133) that currents can be represented by nonstandard C^∞ differential forms, is therefore of considerable interest. We will illustrate this in the simplest case, that of the Dirac δ-function at $0 \in \mathbb{R}$.

Let ϕ be a nonnegative C^∞ function of compact support, such that $\phi(x) = \phi(-x)$, $\phi'(x) \geq 0$ for $x \leq 0$, and $\int\phi = 1$. Then if $\varepsilon > 0$, $\phi_\varepsilon(x) = \phi(x/\varepsilon)/\varepsilon$ has the same properties. Now suppose that ε is a positive infinitesimal, that is, a positive number which is smaller than every positive real number. Then ϕ_ε is a C^∞ function which is infinite at 0 and 0 at every nonzero real number, and whose integral is 1, which is exactly what Dirac envisioned.

The natural reaction of the standard mathematician to the above is to declare that it is nonsense, because infinitesimals don't exist. This is anomalous, because his reaction to any other complaint about the inadequacy of a number system is to construct an extension. In 1960 Robinson introduced nonstandard models of $\mathbb{R}$; these are extensions of $\mathbb{R}$ which possess "all" the properties of $\mathbb{R}$ (in particular, the property of being an ordered field) and also contain both nonzero infinitesimals and infinite numbers.

It is impossible to give here any comprehensive survey of the field of nonstandard analysis and of the various existing and proposed applications.

I will only attempt to close the credibility gap which separates the standard mathematician from the nonstandard practitioner, and to equip the reader to read the nonstandard papers in his favorite subject. For more detailed information on the general theory the reader may consult Robinson [20], Machover and Hirschfeld [17], or Luxemburg [16]. Papers applying nonstandard analysis to a variety of subjects may be found in [9], [14], and [15], and an assortment of additional papers are listed in the bibliography. Illustrations here include only some basic facts about continuity, all from Robinson [20], and the beautiful proofs of the Hilbert and Ruckert Nullstellensätze of S. Kripke* and A. Robinson [23] respectively.

The following notations will be used: $P(X)$ denotes the power set of a set X. If $f: X \to Y$ is a function and if $A \subset X$, $f[A] = \{f(x) | x \in A\}$. If r is a binary relation, that is, a set of ordered pairs, then $D(r) = \{x | \exists y[(x,y) \in r]\}$ and $R(r) = \{y | \exists x[(x,y) \in r]\}$.

1. NONSTANDARD MODELS OF $\mathbb{R}$

By a nonstandard model of $\mathbb{R}$ we mean a proper extension $^*\mathbb{R}$ of $\mathbb{R}$ which, in a sense which we will make precise shortly, has the same properties as $\mathbb{R}$. In particular, $^*\mathbb{R}$ is an ordered field, whose order relation $<$ extends the relation $<$ on $\mathbb{R}$. We introduce some terminology.

Definition: If $a \in {}^*\mathbb{R}$, then a is said to be *infinitesimal* if $|a|$ is less than $|x|$ for every nonzero real number x; and a is *infinite* if $|a|$ is greater than $|x|$ for every $x \in \mathbb{R}$; otherwise a is *finite*. If $a - b$ is infinitesimal, then we say that a is *infinitely close* to b, and we write $a \sim b$. The *monad* of a is $\mu(a) = \{b \in {}^*\mathbb{R} | b \sim a\}$.

Theorem: (1) *Infinitesimals and infinite numbers exist.* (2) *Every finite number* a *is infinitely close to a unique real number* st(a), *called the* standard part *of* a.

Proof: (2) If a is finite, $\mathrm{st}(a) = \sup\{x \in \mathbb{R} | x < a\}$ is the required standard part, as is easily checked. (1) Let $a \in {}^*\mathbb{R} - \mathbb{R}$. If a is infinite, then $1/a$ is infinitesimal. Otherwise, $b = a - \mathrm{st}(a)$

---

* I am grateful to Professor Kripke for his permission to publish his proof of the Hilbert Nullstellensatz.

is infinitesimal and $1/b$ is infinite. Q.E.D.

In the above we have, of course, assumed that a nonstandard model $^*\mathbb{R}$ exists, and that we have made proper use of our assumption that $^*\mathbb{R}$ has "all" the same properties as $\mathbb{R}$. But, as the following shows, there is a restriction on the term "property". We know that every nonempty subset of $\mathbb{R}$ which is bounded above, has a least upper bound. Now $\mathbb{R}$ is a nonempty subset of $^*\mathbb{R}$ which is bounded above; the upper bounds of $\mathbb{R}$ are just the positive infinite numbers. But $\mathbb{R}$ does not have a least upper bound in $^*\mathbb{R}$, because there is no least positive infinite number! In fact the "properties" to which we have referred are the *first order properties*, that is, those in which only quantification over real numbers (and not over sets of real numbers, as in the least upper bound principle) is permitted. More precisely, the first order properties are those which can be expressed in the *first order language* $L(\mathbb{R})$, which we now define.

Let X be a set. The vocabulary of the language $L(X)$ consists of the following:

(1) For each real number x, a *constant symbol* c_x denoting x. We shall usually write just x rather than c_x. Also,

(2) For each set $A \subset \mathbb{R}$, a constant symbol denoting A;

(3) For each binary, ternary, etc. relation or function on $\mathbb{R}$, a constant symbol denoting it;

(4) Individual (real) variables x,y,z, etc;

(5) Connectives: $\wedge$ (and), $\vee$ (or), $\neg$ (not), $\Longrightarrow$ (implies), $\Longleftrightarrow$ (if and only if).

(6) Punctuation: [ ] ( ), and $\in$ (as in "$x \in A$")

(7) Quantifiers: $\forall$, $\exists$.

The grammar of $L(X)$ is well-known. A (well-formed) formula of $L(X)$ is called a *sentence* if it has no free individual variables; otherwise it is a *predicate*. Each sentence is either true or false. The fact that $\mathbb{R}$ is an ordered field can be expressed in $L(\mathbb{R})$. The least upper bound principle cannot be expressed by a single sentence in $L(\mathbb{R})$. However, it can be expressed by an infinite family of sentences: for each $A \subset \mathbb{R}$ we can write in $L(\mathbb{R})$ a sentence stating that if A is nonempty and bounded above then A has a least upper bound. The fact that $\mathbb{R}$ is archimedean *cannot* be formalized in $L(\mathbb{R})$, not even with infinitely many sentences; this follows from the existence, shown below, of nonstandard models of $\mathbb{R}$ (which are nonarchimedean).

We will now give a precise definition of "nonstandard model".

Definition: (1) A *model* for a set S of sentences of a first order language L consists of

(a) a nonempty set $*X$;

(b) for each individual constant symbol x of L, an element $*x$ of $*X$;

(c) for each subset symbol A of L, a subset $*A$ of $*X$;

(d) for each relation (or function) symbol r of L , a relation (or function) $*r$ of the appropriate type on $*X$;

such that each sentence in S, when construed as a statement about $*X$, is true. A model $*X$ of X is a model of the set of true sentences of $L(X)$; $*X$ is *nonstandard* if it is not isomorphic to X . Note that if $*X$ is a nonstandard model of X then we have a map $*: X \to *X$ which extends to a map from the set of all subsets, relations, and functions on X to the set of all subsets, relations, and functions on $*X$. $*$ is one-to-one since if $x \neq y$, then the sentence "$x \neq y$" must be true in $*X$, i.e. , $*x \neq *y$.

2. EXISTENCE OF NONSTANDARD MODELS

The existence of a nonstandard model $*X$ of any infinite set X is deduced from the Compactness Theorem of logic; this theorem states that if every finite subset of a set S of sentences of a first-order language L has a model, then S has a model. In our case we need only augment the language $L(X)$ by adding a new individual constant symbol γ. Let T denote the set of true sentences of $L(X)$ and set

$$S = T \cup \{"c \neq \gamma" | c \in X\}$$

If S' is a finite subset of S, then only finitely many c's occur in S'; let them be $c_1, \ldots, c_n$. To construct a model of S' we need only assign a denotation in X to the symbol γ, for all other symbols already have denotations. If X is infinite, we can choose for the denotation of γ any element of X other than the c_i's; then we have our model of S'. By the Compactness Theorem, S has a model $*X$ and in it the symbol γ denotes an element of $*X$ which does not correspond to any element of X.

It is worth noting that one may easily force $*X$ to have arbitrarily large cardinality; for instead of one new symbol γ we can add a set Γ of new symbols and set

$$S = T \cup \{"c \neq y" \mid c \in X, \gamma \in \Gamma\} \cup \{"\gamma_1 \neq \gamma_2" \mid \gamma_1 \neq \gamma_2 \in \Gamma\}$$

The Compactness Theorem is in turn an easy consequence of Gödel's theorem on the completeness of the predicate calculus. The latter theorem states that a set of sentences S is consistent if and only if it has a model. So if S does not have a model, a contradiction can be deduced; but a deduction uses only finitely many sentences of S, so already some finite subset of S does not have a model.

The above proof of the Compactness Theorem was the only one known until the mid 50's. In 1955 J. Łoš [12] introduced a new concept, the ultraproduct construction, into model theory, which enabled Morel, Scott and Tarski to give in 1958 [8] a direct proof of the Compactness Theorem. Of course, this also leads to a direct construction of nonstandard models of a set X; we sketch this construction below.

Let I be an index set and let F be an ultrafilter on I. On the set X^I of functions from I to X we define an equivalence relation $\simeq$ as follows:

$$f \simeq g \iff \{i \in I \mid f(i) = g(i)\} \in F$$

Thus, we identify two functions if they agree "almost everywhere", where "almost everywhere" means on a set belonging to F. The set of equivalence classes is denoted by X^I/F, which we take to be $^*X$. The injection $^*: X \to {}^*X$ is defined by letting $^*x$ be the equivalence class of the constant function $x: I \to X$.

Let R be an n - ary relation on X; we define $^*R$ on $^*X$ by

$$(f_1,\ldots,f_n) \in {}^*R \iff \{i \in I \mid (f_1(i),\ldots,f_n(i)) \in R\} \in F.$$

It is now a straightforward but tedious task to verify that $^*X = X^I/F$, together with the relations $^*R$ on $^*X$, constitutes a model of $^*X$; for the details see, for instance [1], p.90. The verification is by induction on the number of symbols in a formula.

We specialize this construction to the case of $\mathbb{R}$ in the simplest way. Take $I = \mathbb{N}$, and let F_0 consist of all subsets of $\mathbb{N}$ having finite complement; F_0 is a filter, and Zorn's lemma implies that F_0 can be extended to an ultrafilter F. Then an element of $^*\mathbb{R}$ is a sequence of real numbers, where two sequences are deemed equivalent if they agree on an element of F. An infinite element of $^*\mathbb{R}$ is a sequence which "approaches" ∞, and an infinitesimal is a sequence which "approaches" 0.

The ultraproduct proof of the Compactness Theorem is as follows: Let S be a set of sentences, every finite subset ν of which has a model M_ν. For the index set, I, take the set of all finite subsets ν of S. Define $d_\nu = \{\mu \in I \mid \nu \subset \mu\}$. Then $F_0 = \{d_\nu \mid \nu \in I\}$ is a filter base. If F is an ultrafilter extending F_0, then $\left(\prod_{\nu \in I} M_\nu\right)/F$ is the required model of S; the reader is invited to supply the details of the construction.

3. SOME APPLICATIONS

Let $A \subset \mathbb{R}$, let $f\colon A \to \mathbb{R}$ be a function, and let $a \in A$.

Theorem: f *is continuous at* a *if and only if* $f(x) \sim f(a)$ *whenever* $x \in {}^*A$ *and* $x \sim a$.

Proof: Suppose f is continuous at a. Then this is expressed by the following sentence in $L(\mathbb{R})$:

$$(\forall \varepsilon > 0)(\exists \delta > 0)(\forall x \in A)[|x - a| < \delta \Longrightarrow |f(x) - f(a)| < \varepsilon]$$

Suppose $x \sim a$. Then $|x - a| < \delta$ for every positive $\delta \in \mathbb{R}$, and so $|f(x) - f(a)| < \varepsilon$ for every positive $\varepsilon \in \mathbb{R}$. This says that $f(x) \sim f(a)$.

Conversely, suppose that $f(x) \sim f(a)$ whenever $x \sim a$, and choose $\varepsilon > 0$. Let δ_1 be any positive infinitesimal. Then

$$(\forall x \in A)[|x - a| < \delta_1 \Longrightarrow |f(x) - f(a)| < \varepsilon]$$

but this sentence does not belong to $L(\mathbb{R})$ because it contains the constant symbol δ_1. But we immediately infer

$$(\exists \delta > 0)(\forall x \in A)[|x - a| < \delta \Longrightarrow |f(x) - f(a)| < \varepsilon]$$

and this sentence *does* belong to $L(\mathbb{R})$, so it is true in $\mathbb{R}$. This completes the proof.

The proof of the following is similar:

Theorem: f *is uniformly continuous on* A *if and only if* $f(x) \sim f(y)$ *whenever* $x \in {}^*A$ *and* $y \in {}^*A$ *and* $x \sim y$.

Theorem: *If* f *is continuous on* $[0,1]$, *it is uniformly continuous.*

Proof: Choose x and y in ${}^*[0,1] = \{z \in {}^*\mathbb{R} \mid 0 \le z \le 1\}$, and assume $x \sim y$. Set $a = st(x)$; then $a \in [0,1]$ and $a = st(y)$ also. Since f is continuous at a, and since $x \sim a$ and $a \sim y$, we have $f(x) \sim f(a)$ and $f(a) \sim f(y)$. So $f(x) \sim f(y)$. Q.E.D.

Note that the key to this theorem is that every number in ${}^*[0,1]$ is *near standard*, i.e. infinitely close to a standard number in $[0,1]$. In fact, the condition "every point of ${}^*A$ is near-standard" characterizes compactness of A for any topologized space A (see [20], p. 93).

We now turn to the proof of Hilbert's Nullstellensatz given by S. Kripke. Let k be an algebraically closed field, and let $\mathfrak{p} = (f_1,\ldots,f_s)$ be a prime ideal in the polynomial ring $k[X_1,\ldots,X_n] \equiv k[X]$. The Nullstellensatz asserts that if $g \in k[X]$ vanishes on the set of common zeros of $\mathfrak{p}$, then $g \in \mathfrak{p}$. Let K be the field of quotients of the domain $k[x]/\mathfrak{p}$, and let ${}^*k$ be a nonstandard model of k having infinite transcendence degree over k. Then ${}^*k$ is algebraically closed, because the fact that k is algebraically closed can be expressed in $L(k)$ by the set of sentences $\{\forall c_1 \ldots \forall c_n \exists x[x^n + c_1 x^{n-1} + \ldots + c_n = 0] \mid n \in \mathbb{N}\}$. Thus, K can be imbedded in ${}^*k$:

$$k \to k[x] \to k[x]/\mathfrak{p} \to K \to {}^*k$$

Let ξ_i be the image in ${}^*k$ of X_i and set $\xi = (\xi_1,\ldots,\xi_n)$. Then the images of f_j and g in ${}^*k$ are $f_j(\xi)$ and $g(\xi)$.

Now assume $g \notin \mathfrak{p}$. Then

$$f_1(\xi) = 0 \wedge \ldots \wedge f_s(\xi) = 0 \wedge g(\xi) \neq 0$$

This sentence does not belong to $L(k)$, but

$$\exists \xi [f_1(\xi) = 0 \wedge \ldots \wedge f_s(\xi) = 0 \wedge g(\xi) \neq 0]$$

does! In k it asserts the existence of a point in the zero set of $\mathfrak{p}$ at which g does not vanish, and the proof is complete.

Let f be a holomorphic function defined in a neighborhood U of the origin $0 \in \mathbb{C}^n$. Then f has an extension which we still call $f: {}^*U \to {}^*\mathbb{C}$. We wish to view $f|\mu(0)$ as the germ of f at 0. Clearly, if two holomorphic functions f and g agree on a neighborhood of $0 \in \mathbb{C}^n$, then $f|\mu(0) = g|\mu(0)$. Conversely, if $f|\mu(0) = g|\mu(0)$, choose any positive infinitesimal ε; then

$$(\forall z \in {}^*\mathbb{C}^n)[|z| < \varepsilon \Longrightarrow f(z) = g(z)]$$

This sentence does not belong to $L(\mathbb{C})$, but

$$(\exists \varepsilon > 0)(\forall z \in {}^*\mathbb{C}^n)[|z| < \varepsilon \Longrightarrow f(z) = g(z)]$$

does; interpreted in $\mathbb{C}$ it says that f and g agree on a neighborhood of 0.

Similarly we interpret the germ of an analytic variety V at $0 \in \mathbb{C}^n$ as $V \cap \mu(0)$.

We denote by $\mathfrak{O}_n$ the ring of germs of analytic functions at $0 \in \mathbb{C}^n$. The Ruckert Nullstellensatz states that if p is a prime ideal in $\mathfrak{O}_n$ and if $g \in \mathfrak{O}_n$ vanishes on the set of common zeros of p, then $g \in p$. Robinson's nonstandard proof of the Nullstellensatz is based on the concept of a generic point for a prime ideal in $\mathfrak{O}_n$.

Definition: A point $\xi \in \mu(0) \subset {}^*\mathbb{C}^n$ is *generic* for the prime ideal $p \in \mathfrak{O}_n$ if for every $f \in \mathfrak{O}_n$, $f \in p$ if and only if $f(\xi) = 0$.

It is easily checked that if $\xi \in \mu(0) \subset {}^*\mathbb{C}$ and $\xi \neq 0$, then ξ is generic for the ideal $(0) \subset \mathfrak{O}_1$. The existence of generic points for the zero ideals of $\mathfrak{O}_n$, $n > 1$, is a matter of careful choice of the nonstandard model ${}^*\mathbb{C}$, and it will be convenient to postpone consideration of this point until after the discussion of higher-order structure in the next section.

It is clear that to prove the Nullstellensatz it suffices to demonstrate the existence of a generic point ξ for each prime ideal p in $\mathfrak{O}_n$. We do this by induction on n, assuming the existence of generic points for zero ideals. If $n = 1$, the only prime ideal other than (0) is the maximal ideal (z), and 0 is generic for this ideal.

Let p be prime in $\mathfrak{O}_n$ and set $p' = \mathfrak{O}_{n-1} \cap p$. Then p' is prime in

$\mathbf{O}_{n-1}$ and we have the diagram

$$\begin{array}{ccc} \mathbf{O}_{n-1} & \longrightarrow & \mathbf{O}_n \\ \big\downarrow \lambda' & & \big\downarrow \lambda \\ \mathbf{O}_{n-1}/p' & \longrightarrow & \mathbf{O}_n/p \end{array}$$

in which the horizontal arrows are injections. By the Weierstrass Preparation Theorem we may without loss of generality assume that p contains a function h of the form

$$h = z_n^k + a_1 z_n^{k-1} + \ldots + a_k$$

where $a_i \in \mathbf{O}_{n-1}$ and $a_i(0) = 0$. Applying λ, we obtain

$$0 = \overline{z}_n^k + \overline{a}_1 \overline{z}_n^{k-1} + \ldots + \overline{a_k}$$

where the bar denotes image under λ. This equation shows that the element $\overline{z}_n \in \mathbf{O}_n/p$ is algebraic over $\mathbf{O}_{n-1}/p'$. If $f \in \mathbf{O}_n$, the Weierstrass Division Theorem states that we may write $f = qh + r$, where $r \in \mathbf{O}_{n-1}[z_n]$. It follows that $\overline{f} = \overline{r}$ and $\overline{f}$ is therefore also algebraic over $\mathbf{O}_{n-1}/p'$.

Now our inductive hypothesis states that p' has a generic point $\xi' \in \mu(0) \cap {}^*\mathbb{C}^{n-1}$; i.e. that the evaluation map $e_{\xi'}: \mathbf{O}_{n-1} \to {}^*\mathbb{C}$, defined by $e_{\xi'}(f) = f(\xi')$, has kernel p'. Denote by τ' the induced injection $\mathbf{O}_{n-1}/p' \to {}^*\mathbb{C}$. Since $\mathbf{O}_n/p$ is algebraic over $\mathbf{O}_{n-1}/p'$ and since ${}^*\mathbb{C}$ is algebraically closed, τ' extends to an injection $\tau: \mathbf{O}_n/p \to {}^*\mathbb{C}$. Set $\xi_n = \tau(\overline{z}_n)$ and $\xi = (\xi', \xi_n) \in {}^*\mathbb{C}^n$. Then one readily checks that $\tau \circ \lambda = e_\xi$ and $\tau \circ \lambda$ has kernel p. Thus ξ is the required generic point for p.

It remains only to check that ξ_n is infinitesimal. But if ξ_n is not infinitesimal, we can write

$$\xi_n = -\tau(\overline{a}_1) - \xi_n^{-1}\tau(\overline{a}_2) - \ldots - \xi_n^{-k+1}\tau(\overline{a}_k)$$

which is infinitesimal. This completes the proof.

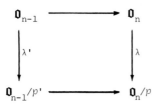

4. HIGHER-ORDER STRUCTURE

The apparatus introduced in Section 2, while already very powerful, is not adequate to handle most "realistic" mathematics. This is illustrated by the following simple nonstandard proof, which is not formalized within the theory so far constructed.

Theorem: *Every continuous function on* $[0,1]$ *attains a maximum.*

"Proof": Let n be an infinite integer and consider the set

$$X = \{f(k/n) \mid k \in {}^*\mathbb{Z},\ 0 \leq k \leq n\}.$$

The following sentence Σ is a true statement about $\mathbb{R}$: "If $n \in \mathbb{N}$ and if h is a function defined on $\{k \in {}^*\mathbb{Z} \mid 0 \leq k \leq n\}$, then the image of h has a largest element." This sentence is true in ${}^*\mathbb{R}$, and shows that X has a largest element $f(k_o/n)$. Set $a = st(k_o/n)$; then $f(a) = st(f(k_o/n))$ since f is continuous. Now let x be any standard number in $[0,1]$; then there is a k and that $x \sim k/n$. Then

$$f(x) \sim f(k/n) \leq f(k_o/n)$$

and $f(x) \leq f(a)$. Q.E.D.

The difficulty with the above argument is that it involves essentially a quantification over subsets of $\mathbb{R}$; in the first order formulation of nonstandard analysis described in Section 1, this procedure is explicitly prohibited. Indeed, the sentence Σ is *false* in ${}^*\mathbb{R}$, for take h to be the function: $h(x) = x$ if x is finite, $h(x) = 0$ if x is infinite!

In order to salvage the above proof, we need a "higher order nonstandard model" of $\mathbb{R}$. Such a model will be equipped with a set ${}^*(\mathbb{R}^{\mathbb{R}})$ of functions from ${}^*\mathbb{R}$ to ${}^*\mathbb{R}$; the elements of ${}^*(\mathbb{R}^{\mathbb{R}})$ are called *internal* functions. ${}^*(\mathbb{R}^{\mathbb{R}})$ will be a proper subset of ${}^*\mathbb{R}^{{}^*\mathbb{R}}$; indeed the function h given above is not internal, but *external*. In the nonstandard model, quantification over functions will extend only to the internal functions.

Similarly, associated to the set $P(\mathbb{R})$ of all subsets of $\mathbb{R}$ will be a set ${}^*P(\mathbb{R})$ of internal subsets of ${}^*\mathbb{R}$; ${}^*P(\mathbb{R})$ is a proper subset of $P({}^*\mathbb{R})$. Let $A \subset \mathbb{R}$, then $A \in P(\mathbb{R})$ and ${}^*A \in {}^*P({}^*\mathbb{R})$. The internal sets of the form ${}^*A$, $A \subset \mathbb{R}$, are the *standard* subsets of $\mathbb{R}$. If ${}^*[P(\mathbb{R})]$ denotes the set of standard sets, we have the inclusions:

$$*[P(\mathbb{R})] \subsetneq *P(\mathbb{R}) \subsetneq P(*\mathbb{R}) .$$

The following development of higher-order theory, due to Robinson and Zakon [24], is based on the observation that in any mathematical discussion there are certain objects, which we will call "atoms", that are not treated as sets; and that all sets considered are built up from the atoms. Typically, atoms are "points" or "numbers".

Let X be a set of atoms. The *universe of* discourse $\mathcal{U} = \mathcal{U}_X$ over X is the smallest set $\mathcal{U}$ satisfying the following conditions:

(1) $X \in \mathcal{U}$

(2) $x \in y \wedge y \in \mathcal{U} \Longrightarrow x \in \mathcal{U}$

(3) $x \in \mathcal{U} \Longrightarrow P(x) \in \mathcal{U}$

(4) $x \in \mathcal{U} \wedge y \in \mathcal{U} \Longrightarrow x \cup y \in \mathcal{U}$

Let $^*\mathcal{U}$ be a nonstandard model of $\mathcal{U}$, and let $\mathbf{E}$ be the relation on $^*\mathcal{U}$ corresponding to the relation $\in$ on $\mathcal{U}$. Let #X be the set of $\mathbf{E}$-minimal elements of $^*\mathcal{U}$ other than $*\emptyset$, i.e.

$$\#X = \{x \in {^*\mathcal{U}} \mid x \neq *\emptyset \wedge \neg (\exists y \in {^*\mathcal{U}})[y \,\mathbf{E}\, x]\}.$$

Then $$\#X = \{x \in {^*\mathcal{U}} \mid x \,\mathbf{E}\, *X\}.$$

Without loss of generality we can assume that the map $*: X \to \#X$ is the inclusion, and that the elements of #X are atoms. We would now like to identify elements of $^*\mathcal{U}$ with elements of the universe of discourse, $\tilde{\mathcal{U}}$, over #X. For instance, if $A \in {^*\mathcal{U}}$ and if each x, such that $x \,\mathbf{E}\, A$, is an atom of #X, then we would like to regard A as a subset of #X . This can be done as follows: If $x \in {^*\mathcal{U}}$, define the rank rk(x) to be the supremum of the lengths n of chains $x_0 \,\mathbf{E}\, x_1 \,\mathbf{E} \cdots \mathbf{E}\, x_n$ where $x_n = x$. rk(x) may be infinite. Let $^*_0\mathcal{U}$ consist of the elements of $^*\mathcal{U}$ having finite rank. We now define a map $\phi: {^*_0\mathcal{U}} \to \tilde{\mathcal{U}}$ inductively: if x is an atom, then $\phi(x) = *x = x$, and otherwise $\phi(x) = \{\phi(y) \mid y \,\mathbf{E}\, x\}$. Let $\#\mathcal{U}$ be the image of $^*_0\mathcal{U}$ under ϕ, and set $\# = \phi \circ *$.

We summarize this construction in the diagram:

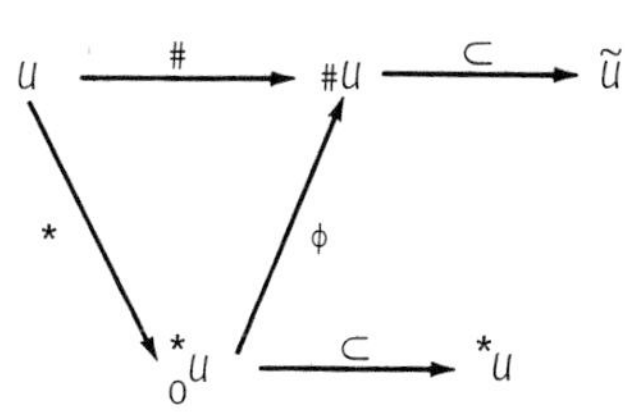

The sets in $\#\mathcal{U}$ are called *internal*, those in $\tilde{\mathcal{U}} - \#\mathcal{U}$ are *external*. Sets in the image of $\#$ are *standard*; they are the sets having names which are also names of elements of $\mathcal{U}$.

We now discard ${}^*\mathcal{U}$ and ${}^*_0\mathcal{U}$ and proceed to regard $\#\mathcal{U}$ as the nonstandard model of the universe of discourse $\mathcal{U}$. However, $\#\mathcal{U}$ is not a model of $\mathcal{U}$ strictly speaking, since we have discarded the sets of ${}^*\mathcal{U}$ having infinite rank! But the situation can be salvaged. Let $L_0(\mathcal{U})$ consist of those sentences of the language $L(\mathcal{U})$ every quantifier of which is restricted to a set in $\mathcal{U}$, that is, is of the form $(\forall x \in A)$ or $(\exists x \in A)$ for some $A \in \mathcal{U}$. Let T_0 be the set of true sentences of $L_0(\mathcal{U})$. Then, as shown in [24], $\#\mathcal{U}$ is a model of T_0.

Let us return now to the proof that every continuous function on $[0,1]$ attains a maximum. For the set of atoms X we take $\mathbb{R}$. The sentence Σ belongs to $L_0(\mathcal{U})$, and interpreted in $\tilde{\mathcal{U}}$ it says: if $n \in {}^*\mathbb{N}$ and if h is an *internal* function defined on $\{k \in {}^*\mathbb{Z} | 0 \le k \le n\}$ then h attains a maximum. The function $h(k) = f(k/n)$ is certainly internal, because f is standard; so Σ may be applied to it.

It is necessary to impose on our nonstandard model a condition that will guarantee that it is sufficiently "large" for all purposes. The required property is that of being an *enlargement*, which we now define.

An element r of $\mathcal{U}$ is a binary relation if it is a set, all of whose elements are ordered pairs; r is said to be *concurrent* if for each set $\{x_1,\ldots,x_n\}$ of finitely many elements of the domain of r there exists a y in the range of r such that $(x_i,y) \in r$ for $i = 1,\ldots,n$. Examples of concurrent relations are the relation $\neq$ on any infinite set A, the relation $<$ on $\mathbb{R}$, and the following relation ρ: $(f,p) \in \rho$ if and only if f is a holomorphic function defined in a neighborhood U_f of $0 \in \mathbb{C}^n$, $f \not\equiv 0$, $f(0) = 0$, $p \in U_f$, $f(p) \neq 0$. A nonstandard model ${}^*\mathcal{U}$ of $\mathcal{U}$ is called an *enlargement* if for every concurrent relation $r \in \mathcal{U}$ there exists a $\gamma_r \in {}^*\mathcal{U}$ such that $({}^*x,\gamma_r) \in {}^*r$ for every x in the domain of r. To see that enlargements exist, use the Compactness Theorem: if T is the set of true sentences of $L(\mathcal{U})$, set

$$T' = T \cup \{\text{"}(x,\gamma_r) \in r\text{"} | r \ \text{concurrent}, \ x \in \text{domain}\ (r)\}$$

where γ_r is a new symbol. One easily sees that every finite subset of T' has a model.

We can now complete the proof of the Ruckert Nullstellensatz. If ${}^*\mathcal{U}$ is an enlargement of $\mathcal{U}$, the concurrence of ρ shows that the ideal (0) in $\mathbf{O}_n$ has a generic point.

5. REPRESENTATION OF DISTRIBUTIONS AND CURRENTS, AND SOME PROBLEMS

Let V be a vector space over $\mathbb{R}$ equipped with a bilinear pairing $\langle v,w\rangle$ from $V \times V$ into $\mathbb{R}$; assume that for each v there is w such that $\langle v,w\rangle \neq 0$, and vice versa. We have in mind, of course, the space of C^∞ functions (or differential forms) with compact support on a manifold M. Let f be an arbitrary linear functional on V, and consider the binary relations "$\langle v,w\rangle = f(v)$" on V. It is easy to see that this relation is concurrent. For let $v_1,\ldots,v_n \in V$ and let W be the subspace spanned by $v_1,\ldots,v_n$. Then since no vector $v \in W$ is in the kernel of all the functionals $v \mapsto \langle v,w\rangle$ for $w \in V$, there exists a $w \in V$ such that $\langle v,w\rangle = f(v)$ for $v \in W$.

Now let us pass to an enlargement; we see that for each linear functional f on V there is a $w \in {}^*V$ such that $\langle v,w\rangle = f(v)$ for every $v \in V$. If we take V to be the space of C^∞ functions with compact support on $\mathbb{R}$, each $\phi \in {}^*V$ is an internal "C^∞" function ${}^*\mathbb{R} \to {}^*\mathbb{R}$, and every distribution can be represented by such a ϕ. The following difficulties immediately arise: First, not every element ϕ of ${}^*V$ represents a linear functional on V; only if $\langle g,\phi\rangle$ is finite for every standard function g can we define a linear functional on V by $g \to \mathrm{st}\langle g,\phi\rangle$. Next, the representation of a linear functional is not unique; the zero functional is represented by every function ϕ such that $\langle g,\phi\rangle$ is infinitesimal for every standard function g. Denote by V^o the $\mathbb{R}$-subspace of ${}^*V$ consisting of the functions which represent functionals, and by V^{oo} the subspace of V^o consisting of the functions which represent the 0 functional. V^o and V^{oo} are external sets. In the notation of Section 1, if ε is an infinitesimal, then $\phi_\varepsilon \in V^o$ and $\varepsilon\phi_\varepsilon \in V^{oo}$. Clearly V^o/V^{oo} is canonically isomorphic to the algebraic dual of V.

It is possible to give reasonable descriptions of V^o and V^{oo} in terms of that weak topology on V defined by the linear functionals $g \to \langle g,\phi\rangle$. Nonstandard models of topological vector spaces in general have been extensively investigated by Henson and Moore [6,7].

Of course, not every $\phi \in V^{oo}$ defines a distribution, for everything said so far ignores the C^∞ topology on V; therein lies the principal obstruction to a satisfactory nonstandard theory of distribution. We conclude by listing some problems: (1) Give necessary and sufficient conditions on $\phi \in {}^*V$ that ϕ represents a distribution; these conditions should refer to the behavior of ϕ as a function from ${}^*\mathbb{R}$ to ${}^*\mathbb{R}$, and not merely to the behavior of ϕ as an element of the abstract vector space V. (2) Give necessary and sufficient conditions on a nonstandard C^∞ differential form that it represents a measure; or a normal current; or a flat current; or a variety.

REFERENCES

1. Bell, J. and A. Slomson, *Models and Ultraproducts*, North-Holland, Amsterdam, 1969.

2. Bernstein, A. and A. Robinson, *Solution of an invariant subspace problem of K. T. Smith and P. R. Halmos*, Pac. J. Math. 16 (1966), 421-431.

3. Fenstad, J. E., *A note on standard vs. nonstandard topology*, Proc. Royal Acad. Sci. Amsterdam 70 (1967), 378-380.

4. Henson, C. W., *On the nonstandard representation of measure*, Trans. Amer. Math. Soc. (to appear).

5. Henson, C. W. and L. C. Moore, Jr., *Invariance of the nonstandard hulls of a uniform space*, Duke Math. J. (to appear).

6. __________, *The nonstandard theory of topological vector spaces*, Trans. Amer. Math. Soc. (to appear).

7. __________, *The nonstandard hulls of locally convex spaces*, Duke Math. J. (to appear).

8. Hurd, A. E., *Nonstandard analysis of dynamical systems. I: Limit notions, stability*, Trans. Amer. Math. Soc. 160 (1971), 1-26.

9. Hurd, A. E. and P. Loeb, ed., *Proceedings of the Victoria, B. C., Symposium on Nonstandard Analysis* (to appear).

10. Kugler, L., *A nonstandard approach to linear functions*, Michigan Math. J. 16 (1969), 157-160.

11. __________, *Nonstandard almost periodic functions on a group*, Proc. Amer. Math. Soc. 22 (1969), 527-533.

12. Loeb, P., *A nonstandard representation of measurable spaces and* L^∞, Bull. Amer. Math. Soc. 77 (1971), 540-544.

13. Łoš, J., *Quelques remarques, théorèmes et problèmes sur les classes*

définissables d'algebrès, in *Mathematical Interpretatations of Formal Systems*, North-Holland, Amsterdam, 1955.

14. Luxemburg, W. A. J., ed., *Applications of Model Theory to Algebra, Analysis, and Probability*, Holt, Rinehart & Winston, New York, 1969.

15. Luxemburg, W. A. J. and A. Robinson, ed., *Contributions to Nonstandard Analysis*, North-Holland, Amsterdam, 1972.

16. Luxemburg, W. A. J., *A general theory of monads*, in [14], 18-86.

17. Machover, M. and J. Hirschfeld, *Lectures on Nonstandard Analysis*, Lecture Notes in Mathematics vol. 94, Springer-Verlag, Berlin, 1969.

18. Morel, A., D. Scott, and A. Tarski, *Reduced products and the compactness theorem*, Notices Amer. Math. Soc. 5 (1958), 674-675.

19. Narens, L., *A nonstandard proof of the Jordan curve theorem*, Pac. J. Math. 36 (1971), 219-229.

20. Robinson, A., *Nonstandard Analysis*, North-Holland, Amsterdam, 1966.

21. __________, *Nonstandard theory of Dedekind rings*, Proc. Royal Acad. Sci. Amsterdam 70 (1967), 442-452.

22. __________, *Compactification of groups and rings and nonstandard analysis*, J. Symbolic Logic 34 (1969), 576-588.

23. __________, *Germs*, in [14], 138-149.

24. Robinson, A. and E. Zakon, *A set-theoretic characterization of enlargements*, in [14], 109-122.

MOIŠEZON SPACES AND THE KODAIRA EMBEDDING THEOREM

R. O. Wells, Jr.*

Rice University

INTRODUCTION

A *Moišezon space* is a compact complex space of complex dimension n whose field of meromorphic functions contains n algebraically independent nonconstant meromorphic functions. Moišezon spaces contain the projective algebraic varieties, but are a more extensive class of complex spaces, as indicated below. In 1954 Kodaira gave a very useful differential geometric characterization of projective manifolds. The purpose of this paper is to discuss the possibility of generalizing Kodaira's theorem to obtain a differential geometric characterization of Moišezon spaces. In particular, we outline one possible approach to achieving this objective, and indicate some of the significant problems which arise.

1. MOIŠEZON SPACES

Let X be a compact complex space of complex dimension n, and let $\mathcal{M}(X)$ be the field of meromorphic functions on X. It is well known that the transcendence degree of $\mathcal{M}(X)$ over $\mathbb{C}$ is at most equal to n (cf. Thimm [20], Siegel [19], Remmert [15]). In other words, there are at most n algebraically independent nonconstant meromorphic functions on X. It has been known for some time that there exist compact complex manifolds (e.g. complex tori) of complex dimension n, where the transcendence degree d of $\mathcal{M}(X)$ over $\mathbb{C}$ is any integer between 0 and n (cf. Siegel [18]).

* Research was partially supported by NSF GP-19011 at Rice University.

Moišezon was the first to systematically study complex n-manifolds X (and more generally spaces) which have precisely n algebraically independent meromorphic functions (cf. Moišezon [10], [11]). On the other hand, complex tori with this property are called *Abelian varieties* and have been an object of study since the time of Abel and Riemann (the period matrix of a Riemann surface of genus g defines an Abelian variety of complex dimension g). Following the current usage, we will call an n-dimensional compact complex space X with n algebraically independent meromorphic functions a *Moišezon space* (or *Moišezon manifold* in case X is nonsingular). A primary example of a Moišezon space is given by projective algebraic subvarieties. To see this we want to prove a simple criterion that a compact complex space be Moišezon.

We recall first the concept of a meromorphic mapping from one compact complex space to another. If X and Y are compact complex spaces, then suppose there exists an analytic subvariety $S \subset X$ and a holomorphic mapping $f: X - S \to Y$, then f is said to be a *meromorphic mapping* from X to Y if the closure of the graph of f in the product space $X \times Y$ is an analytic subvariety (cf. Remmert [14], Whitney [23]). Thus f is extended in an analytic way across S, but points in S may be mapped to sets of points in Y. Meromorphic mappings are generalizations of the classical rational and birational mappings of algebraic geometry. The meromorphic mapping f above is said to be *bimeromorphic* if f maps $X - S$ bihomorphically onto the complement of a subvariety in Y.

Proposition 1: *Let* $f: X \to \mathbb{P}_N(C)$ *be a meromorphic mapping, where* X *is an irreducible complex space of dimension* $\le N$, *and where* f *is a holomorphic mapping on* $X - S$, *for* S *a subvariety of* X. *If there is a point* $x_0 \in X - S$ *such that*

a) x_0 *is not a singular point of* X

b) f *has maximal rank at* x_0 ,

then X *is a Moišezon space.*

Proof: The two conditions a) and b) at x_0 imply that f restricted to a neighborhood U of x_0 is a local embedding. Let $f(U) = Y \subset \mathbb{P}_N$, and assume that we have local coordinates in $\mathbb{P}_N$, $\zeta_1,\ldots,\zeta_N$ at $f(x_0)$ so that $d\zeta_1,\ldots,d\zeta_n$ are cotangent vectors to Y at $f(x_0)$. Then the coordinate functions $\phi_1(\zeta) = \zeta_1$, $\phi_2(\zeta) = \zeta_2,\ldots,\phi_n(\zeta) = \zeta_n$ are n algebraically independent meromorphic functions on $\mathbb{P}_N$. Let $\psi_j = f^*\phi_j$.

Then since f is a meromorphic mapping, it follows that $\psi_1,\ldots,\psi_n$ are meromorphic functions on X (cf. Remmert [14], Whitney [22] for a discussion and definition of the pullback of a meromorphic function). We claim now that $\psi_1,\ldots,\psi_n$ are algebraically independent. By the choice of $\phi_1,\ldots,\phi_n$ and the fact that f has maximal rank at x_0, it follows easily that

$$d\psi_1(x_0) \wedge \ldots \wedge d\psi_n(x_0) \neq 0 . \tag{1}$$

But (1) implies that $\psi_1,\ldots,\psi_n$ are algebraically independent by a result of Remmert ([14], [23]). Basically, if there was a polynomial relation between the ψ's, then differentiating the relation near x_0 would imply a relation between the differentials, which by (1) doesn't exist.

Q.E.D.

A corollary of the above proposition is that any compact subvariety of $\mathbb{P}_N$ is Moišezon, taking f to be the natural injection. It's a well known theorem of Lefschetz that complex tori which are Moišezon are projective algebraic (cf. Siegel [17], Wells [20]). However, the whole point of Moišezon's intensive study of what are now called Moišezon spaces was the existence of Moišezon spaces which are *not* projective algebraic. The first example was given by Nagata in 1957 [12]. Since then there have been numerous examples discovered which illustrate various additional properties (Moišezon [10], Grauert [5], Riemenschneider [16]). In the other direction, there are a variety of theorems of the following type:

Suppose X is a Moišezon space, and, in addition, property P is satisfied, then X is projective algebraic.

Examples of properties P which fit this pattern are:

a) X is 2-dimensional and nonsingular (Chow-Kodaira [3]).

b) X is nonsingular and Kähler (Moišezon [10]).

c) X is a complex torus (Lefschetz; this result was discussed above).

One of the basic theorems of Moišezon asserts that any Moišezon space is bimeromorphically equivalent to a projective algebraic subvariety. Moreover, this bimeromorphic correspondence can be represented in the following manner. Suppose X is a Moišezon space, then there exists a nonsingular projective algebraic manifold $\hat{X}$ and a *proper modification*

$$\pi\colon \hat{X} \to X ,$$

i.e., π is a proper surjective holomorphic mapping which is biholomorphic when restricted to the complement of a subvariety $\hat{S} \subset \hat{X}$ (cf. Whitney [22]).

Moreover, π can be chosen to be a finite sequence of monoidal transformations (cf. Moišezon [11]). Note that, in particular, $\hat{X}$ resolves the singularities of X. Thus a given Moišezon space is, by the above characterization, the "blow down" of a projective algebraic manifold. Most of the examples of non projective Moišezon spaces are explicitly of this form.

The category of Moišezon spaces and meromorphic mappings is closed under bimeromorphic equivalence in the larger category of compact complex spaces. By the above characterization of Moišezon, this is the smallest such category of compact complex spaces which contains the projective algebraic varieties. From the algebraic geometry point of view a similar category of scheme-theoretic spaces (algebraic spaces) has been studied by Artin [1], who shows that the two categories are isomorphic.

2. KODAIRA'S EMBEDDING THEOREM

We now turn to Kodaira's embedding theorem. Let X be a compact complex manifold, then a *Hodge metric* on X is a Kähler metric which represents an integral cohomology class (has integral periods). Kodaira's embedding theorem [9] asserts that any compact complex manifold which admits a Hodge metric has a projective algebraic embedding.* Any integral cohomology class of type (1,1) is the first Chern class of a holomorphic line bundle, and thus the assumption that X carries a Hodge metric is equivalent to the assumption that X carries a holomorphic line bundle with a Hermitian metric h whose *curvature form*

$$\Theta_L(h) = \bar{\partial}\partial \log h_\alpha = \sum_{\mu,\nu} \Theta_{\mu\nu}(z)dz_\mu \wedge d\bar{z}_\nu$$

has a positive definite coefficient matrix (with respect to a local frame e for L, and where $h_\alpha = (e_\alpha, e_\alpha)$ is a positive definite representation for the metric h with respect to the local frame e_α). By definition,

$$c_1(L,h) = \frac{i}{2\pi}\Theta_L(h)$$

---

* This theorem and relevant concepts mentioned here are discussed in detail in [20].

is the *Chern form* of L and is a positive definite differential form on X of type (1,1).

Suppose L is any holomorphic line bundle over a compact complex manifold X then $\mathcal{O}(X,L)$, the vector space of global holomorphic sections of L, is finite dimensional. Let $\{\phi_0,\ldots,\phi_N\}$ be a basis for $\mathcal{O}(X,L)$, and let

$$S = \{x \in X: \phi_0(x) = \ldots = \phi_N(x) = 0\},$$

the set of *base points* of L. It's clear that S is a proper subvariety of X unless there are no nonvanishing sections of L. We assume that there is at least one nonvanishing section of L in order for the following discussion to make sense. The sections $\{\phi_0,\ldots,\phi_N\}$ define a meromorphic mapping

$$\Phi: X \to \mathbb{P}_N \tag{2}$$

which is defined on X - S by the mapping

$$x \to [\phi_0(x),\ldots,\phi_N(x)] \in \mathbb{P}_N$$

where [,...,] denotes homogeneous coordinates in $\mathbb{P}_N$ (cf. Kodaira [9]).

The proof of Kodaira's embedding theorem depends on showing that there are sufficiently many "independent" sections such that the meromorphic mapping (2) is actually a holomorphic embedding. The obstructions to this being the case can be shown to lie in a certain cohomology group. Thus the embedding problem can be reduced to proving a *vanishing theorem* for the space of the obstructions. The theorem in question is the well-known Kodaira Vanishing Theorem, which can be formulated as follows. Let $K_X = \Lambda^n T^*(X)$ $(n = \dim_{\mathbb{C}} X)$ be the *canonical bundle* of X.

Theorem 2 (Kodaira [8]): *Suppose* L *is a positive line bundle over a compact complex manifold* X, *then*

$$H^1(X,\mathcal{O}(L \otimes K_X)) = 0. \tag{3}$$

Kodaira's positivity assumption (in the above form) at each point of X insures a global embedding. To deduce from the meromorphic mapping Φ that X is Moišezon, it suffices, by Proposition 1, to show that Φ is a *local* embedding near a single point $x_0 \in X$. So it makes sense to ask whether a weakening of Kodaira's positivity condition will still insure a local embedding. Note that the positivity condition involves the interaction of the bundle L and the canonical bundle of X.

Riemenschneider has shown that in certain cases a weakening of the positivity condition is quite possible. A holomorphic line bundle $L \to X$ is called *semipositive* if there is a Hermitian metric h on L such that $c_1(L,h)$ is positive semidefinite at each point of X. Riemenschneider has shown that if L is a semipositive holomorphic line bundle over X, and if L is positive near a point x_0 in X, *and* if X is Kähler, then X is Moišezon. The proof depends on using a localization of Kodaira's arguments near the point x_0 (blowing up X at x_0 and inducing a semipositive bundle on the blown up manifold which is positive near the blow up of x_0) as well as showing that the vanishing result (3) above is valid under these weakened positivity assumptions [16].

In Riemenschneider's result it is very important that the Kähler assumption be made. In the original Kodaira theorem the Kählerness of a complex manifold with a positive line bundle L was an automatic consequence of the fact that $c_1(L)$ contained a Kähler form. Moreover, as Riemenschneider points out, the assumption of Kählerness implies that the resulting Moišezon manifold is in fact projective algebraic. Thus this result, which still depends on the standard Kähler geometry, fails to give a sufficient weakening of Kodaira's positivity assumption to account for the non projective algebraic Moišezon manifolds. In fact the proof of the vanishing theorem under the semi-positivity condition in Riemenschneider's paper depends on showing that the *Nakano inequalities* for harmonic (p,q)-forms on a Kähler manifold are valid under the weakened positivity condition. But the standard Hodge-Kähler identities for the harmonic integrals are still employed. Thus one way to approach a more general result is to attempt to follow the path set out by Riemenschneider, but including some appropriate generalization of the Kähler geometry and the associated harmonic integral representation of Dolbeault groups.

3. DIFFERENTIAL GEOMETRY ON MOIŠEZON SPACES

As discussed above, the most general Moišezon space X has a proper modification $\pi: \hat{X} \to X$ where $\hat{X}$ is a projective algebraic manifold. Let $\hat{S}$ be the exceptional subvariety in $\hat{X}$, so that if $S = \pi(\hat{S})$, we have $\pi\colon \hat{X}-\hat{S} \to X-S$ is biholomorphic and moreover, $\operatorname{codim}_{\mathbb{C}} S \geq 2$. Thus X is equipped with a positive line bundle $\hat{L}$, a high power of which gives a projective algebraic embedding of X. Let $\hat{h}$ be a Hermitian metric on $\hat{X}$ so that $c_1(\hat{L},\hat{h}) = \hat{\omega}$ is a positive Chern form for $\hat{L}$. One of the main objects in this section is to decide to what extent X inherits the structure of a positive line bundle.

We want to consider the natural images of $\hat{L}$ and $\hat{\omega}$ under the mapping π above. We define*

$$L = \pi_* \hat{L} \qquad \text{(direct image of sheaves)} \tag{4}$$

$$\omega = \pi_* \hat{\omega} \qquad \text{(push forward of currents)} . \tag{5}$$

Here the direct image sheaf L is the sheaf generated by the presheaf $U \to \Gamma(\pi^{-1}(U),\hat{L})$. In some cases the direct image sheaf may be locally free, which is equivalent to L being the sheaf of sections of a holomorphic line bundle. This is the case, for instance, if $L = K_X$, whereby $L = \mathcal{O}(K_X)$. This follows from the fact that the image S of the exceptional set $\hat{S}$ of points in $\hat{X}$ has complex codimension 2 in X. Moreover, a local section of $\pi_*(\mathcal{O}(K_{\hat{X}}))$ defined outside of S is a holomorphic n-form ($\pi\colon \hat{X} - \hat{S} \to X - S$ is biholomorphic). By the removable singularities theorem for holomorphic functions across subvarieties of codimension ≥ 2, we see that the local sections of $\pi_*(\mathcal{O}(K_{\hat{X}}))$ are precisely local holomorphic n-forms on X. In general, however, the direct image of a locally free sheaf is not locally free.

Letting $\mathcal{D}^{r,s}(\hat{X})$ be the smooth (r,s)-forms with compact support, the interpretation of (5) is given by letting $K^{p,q}(\hat{X})$ be the topological dual of $\mathcal{D}^{n-p,n-q}(\hat{X})$ (with the usual topology), where $\dim_{\mathbb{C}} X = \dim_{\mathbb{C}} \hat{X} = n$. Thus

$$\pi_*\colon K^{p,q}(\hat{X}) \to K^{p,q}(X)$$

is induced by duality from the pullback mapping π^* on smooth forms. Pushforward of currents is the natural generalization of images of chains and cycles under continuous maps (cf. de Rham [4]). Moreover, any smooth form can be considered as a current, and its pushforward as a current is well defined. It's clear that $\pi_*\hat{\omega}$ is smooth outside X, since

---

* $\hat{L} = \mathcal{O}(\hat{L})$ is the sheaf of germs of holomorphic sections of $\hat{L}$.

$$(\pi_*\hat{\omega})|_{X-S} = (\pi^{-1})^*(\hat{\omega}|_{\hat{X}-\hat{S}}),$$

because π is biholomorphic on $\hat{X} - \hat{S}$. Recall that currents of type (p,q) on X can also be considered as (p,q)-forms with distribution coefficients. It is easy to verify that $\pi_*\hat{\omega} = \omega$ has measure coefficients in this case. More, however, is true, and we formulate this in general.*

Proposition 3: *Let* $\phi \in E^{p,q}(\hat{X})$ *be a smooth differential form on* $\hat{X}$, *then* $\pi_*\phi \in E^{p,q}(X-S)$ *and the coefficients of* $\pi_*\phi$ *are (locally)* L^1 *functions on* X.

Proof: First, one checks easily that the coefficients of $\pi_*\phi$ are measures, i.e., the action of $\pi_*\phi$ on (compactly supported) smooth forms extends to (compactly supported) differential forms with continuous coefficients. The question of whether the coefficients of $\pi_*\phi|_{X-S}$ are integrable is equivalent to knowing whether the measure coefficients of $\pi_*\phi$ in X are absolutely continuous with respect to Lebesgue measure (in local coordinates). It follows from a theorem of Stein (cf. Remmert [14]) that there is a fundamental system of neighborhoods of $\hat{S}$ which map to a fundamental system of neighborhoods of S. Using these fundamental neighborhood systems and the monotone convergence theorem, one can deduce the absolute continuity of a local coefficient of $\pi_*\phi$. Q.E.D.

Let us give one simple example of the phenomenon of the mapping (5) giving rise to a differential form with L^1 coefficients on the singular set S. For simplicity we consider a local example. Let $\hat{\mathbb{C}}^2$ be the Hopf blow-up of the origin in $\mathbb{C}^2$. That is, $\hat{\mathbb{C}}^2$ is the subvariety of $\mathbb{C}^2 \times \mathbb{P}_1$ defined by the equation $z_1t_2 = z_2t_1$ where $(z_1,z_2) \in \mathbb{C}^2$, and $[t_1,t_2] \in \mathbb{P}_1$, are affine and homogeneous coordinates, respectively. The natural Kähler form on $\mathbb{C}^2 \times \mathbb{P}_1$ is given by

$$\hat{\omega} = i\partial\bar{\partial}(|z_1|^2 + |z_2|^2) + i\partial\bar{\partial} \log (|t_1|^2 + |t_2|^2)$$

which when restricted to $\hat{\mathbb{C}}^2$ is of the form (in the local coordinates in

* As was pointed out to me by Reese Harvey.

$\hat{\mathbb{C}}^2$ given by $t_1 \neq 0$)

$$\hat{\omega} = i\partial\bar{\partial}(|z_1|^2 + |z_2|^2) + i\partial\bar{\partial} \log (1 + |\frac{z_2}{z_1}|^2)$$

where we have set $t_2/t_1 = z_2/z_1$. Thus $\pi_*\hat{\omega} = \omega$ is given by

$$i\partial\bar{\partial}(|z_1|^2 + |z_2|^2) + i\partial\bar{\partial} \log (|z_1|^2 + |z_2|^2) - i\partial\bar{\partial} \log |z_1|^2.$$

But $\partial\bar{\partial} \log |z_1|^2 \equiv 0$, since $\log|z_1|^2$ is pluriharmonic. Thus we obtain that $\pi_*\hat{\omega}$ is the usual Euclidean Kähler form on $\mathbb{C}^2$ plus the term $i\partial\bar{\partial} \log |z|^2$ which is an L^1 function in $\mathbb{C}^2$ (but note that it is not in L^2).

Returning to the relations (4) and (5), we consider the special case where L is a locally free sheaf (e.g., if $L = K_{\hat{X}}$). Suppose that $L = \mathcal{O}(L)$, where L is a holomorphic line bundle on X. Then the current ω is a generalized Kähler form on X which is smooth outside of S, and has L^1 coefficients at all points of X. In some sense, the pair (L,ω) and more generally (L,ω) is the naturally occurring "positive" differential geometric object on a Moišezon space. There are numerous questions concerning L and ω. Two immediate ones are: What can one say about the behavior of the metric h defining the Chern form ω on the fibres of $L|_{X-S}$ as the fibres of L approach fibres of L over S? Also, is ω a current representative for the Chern class of L?

Another closely related pair of geometric objects on X is given by Riemenschneider [16], who considers an *ample* line bundle $\hat{L} \to \hat{X}$, inducing a surjection

$$\mathcal{O}^q_{\hat{X}} \to \hat{L} \to 0 \tag{6}$$

which induces the injection of line bundles

$$0 \to \hat{L}^* \to \hat{X} \times \mathbb{C}^q. \tag{7}$$

By the assumption of ampleness (Griffiths [7]) the trivial metric on $\hat{X} \times \mathbb{C}^q$ induces a *strictly negative* curvature on the line bundle $\hat{L}^*$. Then by taking the direct image of (6)

$$\pi_*\mathcal{O}^q_{\hat{X}} = \mathcal{O}^q_X \qquad \text{(since } S \text{ has } \operatorname{codim}_{\mathbb{C}} \geq 2).$$

We let

$$L = \operatorname{im}(\pi_* : \mathcal{O}^q_{\hat{X}} \to \pi_*\hat{L})$$
$$= \operatorname{im}(\mathcal{O}^q_X \to L),$$

which is a coherent sheaf of rank one on X. The associated linear space (cf. Grauert [5])

$$L(L) \longmapsto X \times \mathbb{C}^q$$

is embedded in $X \times \mathbb{C}^q$. The trivial metric on $X \times \mathbb{C}^q$ induces a metric h on the fibres of $L(L)$ and $L(L) \cong L^*$ off of S since π is biholomorphic there. Thus the Chern form of $L(L)$ computed outside of S agrees with $-\omega = -\pi_*\hat{\omega}$ (if we let $\hat{\omega}$ be the Chern form on L induced by (7) above). Riemenschneider calls such a sheaf L an *almost positive coherent sheaf of rank* 1. It has associated with it a positive Chern form defined outside of the singular set S. The triple $(L,h,-\omega|_{X-S})$ is then a globally defined metric h on L which has a negative Chern form outside of S. The main difference between this construction and the previous one is that $L(L)$ has a metric defined in each fibre (recall that the fibre dimension in $L(L)$ may increase at points off S). However this metric is used to compute the curvature (Chern form) only outside of S. In fact the form ω will coincide with the Chern form discussed above which was shown to have L^1 coefficients. Moreover, a high power m of a positive line bundle on X will be ample and the curvature ω will be changed only by multiplication by m, which doesn't affect the nature of the singularities on S.

The basic problem can thus be stated: given a rank 1 coherent sheaf L on a complex manifold X, and suppose that $L = \mathcal{O}(L)$, where L is a holomorphic line bundle on $X - S$, and moreover, suppose h is a metric on $L|_{X-S}$ and that $c_1(L,h) = \omega \in E^{1,1}(X-S)$ is a Kähler form on $X - S$ which has L^1 coefficients on X, and thus defines a global current on X. Then is X necessarily Moišezon?

This question was considered by Riemenschneider who conjectured in [16] and [17] that X is Moišezon under the stronger condition that the metric h be defined on *all* of L (in an appropriate sense). The conjecture was verified in the case that S consists of a discrete set.

To show that a complex manifold satisfying the above differential geometric condition is Moišezon would depend, in view of Proposition 1 and the discussion in §2, on proving that

$$H^1(X,\mathcal{O}(L \otimes K_X)) = 0, \tag{8}$$

under the given positivity hypothesis on L. To show (8) in the case where L has a metric defined in all of X with positive Chern form, the method initiated by Kodaira was to represent the Dolbeault group by harmonic forms (cf. [20], where this is discussed in some detail); for an alternative approach using pseudoconvexity in the total space of L see Grauert [5].

To use the approach of harmonic integrals, what is necessary is some sort of harmonic integral representation for the Dolbeault group $H^1(X, (L \otimes K_X))$ with respect to the metric h and the Chern form ω which both become singular in some prescribed manner along S. Given such a representation then one would like to investigate whether a version of Nakano's inequalities and the resulting vanishing theorem of Kodaira are valid.

There is some evidence that such a representation and vanishing theorem is possible. Namely, suppose we have the situation considered at the beginning of this section: $\pi:\hat{X} \to X$, where $\hat{X}$ and X are compact complex manifolds, π is a proper modification, and $\hat{X}$ is projective algebraic. Suppose that $\hat{L}$ is a positive line bundle in $\hat{X}$ with metric $\hat{h}$, and positive Chern form $\hat{\omega}$, and consider $\hat{\omega}$ as a Kähler metric on $\hat{X}$. Suppose again that $\pi_*(\mathcal{O}(\hat{L})) = \mathcal{L}$ is locally free and that $\mathcal{L} = \mathcal{O}(L)$, where L is a holomorphic line bundle on X. Let $\mathcal{H}^{p,q}(\hat{X},\hat{L})$ denote the harmonic (p,q) forms on $\hat{X}$ with coefficients in $\hat{L}$ with respect to the metrics $\hat{h}$ and $\hat{\omega}$ (cf. [20]). Letting $\mathcal{K}^{p,q}(X,L)$ denote the currents of type (p,q) on X with coefficients in L (cf. [21]), we see that there is a natural mapping

$$\pi_*: \mathcal{H}^{p,q}(\hat{X},\hat{L}) \to \mathcal{K}^{p,q}(X,L), \tag{9}$$

considering the harmonic forms $\mathcal{H}^{p,q}(\hat{X},\hat{L})$ as currents on $\hat{X}$. Let us denote the *image* of $\mathcal{H}^{p,q}(\hat{X},\hat{L})$ in (9) by $\mathcal{H}^{p,q}(X,L)$. Then we have the following diagram:

$$\begin{array}{ccc} H^q(\hat{X},\Omega^p(\hat{L})) & \xrightarrow{\ \hat{\alpha}\ } & \mathcal{H}^{p,q}(\hat{X},\hat{L}) \\ \uparrow \pi^* & & \downarrow \pi_* \\ H^q(X,\Omega^p(L)) & \xrightarrow{\ \alpha\ } & \mathcal{H}^{p,q}(X,L) \end{array} , \tag{10}$$

where π^* is the mapping on Dolbeault groups induced by the pullback of differential forms, $\hat{\alpha}$ is the *isomorphism* of Dolbeault groups to harmonic forms given by Hodge theory, and α is the induced mapping.

Theorem 4. *The mapping*

$$\alpha\colon H^q(X,\Omega^p(L) \to \mathcal{H}^{p,q}(X,L) \subset \mathcal{K}^{p,q}(X,L)$$

is an injection.

Proof: First, we note that if $\pi_*\phi = 0$ in $\mathcal{K}^{p,q}(X,L)$, then $\pi_*\phi$ vanishes as a current, and then as a smooth differential form identically outside of S. Then ϕ vanishes identically outside of $\hat{S}$, and since ϕ is smooth, we obtain $\phi \equiv 0$. Second, it is shown in [21] that a proper, surjective holomorphic mapping always induces an injective mapping on Dolbeault (and de Rham) groups. Thus, in particular, the proper modification π above induces an injection π^* in (10). Hence α is injective.

Q.E.D.

Remark: Simple examples show that π^* need not be an isomorphism, thus α need *not* be surjective. Then if we consider $\mathcal{H}^{p,q}(X,L)$ as being generalized harmonic forms with singularities along S which represent the Dolbeault group $H^q(X,\Omega^p(L))$, there are more harmonic forms in general than Dolbeault classes. This is due in this case to the singularities in the mapping π. Compare this situation with that encountered by Bloom and Herrera [2] who represented singular cohomology with de Rham groups on an analytic space, the naturally occurring de Rham groups being in general *larger* than the singular cohomology groups.

Corollary 5. $$H^1(X,\mathcal{O}(L \otimes K_X)) = 0.$$

Proof: By the positivity of $\hat{L}$, we obtain that $\mathcal{H}^{0,1}(\hat{X},\hat{L}) \cong \mathcal{H}^{0,1}(X,L) = 0$, and then by the injection α, we see that the corollary follows.

Q.E.D.

Therefore, the problem becomes: give an *intrinsic* characterization of $\mathcal{H}^{p,q}(X,L)$ in terms of L, ω and the associated differential operators of the standard Kähler theory. The "pushforward" of the standard Kähler operators from $\hat{X}$ to X have singularities along S of a yet-to-be-made-precise type. The "generalized harmonic forms" $\mathcal{H}^{p,q}(X,L)$ are smooth forms outside of S which are integrable across S, and so it may be possible to define these currents as the kernel of a Laplacian (with coefficients singular along S) acting on currents of type (p,q) with, say, L^1 coefficients.

This leads to the general (and specific) problem of developing an appropriate Hodge theory "with singularities". For the case considered above the candidate for the "harmonic forms" is there, and any good theory should be somehow consistent with the diagram (10). For a Hodge theory on an analytic space (with singularities) one could consider an analogous diagram, given by a resolution of singularities, but (10) should be a priori easier, since we still have a differentiable manifold structure on X to work with. It remains to find the right interpretation of the differential operators, appropriate Hilbert spaces, appropriate finiteness theorems, etc. The problem seems to be a difficult one, but worthy of pursuit.

REFERENCES

1. Artin, M., *Algebraization of formal moduli: II, Existence of modifications,* Ann. of Math. 91 (1970), 88-135.

2. Bloom, T. and M. Herrera, *De Rham cohomology of an analytic space,* Invent. Math. 7 (1969), 275-296.

3. Chow, W. L. and K. Kodaira, *On analytic surfaces with two independent meromorphic functions,* Proc. Nat. Acad. Sci. U.S.A., 38 (1952), 319-325.

4. de Rham, G., *Variétés Differentiables,* Hermann, Paris, 1955.

5. Grauert, H., *Über Modifikationen und exzeptionelle analytische Mengen,* Math. Ann., 146 (1962), 331-368.

6. __________, and O. Riemenschneider, *Verschwindungssätze für analytische Kohomologiegruppen auf komplexen Räumen,* Invent. Math. 11 (1970), 263-292.

7. Griffiths, P. A., *Hermitian differential geometry, Chern classes, and positive vector bundles,* in *Global Analysis,* Princeton Univ. Press, 1969, pp. 185-251.

8. Kodaira, K., *On a differential geometric method in the theory of analytic stacks,* Proc. Nat. Acad. Sci. U.S.A. 39 (1953), 1268-1273.

9. __________, *On Kähler varieties of restricted type,* Ann. of Math. 60 (1954), 28-48.

10. Moišezon, B. G., *On n-dimensional compact varieties with n algebraically independent meromorphic functions, I, II, III,* Amer. Math. Soc. Translat. 63(2), 1967, 51-177 (Izvest. Akad. SSSR, Ser. Mat. 30, 133-174; 345-386; 621-656 (1966)).

11. __________, *Resolution theorems for compact complex spaces with a sufficiently large field of meromorphic functions*, Math. USSR - Izvestia 1 (1967), 1331-1356 (Izvest. Akad. Nauk. SSSR, Ser. Mat. 31 (1967), 1385-1414).

12. Nagata, M., *Existence theorems for nonprojective complete algebraic varieties*, Ill. J. Math. 2 (1958), 490-498.

13. Remmert, R., *Holomorphe and meromorphe Abbildungen komplexer Räume*, Math. Ann. 133 (1957), 328-370.

14. __________, *Analytic and algebraic dependence of meromorphic functions*, Amer. J. Math., 82 (1960), 891-899.

15. Riemenschneider, O., *Characterizing Moišezon spaces by almost coherent analytic sheaves*, Math. Zeit., 123 (1971), 263-284.

16. __________, *A generalization of Kodaira's embedding theorem*, Math. Ann. 200 (1973), 99-102.

17. Siegel, C. L., *Analytic Functions of Several Complex Variables*, Inst. for Advanced Study, 1962.

18. __________, *Meromorphic Funktionen auf kompakten analytischen Mannigfaltigkeiten*, Nachr. Akad. Wiss., Göttingen, 1955, 71-77.

19. Thimm, W., *Über die algebraischen Relationen zwischen meromorphen Funktionen in abgeschlossenen Raümen*, Dissertation, Königsberg, 1939.

20. Wells, R. O. Jr., *Differential Analysis on Complex Manifolds*, Prentice-Hall, Inc., Englewood Cliffs, N. J., 1973.

21. __________, *Comparison of Dolbeault and de Rham groups for a proper surjective mapping*, (to appear).

22. Whitney, Hassler, *Complex Analytic Varieties*, Addison-Wesley, Reading, Mass., 1973.

POSITIVE (p,p) FORMS, WIRTINGER'S INEQUALITY, AND CURRENTS

Reese Harvey and A. W. Knapp*

Rice University and Cornell University

The purpose of this paper is to discuss positivity of (p,p) forms, to generalize Wirtinger's Inequality [2], and to indicate a relationship among the generalization, the results of [7], [4], and [5], and the Hodge Conjecture. In the first section three notions of positivity (weak, regular, and strong) for (p,p) forms over a complex vector space are examined, and a canonical form for positive (p,p) forms is described. The canonical form implies one of the two generalizations of Wirtinger's Inequality that we obtain, and it allows us to answer negatively two questions of Lelong [10] concerning weak positivity. In the second section a conjecture (I) concerning positive Plateau problems on a complex projective manifold is stated. This conjecture is equivalent to another conjecture (II) proposing a sufficient condition for a cohomology class to be determined by a rational positive analytic cycle. The conjecture (II) easily implies the Hodge Conjecture (III).

It is our pleasure to thank Bob Kujala, John Polking, Bernie Shiffman, and Al Vitter for helpful conversations, and David Mumford for providing us with an example of a strongly positive integral class which does not contain a positive analytic cycle.

1. POSITIVE (p,p) FORMS

Suppose E is a complex vector space of complex dimension n with i

* Research of the first author partially supported by a Sloan Fellowship and NSF GP-19011; research of the second author partially supported by NSF GP-28251.

the operation of multiplication by $\sqrt{-1}$. See Weil [13] for a more complete discussion of the complex exterior algebra described below. Let $\Lambda^r_{\mathbb{R}}E$ denote the space of exterior r-vectors over E considered as a real vector space of real dimension 2n, and let $\Lambda^r_{\mathbb{C}}E$ denote the complexified space of exterior r-vectors over E or, equivalently, the space of exterior r-vectors over $E_{\mathbb{C}}$, the complexification of E. The notion of conjugation on $E_{\mathbb{C}}$ extends to $\Lambda^r_{\mathbb{C}}E$ and will be denoted by bar. The operation i extends to $\Lambda^1_{\mathbb{C}}E$ with eigenvalues $\pm i$. Let $\Lambda^{1,0}E$ denote the eigenspace corresponding to i and $\Lambda^{0,1}E$ the eigenspace corresponding to $-i$. Then $\Lambda^1_{\mathbb{C}}E = \Lambda^{1,0}E \oplus \Lambda^{0,1}E$. This decomposition induces a natural direct sum decomposition $\Lambda^r E = \Sigma\, \Lambda^{p,q}E$, where $\Lambda^{p,q}$ denotes the space of r-vectors of type (p,q) and the summation extends over all (p,q) where $p + q = r$. The map of E into $E_{\mathbb{C}}$ followed by projection on $\Lambda^{1,0}E$ is a natural complex isomorphism.

Let E' denote the space of real-valued linear functionals on E considered as a real vector space, and let E* denote the space of complex linear functionals on the complex space E. The operation on E' dual to the operation i on E will also be denoted i. Note that $E^* = \Lambda^{1,0}E'$. It is customary to identify E' with $\Lambda^{1,0}E'$ by taking the projection on $\Lambda^{1,0}E'$ of *twice* the natural injection of E' into its complexification $E'_{\mathbb{C}}$, since this identification agrees with the usual isomorphism $E' \cong E^*$ defined by sending F(x) into $F(x) - iF(ix)$. The natural isomorphism $\Lambda^{p,p}_{\mathbb{R}}E' = (\Lambda^{p,p}_{\mathbb{R}}E)'$ will be used below.

Let I be a multi-index, say $I = (i_1,\ldots,i_p)$. If $\{e_1,\ldots\}$ is a set in $E^* = \Lambda^{1,0}E'$, let $e^I = e_{i_1} \wedge \ldots \wedge e_{i_p}$. If $\{e_1,\ldots,e_n\}$ is a basis for $E^* = \Lambda^{1,0}E'$, then

$$\{e^I \wedge \bar{e}^J:\ I \text{ and } J \text{ strictly increasing},\ |I| = p,\ |J| = q\}$$

is a basis for $\Lambda^{p,q}E'$. Therefore each A in $\Lambda^{p,q}E'$ may be uniquely expressed as $\Sigma'\, a_{IJ} e^I \wedge \bar{e}^J$ with the coefficients in $\mathbb{C}$, with Σ' denoting summation over strictly increasing multi-indices, and with $|I| = p$ and $|J| = q$. In such a basis, $\bar{A} = \Sigma'\, \bar{a}_{IJ} \bar{e}^I \wedge e^J$.

Let $\sigma_k = 2^{-k}$ if k is even and $\sigma_k = i2^{-k}$ if k is odd. For For $e_1,\ldots,e_k$ in $E^* = \Lambda^{1,0}E'$ and $I = (1,\ldots,k)$, we have

$$\frac{i}{2} e_1 \wedge \bar{e}_1 \wedge \frac{i}{2} e_2 \wedge \bar{e}_2 \wedge \ldots \wedge \frac{i}{2} e_k \wedge \bar{e}_k = \sigma_k e^I \wedge \bar{e}^I.$$

If ξ and η are in $\Lambda^{k,0}E$, then the conjugate of $\sigma_k \xi \wedge \bar{\eta}$ is

$\sigma_k \eta \wedge \overline{\xi}$.

A form A in $\Lambda^r_{\mathbb{C}} E'$ is said to be *real* if $A = \overline{A}$ or equivalently if A is in $\Lambda^r_{\mathbb{R}} E'$. Let Herm E be the real vector space of complex-valued functions $H(u,v)$ on $E \times E$ that are i-linear in u and satisfy $\overline{H(u,v)} = H(v,u)$. Passage from H to minus its imaginary part gives us a real-linear isomorphism

$$\text{Herm } E \to \Lambda^{1,1}_{\mathbb{R}} E', \tag{1.1}$$

expressed in terms of a basis $\{e_1,\ldots,e_n\}$ for $E^* = \Lambda^{1,0}E'$ as $\Sigma\, a_{ij} e_i \otimes \overline{e}_j \to \frac{i}{2} \Sigma\, a_{ij} e_i \wedge \overline{e}_j$ with the matrix (a_{ij}) Hermitian. In addition, we have a real-linear isomorphism

$$\Lambda^{p,p}_{\mathbb{R}} E' \to \text{Herm}(\Lambda^{p,0}E) \tag{1.2}$$

given by sending A into the Hermitian form H with $H(\xi,\eta) = A(\sigma_p^{-1} \xi \wedge \overline{\eta})$. This map is given in terms of a basis $\{e_1,\ldots,e_n\}$ for $E^* = \Lambda^{1,0}E'$ and its dual basis $\{g_1,\ldots,g_n\}$ for $\Lambda^{1,0}E$ by sending an element $A = \Sigma'\, a_{IJ} \sigma_p e^I \wedge \overline{e}^J$ of $\Lambda^{p,p}E'$ into $H \in \text{Herm}(\Lambda^{p,0}E)$ defined by $H(\Sigma'\, \zeta_I g^I, \Sigma'\, \eta_J g^J) = \Sigma'\, a_{IJ} \zeta_I \overline{\eta}_J$. Combining (1.2) and (1.1), we obtain a real-linear isomorphism

$$\Lambda^{p,p}_{\mathbb{R}} E' \to \Lambda^{1,1}_{\mathbb{R}} (\Lambda^{p,0}E)'. \tag{1.3}$$

An exterior $(p,0)$ form that can be expressed as e^I with $e_{i_1},\ldots,e_{i_p}$ in $E^* = \Lambda^{1,0}E'$ is called *decomposable*. An exterior (p,p) form will be called *elementary* if it can be expressed as $c\zeta \wedge \overline{\zeta}$ with c in $\mathbb{C}$ and ζ in $\Lambda^{p,0}E'$, and it will be called *decomposable* if in addition ζ can be chosen decomposable.

A form $A \in \Lambda^{2n}E' = \Lambda^{n,n}E'$ is said to be a *positive volume form* if $A = cie_1 \wedge \overline{e}_1 \wedge \ldots \wedge ie_n \wedge \overline{e}_n$, with $c \geq 0$. This notion of positive volume form is of course independent of the choice of basis $\{e_1,\ldots,e_n\}$.

Definition 1.1: A form $A \in \Lambda^{p,p}E'$ is *weakly positive* if

$$\begin{aligned} &A \wedge \sigma_k e^I \wedge \overline{e}^I \text{ is a positive volume form for all} \\ &e_1,\ldots,e_k \in \Lambda^{1,0}E' = E^* \text{ with } I = (1,\ldots,k) \text{ and} \\ &p + k = n. \end{aligned} \tag{1.4}$$

A form $A \in \Lambda^{p,p}E'$ is *positive* if

$A \wedge \sigma_k \zeta \wedge \bar{\zeta}$ is a positive volume form for all $\zeta \in \Lambda^{k,0}E'$ with $p + k = n$. (1.5)

A form $A \in \Lambda^{p,p}E'$ is *strongly positive* if

A can be expressed as $\Sigma\, \sigma_p \zeta_j \wedge \bar{\zeta}_j$ with each $\zeta_j \in \Lambda^{p,0}E'$ decomposable. (1.6)

Let WP^p, P^p and SP^p denote the cones in $\Lambda^{p,p}_{\mathbb{R}}E'$ defined by (1.4), (1.5), and (1.6) respectively. If ζ is any element of $\Lambda^{p,0}E'$, then $\sigma_p \zeta \wedge \bar{\zeta}$ is in P^p, since $\sigma_p \zeta \wedge \bar{\zeta} \wedge \sigma_k \eta \wedge \bar{\eta}$ can be checked to be equal to the positive volume form $\sigma_n \zeta \wedge \eta \wedge \bar{\zeta} \wedge \bar{\eta}$ for an arbitrary η in $\Lambda^{k,0}E'$. Therefore $SP^p \subset P^p \subset WP^p$. If $p = 1$ or $n - 1$, then $P^p = WP^p$ since every form in $\Lambda^{1,0}E'$ or $\Lambda^{n-1,0}E'$ is decomposable. An immediate consequence of Theorem 1.2 below (for $p = 1$ and $p = n - 1$) is that $SP^p = P^p = WP^p$ for $p = 1$ or $n - 1$.

The condition (1.4) may be equivalently stated as follows: If $L\colon E_1 \to E_2$ is a complex linear map let $L^*\colon \Lambda^{p,q}E_2' \to \Lambda^{p,q}E_1'$ denote the usual pull-back of (p,q) forms from E_2 to E_1.

L^*A is a positive volume form for all complex linear maps $L\colon F \to E$ with $\dim_{\mathbb{C}} F = p$. (1.4')

To check that (1.4) and (1.4)' are equivalent consider the case where F is a subspace of E and L is inclusion. Let $\{e_1, \ldots, e_n\}$ be a basis for $\Lambda^{1,0}E' = E^*$ with $\{e_1, \ldots, e_k\}$ vanishing on F. Then $L^*A \wedge \sigma_k e^I \wedge \bar{e}^I = A \wedge \sigma_k e^I \wedge \bar{e}^I$ where $I = (1, \ldots, k)$; and L^*A is a positive volume form if and only if $L^*A \wedge \sigma_k e^I \wedge \bar{e}^I$ is a positive volume form.

A weakly positive (p,p) form A is said to be *non-degenerate* or *definite* if $A \wedge \sigma_k e^I \wedge \bar{e}^I$ in (1.4) is greater than zero for all $\{e_1, \ldots, e_k\}$ linearly independent. Similarly a positive (p,p) form A is said to be *non-degenerate* or *definite* if $A \wedge \sigma_k \zeta \wedge \bar{\zeta}$ in (1.5) is greater than zero for all $\zeta \in \Lambda^{k,0}E'$, $\zeta \neq 0$. Note that the set of (weakly) positive (p,p) forms that are non-degenerate is the interior of (WP^p) P^p.

Suppose E is furnished with a Hermitian inner product. In terms of the extension of this inner product to $\Lambda^r_{\mathbb{C}}E'$, we obtain a canonical form for real (p,p) forms and for positive (p,p) forms. This theorem is well known for $p = 1$ but we include the proof for the sake of completeness. One can easily show that for ζ in $\Lambda^{p,0}E'$, $|\zeta|^2 = 2^p$ if and only if $|\sigma_p \zeta \wedge \overline{\zeta}| = 1$.

Theorem 1.2: *Let* $N = \binom{n}{p}$. *If* A *is in* $\Lambda_{\mathbb{R}}^{p,p}E'$, *there exist real numbers* $\{\lambda_1,\dots,\lambda_N\}$ *and a set* $\{\zeta_1,\dots,\zeta_N\}$ *of mutually orthogonal vectors with* $|\zeta_j|^2 = 2^p$ *in* $\Lambda^{p,0}E'$ *such that* $A = \sum_{j=1}^{N} \lambda_j \sigma_p \zeta_j \wedge \overline{\zeta}_j$. *The numbers* λ_j *and their multiplicities are unique, and the subspaces spanned by* $\{\zeta_j : \lambda_j = \lambda\}$ *are unique. In addition,* A *is positive if and only if each* λ_j *is* ≥ 0.

Proof (for $p = 1$): Choose a basis $\{e_1,\dots,e_n\}$ of $E^* = \Lambda^{1,0}E'$ with $\langle e_i, e_j \rangle = 0$ for $i \neq j$ and $|e_i|^2 = 2$. Then $|\frac{i}{2} e_i \wedge \overline{e}_j| = 1$. Write $A = \frac{i}{2} \Sigma\, a_{ij} e_i \wedge \overline{e}_j$ with (a_{ij}) Hermitian. Diagonalize (a_{ij}) by a unitary matrix and change the e's accordingly. The existence of the decomposition follows. Any different choice of orthogonal basis $\{e_i\}$ with $|e_i|^2 = 2$ (in particular, a second choice for the ζ_j) leads to a matrix (b_{ij}) that is conjugate to (a_{ij}) by a unitary matrix. Consequently the uniqueness follows from standard linear algebra. From (1.4') A is weakly positive if and only if (a_{ij}) is positive semidefinite and hence if and only if all λ_j are ≥ 0. Q.E.D.

Remark: Since $\frac{i}{2} \zeta \wedge \overline{\zeta}$ with $\zeta \in \Lambda^{1,0}E'$ is positive, this argument proves that weakly positive implies positive for $p = 1$ (and for $p = n - 1$ by duality).

Proof (for general p): Let $L: \Lambda_{\mathbb{R}}^{p,p}E' \to \Lambda_{\mathbb{R}}^{1,1}(\Lambda^{p,0}E')$ be the isomorphism of (1.3). If A is in $\Lambda_{\mathbb{R}}^{p,p}E'$, A can be expanded in terms of a basis as $\sigma_p \Sigma'\, a_{IJ} e^I \wedge \overline{e}^J$ with (a_{IJ}) Hermitian. Then $L(A)$ is positive (weakly positive = positive for $L(A)$) if and only if (a_{IJ}) is positive semidefinite, and (a_{IJ}) is positive semidefinite if and only if A is positive. Hence the result for general p follows from the case $p = 1$.

Q.E.D.

Remark: The notion of positivity does not depend upon the Hermitian inner product on E, and hence the theorem shows that all the λ_j are ≥ 0

for one inner product if and only if they are ≥ 0 for any other inner product. More generally the numbers of λ_j that are respectively > 0, < 0, and $= 0$ are independent of the inner product.

The Hermitian inner product on E induces an operator on $\Lambda_{\mathbb{C}}E'$ usually denoted $*$. If K is a cone in $\Lambda_{\mathbb{R}}^{p,p}E'$, the dual cone K^0 is by definition the set of all $B \in \Lambda_{\mathbb{R}}^{k,k}E'$ $(p + k = n)$ such that $A \wedge B$ is a positive volume form for all $A \in K$.

Corollary 1.3:

(a) *If* A *and* B *are positive then* $A \wedge B$ *is positive.*

(b) *If* A *is positive then* $*A$ *is positive.*

(c) *If* A *is positive and* $L: F \to E$ *then* L^*A *is positive.*

(d) *The dual cone* $(P^p)^0$ *equals* P^k *with* $p + k = n$.

Proof: (a) By Theorem 1.2 it suffices to check the result for $A = \sigma_p \zeta \wedge \bar{\zeta}$ and $B = \sigma_q \eta \wedge \bar{\eta}$. Then $A \wedge B = \sigma_p \zeta \wedge \bar{\zeta} \wedge \sigma_q \eta \wedge \bar{\eta} = \sigma_{p+q} \zeta \wedge \eta \wedge \bar{\zeta} \wedge \bar{\eta}$, which is a positive volume form. Similarly (b) and (c) need only be checked for $A = \sigma_p \zeta \wedge \bar{\zeta}$ with $\zeta \in \Lambda^{p,0}E'$ because of Theorem 1.2; hence (b) and (c) follow. Part (d) is an immediate consequence of Theorem 1.2. Q.E.D.

Corollary 1.4: *The set of extreme rays in the cone of positive* (p,p) *forms is the set of rays determined by positive elementary* (p,p) *forms* (i.e., $\{\sigma_p \zeta \wedge \bar{\zeta}: \zeta \in \Lambda^{p,0}E'\}$).

Proof: First consider the case $p = 1$. Suppose $A = \sum_{j=1}^{n} \lambda_j \frac{i}{2} e_j \wedge \bar{e}_j$ determines an extreme ray in $P^1 \subset \Lambda_{\mathbb{R}}^{1,1}E'$. Then obviously all λ_j must vanish but one. Hence each extreme ray is determined by a decomposable vector $\frac{i}{2} e \wedge \bar{e}$ for some e in $E^* = \Lambda^{1,0}E'$. Conversely, each decomposable vector $\frac{i}{2} e \wedge \bar{e}$ determines an extreme ray in P^1. Suppose $\frac{i}{2} e \wedge \bar{e}$ is expressed as the sum of two positive elements $\alpha, \beta \in P^1$. Then α and β must be dependent since otherwise the rank of $\frac{i}{2} e \wedge \bar{e} = \alpha + \beta$ would be at least 2. This proves Corollary 1.4 for $p = 1$. The case $p \geq 1$ easily follows from the case $p = 1$ by using the isomorphism (1.3). Q.E.D.

In fact the cone P of positive (p,p) forms is nicely stratified with strata S_j equal to the set of A in P with exactly j positive eigenvalues. The lowest stratum S_1, which is the set of extreme rays, is $G(k,n,\mathbb{C}) \times \mathbb{R}^+$, where $G(k,n,\mathbb{C})$ denotes the Grassmannian of complex k-planes in complex n-space.

Our notion of weak positivity coincides with the notion of positivity occurring in Lelong [10]. Lelong ([10], p. 60) asked two questions, which translate into (A) and (B) below:

(A) *Is every weakly positive* (p,p) *form strongly positive?*

(B) *If* A *is in* $\Lambda^{p,p}_{\mathbb{R}}E'$ *and* B *is in* $\Lambda^{q,q}_{\mathbb{R}}E'$ *and both are weakly positive, is* $A \wedge B$ *weakly positive?*

Obviously (A) implies (B). In addition (B) for $q = k = n - p$ implies (A) as follows. Since SP^p is the cone on the convex hull of the compact set $G(k,n,\mathbb{C})$ in $\Lambda^{p,p}_{\mathbb{R}}E'$, it is a closed cone. By definition $(SP^p)^0 = WP^k$ and hence $SP^p = (WP^k)^0$. That is,

$$\text{the cones } SP^p \text{ and } WP^k \text{ are dual.} \tag{1.7}$$

If (B) is true for $q = k = n - p$ then $WP^p \subset (WP^k)^0$. Since $(WP^k)^0 = SP^p$ this proves that (B) implies (A). Now there is a variety of ways of showing (A) and (B) are false for $2 \le p \le n - 2$.

For example if (A) were true then certainly the notions of positive and strongly positive would have to agree. However the next result shows that if $\zeta \in \Lambda^{p,0}E'$ is not decomposable, then although $\sigma_p \zeta \wedge \bar{\zeta}$ is positive it is not strongly positive. (If $2 \le p \le n - 2$, then there exists $\zeta \in \Lambda^{p,0}E'$ not decomposable.)

Proposition 1.5: *An elementary vector* $A = \sigma_p \zeta \wedge \bar{\zeta}$ *with* $\zeta \in \Lambda^{p,0}E'$ *is strongly positive if and only if* A *is decomposable (i.e.,* ζ *is a decomposable* (p,0) *form).*

Proof: If A is strongly positive, then by Theorem 1.2 $A = \Sigma\, \sigma_p \zeta_j \wedge \bar{\zeta}_j$ with each ζ_j a decomposable (p,0) form. Since $A = \sigma_p \zeta \wedge \bar{\zeta}$ determines an extreme ray in $P^p \subset \Lambda^{p,p}_{\mathbb{R}}E$ by Corollary 1.4, it must be a positive multiple of $\sigma_p \zeta_j \wedge \bar{\zeta}_j$ for some j. This implies ζ is a multiple of ζ_j and hence decomposable. Q.E.D.

In particular, if $\{e_1, e_2, e_3, e_4\}$ is a basis for $\mathbb{C}^4$ then $A = (e_1 \wedge e_2 + e_3 \wedge e_4) \wedge (\bar{e}_1 \wedge \bar{e}_2 + \bar{e}_3 \wedge \bar{e}_4)$ is a positive (2,2) form on $\mathbb{C}^4$ which cannot be expressed as the sum of positive decomposable (2,2) forms, since $e_1 \wedge e_2 + e_3 \wedge e_4$ is not a decomposable (2,0) form.

Corollary 1.6: *For* $2 \le p \le n-2$ *the inclusions* $SP^p \subset P^p \subset WP^p$ *are proper.*

Proof: $SP^p \ne P^p$ from above, and $P^p \ne WP^p$ by duality since $P^k \ne SP^k$ with $p + k = n$. Q.E.D.

To explicitly construct a weakly positive (p,p) form which is not positive, suppose $\zeta_1^* \in \Lambda^{k,0}E'$ is not decomposable and $|\sigma_k \zeta_1^* \wedge \bar{\zeta}_1^*| = 1$. Extend ζ_1^* to an orthogonal basis $\{\zeta_1^*,\ldots,\zeta_N^*\}$ for $\Lambda^{k,0}E'$. Let $\{\zeta_1,\ldots,\zeta_N\}$ denote the dual basis for $\Lambda^{p,0}E'$, and set $A_1 = \sum_{j=2}^{N} \lambda_j \sigma_p \zeta_j \wedge \bar{\zeta}_j$ with each $\lambda_j > 0$. Let $A = A_1 - \varepsilon\sigma_p \zeta_1 \wedge \bar{\zeta}_1$ with $\varepsilon > 0$ to be chosen. $A_1 \wedge \sigma_k \eta \wedge \bar{\eta}$ is ≥ 0 for all $\eta \in \Lambda^{k,0}E'$ and equal to zero if and only if $\eta = c\zeta_1^*$. Since ζ_1^* is not decomposable, $A_1 \wedge \sigma_k \eta \wedge \bar{\eta}$ is > 0 for each $\eta \in \Lambda^{k,0}E'$ with η decomposable and nonzero. Choose $\varepsilon > 0$ strictly less than the minimum of $A_1 \wedge \sigma_k \eta \wedge \bar{\eta}$ over all decomposable η with $|\sigma_k \eta \wedge \bar{\eta}| = 1$. Then A is weakly positive (in fact non-degenerate) since

$$A_1 \wedge \sigma_k \eta \wedge \bar{\eta} - \varepsilon\sigma_n(\zeta_1 \wedge \eta) \wedge (\bar{\zeta}_1 \wedge \bar{\eta}) > 0$$

for all decomposable nonzero η. Since $A \wedge \sigma_k \zeta_1^* \wedge \bar{\zeta}_1^*$ equals $-\varepsilon$ times a positive volume form, A cannot be strongly positive. The above is just a matter of finding a hyperplane (linear functional) through zero in $\Lambda^{k,k}_{\mathbb{R}}E'$ separating $\sigma_k \zeta_1^* \wedge \bar{\zeta}_1^*$ and SP^k. Also note that although A is in WP^p and $B = \sigma_k \zeta_1^* \wedge \bar{\zeta}_1^*$ is in SP^k (with $k + p = n$), the product $A \wedge B$ is a negative volume form (cf. (B) and Corollary 1.3 (a); of course $A, B \in SP$ imply $A \wedge B \in SP$).

If $A = \sigma_p \zeta \wedge \bar{\zeta}$ with ζ decomposable, say $\zeta = e_1 \wedge \ldots \wedge e_p$ with $\{e_1,\ldots,e_n\}$ an orthogonal basis for $E^* = \Lambda^{1,0}E'$ and $|e_j|^2 = 2$, then $*A = \sigma_k \eta \wedge \bar{\eta}$ where $\eta = e_{p+1} \wedge \ldots \wedge e_n$. Therefore:

$*$ is an isomorphism of SP^p onto SP^k and hence by duality an isomorphism of WP^p onto WP^k (where $p + k = n$). (1.8)

Also by (1.4') and an elementary calculation

part (c) of Corollary 1.3 remains valid with P replaced by WP or SP. (1.9)

The statements (1.7), (1.8) and (1.9) above provide the analogue of Corollary 1.3 for WP and SP.

Next we generalize Wirtinger's Inequality. We continue to assume E has a Hermitian inner product, and we consider the induced $\star$ operator mapping $\Lambda^{p,q}E'$ isometrically onto $\Lambda^{n-p,n-q}E'$. If A and B are in $\Lambda^r_{\mathbb{C}}E'$, then $(A,B) = \star(A \wedge \overline{\star B})$ defines the Hermitian inner product on $\Lambda^r_{\mathbb{C}}E'$. Let $|\cdot|_2$ denote the norm associated with this inner product. Let $\omega \in \Lambda^{1,1}_{\mathbb{R}}E'$ denote the image of the given element of Herm(E) under the map defined by (1.1). This positive (1,1) form ω is called the standard Kähler form on the Hermitian space E. In terms of an orthogonal basis $\{e_1,\ldots,e_n\}$ for E* with $|e_j|_2^2 = 2$, one has $\omega = \frac{i}{2}\sum_{j=1}^{n} e_j \wedge \overline{e}_j$. Let ω^k denote $\omega\wedge\ldots\wedge\omega$ taken k times.

The classical Wirtinger Inequality says that for a decomposable form $A \in \Lambda^{2p}_{\mathbb{R}}E'$ and $p + k = n$

$\star(A \wedge \frac{1}{k!}\omega^k) \le |A|_2$ with equality if and only if A is a positive decomposable (p,p) form. (1.10)

See Federer [2], p. 40, for a nice proof. Before generalizing (1.10) to forms A that are not necessarily decomposable, we must discuss some other norms on $\Lambda_{\mathbb{C}}E'$ that agree with $|\cdot|_2$ on decomposable 2p-forms.

Definition 1.7: (1) Let $|\cdot|_1$ (called the *mass norm*) denote the norm on $\Lambda^{2p}_{\mathbb{R}}E'$ whose closed unit ball is the convex hull of $\mathcal{D} = \{A \in \Lambda^{2p}_{\mathbb{R}}E' : A \text{ is decomposable and } |A|_2 = 1\}$.

(2) Let $\|\cdot\|_1$ denote the norm on $\Lambda^{2p}_{\mathbb{R}}E'$ whose closed unit ball is the convex hull of $\mathcal{D} \cup \mathcal{E}$ where

$$\mathcal{E} = \{A \in \Lambda^{p,p}_{\mathbb{R}}E' : A = \pm\sigma_p \zeta \wedge \overline{\zeta} \text{ with } \zeta \in \Lambda^{p,0}E' \text{ and } |A|_2 = 1\}.$$

Note that $\mathcal{D} \subset \mathcal{D}\cup\mathcal{E} \subset \{A: |A|_2 = 1\}$ implies that $|A|_2 \le \|A\|_1 \le |A|_1$ for all $A \in \Lambda^{2p}_{\mathbb{R}}E'$.

Theorem 1.8 (Generalized Wirtinger Inequality): *Suppose* A *is in* $\Lambda^{2p}_{\mathbb{R}}E'$

(and $p + k = n$). *Then*

(a) $*(A \wedge \frac{1}{k!}\omega^k) \le |A|_1$ *with equality if and only if* $A \in SP^p$.

(b) $*(A \wedge \frac{1}{k!}\omega^k) \le \|A\|_1$ *with equality if and only if* $A \in P^p$.

Of course the inequality in (b) is stronger than the inequality in (a) since $\|A\|_1 \le |A|_1$. Before proving Theorem 1.8 we give some alternate descriptions of the norms $|\cdot|_1$ and $\|\cdot\|_1$.

If A is in $\mathcal{D}$, then A belongs to the convex hull of $\mathcal{D}$ so that $|A|_1 \le 1$. Also $1 = |A|_2 \le |A|_1$ for $A \in \mathcal{D}$. That is,

$$|A|_1 = |A|_2 \quad \text{for} \quad A \in \mathcal{D}. \tag{1.11}$$

(In fact $|A|_1 = |A|_2$ if and only if a multiple of A belongs to $\mathcal{D}$, since every point of $\{A: |A|_2 = 1\}$ is extreme.) Since for arbitrary A, $A/|A|_1$ belongs to the convex hull of $\mathcal{D}$, there exist $0 \le t_j \le 1$ with $\Sigma_1^N t_j \le 1$ and $B_j \in \mathcal{D}$ such that $A/|A|_1 = \Sigma\, t_j B_j$. Let $A_j = t_j |A|_1 B_j$, so that $A = \Sigma\, A_j$. Note that $|A_j|_1 = t_j|A|_1$ since, by (1.11), $|B_j|_1 = 1$. Therefore $\Sigma\, |A_j|_1 \le |A|_1$ and $|A|_1 = \Sigma\, |A_j|_1$. This proves

(1') $|A|_1$ equals the infimum of $\Sigma\, |A_j|_2$ taken over all collections A_j with a multiple of each A_j belonging to $\mathcal{D}$ and $A = \Sigma\, A_j$.

Similarly, one can show that a multiple of A belongs to $\mathcal{D} \cup \mathcal{E}$ if and only if $\|A\|_1 = |A|_2$ and that

(2') $\|A\|_1$ equals the infimum of $\Sigma\, |A_j|_2$ taken over all collections A_j with a multiple of each A_j belonging to $\mathcal{D} \cup \mathcal{E}$ and $A = \Sigma\, A_j$.

Let $|B|_\infty = \sup\{|A \wedge B|_2 : A \in \mathcal{D}\}$. This defines a norm (the *comass norm*) on $\Lambda^{2k}_{\mathbb{R}} E'$ $(p + k = n)$ dual to $|\cdot|_1$.

(1") $|A|_1 = \sup |A \wedge B|_2$ taken over all $B \in \Lambda^{2k}_{\mathbb{R}} E'$ with $|B|_\infty \le 1$.

Similarly $\|B\|_\infty = \sup\{|A \wedge B|_2 : A \in \mathcal{D} \cup \mathcal{E}\}$ defines a norm on $\Lambda^{2k}_{\mathbb{R}} E'$ dual to $\|\cdot\|_1$.

(2") $\|A\|_1 = \sup |A \wedge B|_2$ taken over all $B \in \Lambda^{2k}_{\mathbb{R}} E'$ with $\|B\|_\infty \le 1$.

Proof of the Generalized Wirtinger Inequality:

(a) Choose A_j with $A_j/|A_j|_1 \in \mathcal{D}$ and $A = \Sigma A_j$ so that $|A|_1 = \Sigma |A_j|_2$. By (1.10)

$$*(A \wedge \frac{1}{k!}\omega^k) = \Sigma *(A_j \wedge \frac{1}{k!}\omega^k) \leq \Sigma |A_j|_1 = |A|_1.$$

This proves the inequality part of (a). Equality holds if and only if $*(A_j \wedge \frac{1}{k!}\omega^k) = |A_j|_1$ for each j. By the equality part of (1.10) if these equalities hold for all j, then each A_j is a positive decomposable (p,p) form and hence $A = \Sigma A_j$ is in SP^p. Conversely, if $A = \Sigma A_j$ is in SP with each A_j positive and decomposable, then equality in (1.10) for each A_j implies that $*(A \wedge \frac{1}{k!}\omega^k) = \Sigma |A_j|_1 \geq |A|_1$, which proves equality in (a).

(b) Suppose $A = \Sigma' a_{IJ}\sigma_p e^I \wedge \bar{e}^J$ belongs to $\Lambda^{p,p}_{\mathbb{R}}E'$ where $\{e_1,\ldots,e_n\}$ is an orthogonal basis for $E^* = \Lambda^{1,0}E'$ with $|e_j|_2^2 = 2$. Since $\frac{1}{k!}\omega^k = \sum'_{|J|=k} \sigma_k e^J \wedge \bar{e}^J$, $*(A \wedge \frac{1}{k!}\omega^k)$ equals $\sum'_{|I|=p} a_{II}$, the trace of (a_{IJ}). Let $\{\lambda_1,\ldots,\lambda_N\}$ denote the eigenvalues and $\{\zeta_1,\ldots,\zeta_N\}$ the eigenvectors of A given by Theorem 1.2. Then, since the trace of (a_{IJ}) equals $\Sigma \lambda_j$,

$$*(A \wedge \frac{1}{k!}\omega^k) = \sum_{j=1}^{N} \lambda_j . \tag{1.12}$$

Just as in (1.11), for $A \in \Lambda^{p,p}_{\mathbb{R}}E'$, $\|A\|_1 = |A|_2$ if and only if a multiple of A belongs to $\mathcal{E}$. In particular, for $A_j = \lambda_j\sigma_p\zeta_j \wedge \bar{\zeta}_j$ we have $\|A_j\|_1 = |\lambda_j|$. Therefore if A is positive,

$$*(A \wedge \frac{1}{k!}\omega^k) = \Sigma \lambda_j = \Sigma \|A_j\|_1 \geq \|A\|_1.$$

In particular, if a multiple of A belongs to $\mathcal{E}$, then

$$|*(A \wedge \frac{1}{k!}\omega^k)| = \|A\|_1 \quad \text{and} \quad *(A \wedge \frac{1}{k!}\omega^k) \leq \|A\|_1 \tag{1.13}$$

with equality if and only if A is positive.

The proof of the inequality in (b) proceeds exactly as in (a) except the A_j are chosen so that $A_j/\|A_j\|_1$ is in $\mathcal{D} \cup \mathcal{E}$. Because of (1.13) above and the fact that a multiple of A_j belongs to $\mathcal{E}$ or $\mathcal{D}$, $*(A_j \wedge \frac{1}{k!}\omega^k) \leq \|A_j\|_1$ with equality if and only if A_j is positive. Therefore

$$*(A \wedge \frac{1}{k!}\omega^k) = \Sigma *(A_j \wedge \frac{1}{k!}\omega^k) \le \Sigma \|A_j\|_1 = \|A\|_1 ,$$

with equality implying that each A_j is of the form $\sigma_p \zeta \wedge \overline{\zeta}$ for some $\zeta \in \Lambda^{p,0}E'$. Therefore equality implies that A is in P^p. Q.E.D.

As noted above each positive (p,p) form is not necessarily a positive combination of positive decomposable (p,p) forms. We finish this section by proving that each real (p,p) form is the sum of real decomposable (p,p) forms.

Proposition 1.9: *There exists a basis for* $\Lambda^{p,p}_{\mathbb{R}}E'$ *consisting of positive decomposable* (p,p) *forms.*

Proof: Suppose $\{e_1,\ldots,e_n\}$ is a basis for $E^* = \Lambda^{1,0}E'$. The proof is by induction on p. It suffices to show that $\Lambda^{p,p}E'$ has a basis consisting of positive decomposable (p,p) forms. For $p = 1$ it is true by Theorem 1.2 or because $2\,\mathrm{Re}(\frac{i}{2} e_i \wedge \overline{e}_j) =$

$$= \frac{i}{2} e_i \wedge \overline{e}_j + \frac{i}{2} e_j \wedge \overline{e}_i$$

$$= \frac{i}{2}(e_i + e_j) \wedge (\overline{e}_i + \overline{e}_j) - \frac{i}{2} e_i \wedge \overline{e}_i - \frac{i}{2} e_j \wedge \overline{e}_j ,$$

and $2\,\mathrm{Im}(\frac{i}{2} e_i \wedge \overline{e}_j) =$

$$= \frac{1}{2} e_i \wedge \overline{e}_j - \frac{1}{2} e_j \wedge \overline{e}_i$$

$$= \frac{i}{2}(e_i + ie_j) \wedge (\overline{e}_i - i\overline{e}_j) - \frac{i}{2} e_i \wedge \overline{e}_i - \frac{i}{2} e_j \wedge \overline{e}_j .$$

Consider $e^I \wedge \overline{e}^J$. Let $I' = (i_2,\ldots,i_p)$ and $J' = (j_2,\ldots,j_p)$.

Then $$\sigma_p e^I \wedge \overline{e}^J = (\frac{i}{2} e_{i_1} \wedge \overline{e}_{j_1}) \wedge (\sigma_{p-1} e^{I'} \wedge \overline{e}^{J'}) ;$$

and by the induction hypothesis for 1 and $p - 1$, respectively, the two factors on the right can be expressed as linear combinations of positive decomposable forms. Q.E.D.

Corollary 1.10: *The* (p,p) *form* $\frac{1}{p!}\omega^p$ *belongs to the interior of the cone* SP^p *(i.e., is strongly positive non-degenerate).*

Proof: It suffices to show that $*(\frac{1}{p!}\omega^p \wedge B) > 0$ for all non-zero

$B \in WP^k$, by (1.7). Choose a basis $\{A_1,\ldots,A_N\}$ for $\Lambda_{\mathbb{R}}^{p,p}E'$ consisting of positive decomposable (p,p) forms with $|A_j|_2 = 1$. We may choose an orthogonal basis $\{e_1,\ldots,e_n\}$ with $|e_j|_2^2 = 2$ for $E^* = \Lambda^{1,0}E'$ so that $A_1 = \sigma_p e^I \wedge \bar{e}^I$ with $I = (1,\ldots,p)$.

Suppose $\frac{1}{p!}\omega^p \wedge B = 0$ for some $B \in WP^k$. Let $B = \Sigma' b_{IJ} e^I \wedge \bar{e}^J$. As noted before $*(\frac{1}{p!}\omega^p \wedge B) = \Sigma' b_{II}$. If $|I| = k$, $|J| = p$, I and J strictly increasing, and $I \cup J = \{1,\ldots,n\}$, then $b_{II} = *(\sigma_p e^J \wedge \bar{e}^J \wedge B)$; which is ≥ 0 since B is in WP^k. Therefore $*\frac{1}{p!}\omega^p \wedge B = \Sigma' b_{II} = 0$ if and only if each $b_{II} = 0$. In particular, this proves that if $\frac{1}{p!}\omega^p \wedge B = 0$, with $B \in WP^k$, then, for $I = (1,\ldots,p)$, $0 = b_{II} = *(A_1 \wedge B) = (A_1, *B)$. Similarly $(A_j, *B)$ must vanish for $1 \leq j \leq N$ so that $B = 0$. Q.E.D.

2. ANALYTIC CYCLES ON KÄHLER MANIFOLDS

Suppose X is a Kähler manifold of complex dimension n, with Kähler form ω. Suppose $T \in \mathcal{D}'^{2p}(X)$ is a real-valued current on X of degree $2p$ which is representable by integration (i.e., when expressed as a $2p$ form with distribution coefficients, the distributions are measures). Let $\|T\|$ denote the positive measure defined by letting $\|T\|(f) = \sup\{|T(\phi)|: |\phi(z)|_\infty \leq f(z)\}$ for each positive continuous function f on X and let $M(T) = \|T\|(1) = \sup\{|T(\phi)|: |\phi|_\infty \leq 1\}$ (the mass norm of T). Then there exists a $\|T\|$ measurable function $\vec{T}$ from X to $\Lambda^{2p}\mathbb{C}^n$ with $T(\phi) = \int \langle\vec{T},\phi\rangle_x d\|T\|$ (see [2]). A current T defined on an open set Ω contained in $\mathbb{C}^n$ and of degree (p,p) (with $p + k = n$) will be called *weakly positive* if for all $\psi \in C_0^\infty(\Omega)$, $\psi \geq 0$, and all complex linear projections $\pi: \mathbb{C}^n \to \mathbb{C}^k$, the push forward $\pi_*(\psi T)$ is a positive measure. Such a current T will be called *positive* if for all $\zeta \in \Lambda^{k,0}\mathbb{C}^n$, $T \wedge \sigma_k \zeta \wedge \bar{\zeta}$ is a positive measure on Ω. The current T will be called *strongly positive* if for all $B \in \Lambda_{\mathbb{R}}^{k,k}\mathbb{C}^n$ which belong to WP^k, $T \wedge B$ is a positive measure on Ω. A current T of degree (p,p) on X is said to be *weakly positive, positive, or strongly positive* if the restriction of T to each coordinate chart is weakly positive, positive, or strongly positive, respectively. One can show that a current T on X is weakly positive, positive, or strongly positive if and only if T is representable by integration and $\vec{T}(z)$ is weakly positive, positive, or strongly positive, respectively, for almost all z in X (with respect to $||T||$).

For the proof that a weakly positive current T on $\mathbb{C}^n$ is representable by integration see Lelong [10] or use Proposition 1.9 to choose a basis $\{B_1,\ldots,B_N\}$ for $\Lambda^{k,k}_{\mathbb{R}}E'$ consisting of positive decomposable (k,k) forms and let $\{A_1,\ldots,A_n\}$ denote the dual basis for $\Lambda^{p,p}_{\mathbb{R}}E'$ where $p + k = n$. Then expressing T in terms of the basis $A_1,\ldots,A_N$, we see that the coefficients $T \wedge B_j$ are positive measures.

The Generalized Wirtinger's Inequality immediately applies to currents, since $T(\frac{1}{k!}\omega^k) = \int *(\vec{T} \wedge \frac{1}{k!}\omega^k)d||T||$ and $\int d||T|| = M(T)$.

Theorem 2.1: *Suppose* T *is a real-valued current of degree* $2p$ *on* X (*with* $p + k = n$) *which is representable by integration. Then*

(a) $T(\frac{1}{k!}\omega^k) \le M(T)$ *and equality holds if and only if* T *is strongly positive.*

(b) $T(\frac{1}{k!}\omega^k) \le M'(T)$ *and equality holds if and only if* T *is positive, where* $M'(T) = \sup\{|T(\phi)| : ||\phi||_\infty \le 1\}$.

The next three results are well known. First we briefly sketch how locally rectifiable currents can be used to compute cohomology with coefficients in $\mathbb{Z}$ on an oriented manifold. See Federer [2] for similar results concerning homology. Suppose X is a real n-dimensional oriented C^∞ manifold. Let $\mathcal{D}'^p = \mathcal{D}'_k$ $(p + k = n)$ denote the sheaf of germs of currents of degree p or dimension k on X. Let $\mathcal{R}^p = \mathcal{R}_k$ $(p + k = n)$ denote the subsheaf of germs of locally rectifiable currents of degree p or dimension k on X (see [2]). The exterior derivative d operating on smooth forms of degree p extends as a differential operator to currents of degree p, and we have the complex

$$0 \to \mathbb{C} \to \mathcal{D}'^0 \xrightarrow{d} \mathcal{D}'^1 \xrightarrow{d} \cdots \xrightarrow{d} \mathcal{D}'^n \to 0$$

which is an exact resolution of the sheaf $\mathbb{C}$ by fine sheaves $\mathcal{D}'^p$. This proves that

$$H^p(X,\mathbb{C}) \cong \{T \in \mathcal{D}'^p(X) : dT = 0\}/d\mathcal{D}'^{p-1}(X) .$$

A locally rectifiable current is said to be *locally integral* if its exterior derivative is also locally rectifiable. Let $\mathcal{I}^p$ denote the sheaf of germs of locally integral currents of degree p. Then

$$0 \to \mathbb{Z} \to I^0 \xrightarrow{d} I^1 \xrightarrow{d} \cdots \to I^n \to 0 \tag{2.1}$$

is an exact resolution of the sheaf $\mathbb{Z}$ (if X is not oriented the sequence is an exact resolution of the orientation sheaf). The usual cone construction provides a proof of local exactness of (2.1) (see [2]). Each sheaf I^p is soft. That is given a locally integral current T on a neighborhood U of a closed subset F of X, there exists a locally integral current S on X with S equal to T in a neighborhood of F. Choose $f \in C^\infty(X)$ with $f \equiv 1$ on a neighborhood of $X - U$ and $f \equiv 0$ on a neighborhood of F, and $0 \le f \le 1$. Let χ_ε denote the characteristic function of the set $\{x \in X: f(x) < \varepsilon\}$. Then it follows easily from Federer's theory of slicing [2] that $\chi_\varepsilon T$ is a locally integral current for almost all ε. Now take $S = \chi_\varepsilon T$. This proves each I^p is soft. The fact that (2.1) is an exact resolution of the sheaf $\mathbb{Z}$ by soft sheaves I^p implies the following theorem, on a smooth oriented manifold X.

Theorem 2.2: $H^p(X,\mathbb{Z}) = \{T \in R^p_{loc}(X): dT = 0\}/dI^{p-1}_{loc}(X)$.

Note that the natural map of $H^p(X,\mathbb{Z})$ into $H^p(X,\mathbb{C})$ is induced by the inclusion map of $R^p_{loc}(X)$ into $\mathcal{D}'^p(X)$.

Now assume that X is a complex manifold of complex dimension n. Integration over a subvariety defines a current on X (see [10], [7], and [5]). Let $\operatorname{Reg} V$ denote the regular points of a subvariety V. The second result is the following:

Proposition 2.3: *If* V *is a pure* k *dimensional subvariety of* X *then* $[V](\phi)$ *defined by integrating* $\phi \in \mathcal{D}^{2k}(X)$ *over* $\operatorname{Reg} V$ *is a locally rectifiable current on* X *which is* d*-closed and positive of type* (p,p) $(p + k = n)$.

Remark: Let $T = [V]$. Since $\vec{T}(z)$ is decomposable for $z \in \operatorname{Reg} V$, the three notions weakly positive, positive, and strongly positive all agree for T.

An *analytic cycle of dimension* k (or *holomorphic* k*-chain*) on X is a current T of type (p,p) $(p + k = n)$ which can be expressed as $\Sigma n_j[V_j]$ where each $n_j \in \mathbb{Z}$ and $V = \cup V_j$ is a pure k-dimensional subvariety of X with irreducible components $\{V_j\}$. An analytic cycle is positive if and only if each $n_j \ge 0$. Let $Z_k(X)$ denote the group of all analytic cycles of dimension k on X and let $Z_k^+(X)$ denote the set of positive analytic k-cycles. If T can be expressed as $\Sigma\, r_j[V_j]$ with each r_j a rational number then T is called a *rational analytic* k-*cycle*. The third result is a slight generalization of Proposition 2.3.

Corollary 2.4: *Each analytic* k-*cycle is a locally rectifiable,* d-*closed current of type* (p,p) $(p + k = n)$.

Using Theorem 2.1, we see that each analytic k-cycle $\Sigma\, n_j[V_j]$ determines a cohomology class in $H^{2p}(X,\mathbb{Z})$ where $p + k = n$. Let $\pi\colon H^{2p}(X,\mathbb{Z}) \to H^{2p}(X,\mathbb{C})$ denote the natural inclusion map.

Now assume that X is a compact Kähler manifold. Then each $H^{2r}(X,\mathbb{C})$ can be expressed as the direct sum $\Sigma\, H^{s,t}(X)$ where $s + t = r$ and $H^{s,t}(X)$ denotes the space of harmonic forms of type (s,t). Suppose $T \in Z_k(X)$; then T determines a cohomology class in $H^{2p}(X,\mathbb{C})$ $(p+k=n)$. This cohomology class must belong to $H^{p,p}(X)$ since $\int_T \phi = 0$ for all $\phi \in H^{s,t}(X)$ with $s + t = k$, $s \neq t$. That is, π maps the subspace of $H^{2p}(X,\mathbb{Z})$ determined by $Z_k(X)$ into $H^{p,p}(X)$.

Now assume that X is a complex submanifold of some complex projective space. The original conjecture of Hodge [6] was that

III (*over* $\mathbb{Z}$): *Each class* $c \in H^{2p}(X,\mathbb{Z})$ *with* $\pi c \in H^{p,p}(X)$ *is determined by an analytic* k-*cycle* $(p + k = n)$.

While this is true for $p = 1$ (Kodaira-Spencer [9]), it is false for $p > 1$ (Atiyah-Hirzebruch [1]). Conjecture III has been reformulated for general p over the rationals $\mathbb{Q}$.

III (*over* $\mathbb{Q}$): *Each class* $c \in H^{p,p}(X) \cap \pi H^2(X,\mathbb{Z}) \subset H^{2p}(X,\mathbb{C})$ *is determined by a rational analytic* k-*cycle,* $\frac{1}{m}T$ $(T \in Z_k(X),\ m \in \mathbb{Z}$ *and* $p + k = n)$.

A smooth form ϕ of type (p,p) is strongly positive definite if ϕ remains positive under small perturbations or equivalently if (for each point $z \in X$) each of the eigenvalues λ_j is positive for j from 1 to $\binom{n}{p}$, or equivalently if for each point $z \in X$, $\phi(z)$ belongs to the interior of $SP \subset \Lambda^{p,p}_{\mathbb{R}} T'_z(X)$.

Consider the following two "Plateau problems" on X (instead of fixing a boundary, we fix a cohomology class). Suppose a class $c \in \pi H^{2p}(X,\mathbb{Z})$

is given. Let $M(c)$ denote the infimum of $\{M(T)\}$ taken over all $T \in \mathcal{R}^{2p}(X)$ with $T \in c$, and let $m(c)$ denote the infimum of $\{M(\psi)\}$ taken over all $\psi \in \mathcal{D}'^{2p}(X)$ with $\psi \in c$. Find a rectifiable current $T \in \mathcal{R}^{2p}(X)$ with $T \in c$ such that

$$(1) \quad M(T) = M(c) \quad \text{or} \quad (2) \quad M(T) = m(c) .$$

It follows easily from Federer [2] that problem (1) always has a solution T; however, it is not always true that this solution T is also a minimum among the larger class of competitors in (2) (cf. Mumford's example at the end of this section). Let T_k denote a solution to problem (1) for the class kc, where $k \in \mathbb{Z}^+$ (i.e., $M(T_k) = M(kc)$). Then $m(c) \leq \leq \frac{1}{k} M(T_k) \leq M(c)$, since $M(kc) \leq kM(c)$. Federer has shown that $\lim_{k\to\infty} \frac{1}{k} M(T_k) = m(c)$.

Next we state a conjecture about "positive Plateau problems", which concludes that equality holds in the above limit for some finite number k.

I *Suppose* $c \in H^{p,p}(X) \cap \pi H^{2p}(X,\mathbb{Z}) \subset H^{2p}(X,\mathbb{C})$ *contains a strongly positive definite* (p,p) *form* ϕ. *Then there exists an integer* $m \in \mathbb{Z}^+$ *such that for the class* mc *a solution* $T \in \mathcal{R}^{2p}(X)$ *to* (1) *above also satisfies* (2).

The following conjecture will be shown to be equivalent to I in the next theorem:

II *Suppose* $c \in H^{p,p}(X) \cap \pi H^{2p}(X,\mathbb{Z}) \subset H^{2p}(X,\mathbb{C})$ *contains a strongly positive definite* (p,p) *form* ϕ. *Then there exists an integer* $m \in \mathbb{Z}^+$ *and a positive analytic* k-*cycle* $T \in Z_k^+(X)$ *such that* $\frac{1}{m} T \in c$ $(p + k = n)$.

Theorem 2.5: *For a given compact Kähler manifold* X, I *is true if and only if* II *is true.*

Proof: First assume that I is true. Let $T \in \mathcal{R}^{2p}(X)$ with $T \in mc$, and let $\phi' = m\phi$. Then by Theorem 2.1

$$M(\phi') = \int \phi' \wedge \frac{1}{k!} \omega^k \leq M(T) \tag{2.2}$$

(the first equality uses the Generalized Wirtinger Inequality, as opposed to the usual Wirtinger Inequality). If T is a solution to (2) then equality must hold in (2.2). By Theorem 2.1 this implies that T is positive.

By the structure theorem of [7] (cf. [4] and [5]) T must be a positive analytic k-cycle. This proves II. Conversely, suppose $S \in Z_k^+(X)$ and $S \in mc$. Then by Theorem 2.1

$$M(S) = \int S \wedge \frac{1}{k!}\omega^k = \int \psi \wedge \frac{1}{k!}\omega^k \le M(\psi)$$

for each $\psi \in c$. This proves (2) for $S \in Z_k^+(X)$. However if (1) is true for $T \in mc$ and $S \in Z_k^+(X) \cap c$ then T must belong to $Z_k^+(X)$ (see (2.2) with $\phi' = S$ to conclude T is positive of type (p,p)). Therefore II implies I. Q.E.D.

Next we note that II implies the Hodge Conjecture III. Assume X is a submanifold of $\mathbb{P}^N$ and let ω denote the Fubini-Study (1,1) form on $\mathbb{P}^N$ restricted to X. Assume II is true and let $c \in H^{p,p}(X) \cap H^{2p}(X,\mathbb{Z})$. Then, given a smooth representative $\psi \in c$ of type (p,p), we can find a positive integer q such that $\phi = \psi + q\frac{1}{p!}\omega^p$ is strongly positive definite because of Corollary 1.10. By assumption there exist $T \in Z_k^+(X)$ and $m \in \mathbb{Z}$ with $[T] = m[\phi]$. Since $\frac{1}{p!}\omega^p$ determines the same class as a k-linear section S of X in $\mathbb{P}^N$ $(S \in Z_k^+(X))$, $T - mqS \in mc$ (and $T - mqS \in Z_k(X)$).

Remark: It follows from Theorem 5.8 in Kleiman [14] that if the Hodge Conjecture III is true then Conjecture II (and hence I) is true for (using the above notation) classes c of the form $[\psi + q\frac{1}{p!}\omega^p]$, if q is chosen sufficiently large depending on ψ. Therefore, Conjecture I (a "Plateau problem") for classes c of the form $[\psi + q\frac{1}{p!}\omega^p]$ with q large is equivalent to the Hodge Conjecture III.

The statement II is of course true for $p = 1$ by the Kodaira Embedding Theorem as follows: There exists a line bundle L over X with first Chern form ϕ. By Kodaira [8] there exists an integer m such that the mapping $X \xrightarrow{f} P(H^0(X,(L^{-1})^m))$ is an embedding. Therefore $m\phi - f^*([H]) = d\psi$ where $[H]$ is a hyperplane section of $f(X)$.

One might conjecture that statement II (or equivalently I) is valid modulo torsion. That is, if $c \in H^{p,p}(X) \cap \pi H^{2p}(X,\mathbb{Z})$ contains a strongly positive definite (p,p) form ϕ then c contains a positive analytic k-cycle $T \in Z_k^+(X)$. David Mumford's example, mentioned in the introduction, shows that this is false in the simplest case $p = 1$ and dimension $X = 2$. See [11] and [12] for the results about ruled surfaces needed below. Suppose $X \xrightarrow{\pi} C$ is a $\mathbb{P}^1$ bundle over an algebraic curve of genus $g \geq 4$. Choose X homeomorphic to $C \times \mathbb{P}^1$ and generic. Then $H^2(X,\mathbb{Z}) = \mathbb{Z} \cdot e \oplus \mathbb{Z} \cdot f$ where e contains $C \times \{z\}$, $z \in \mathbb{P}^1$ and f contains $\{w\} \times \mathbb{P}^1$, $w \in C$. Briefly, one can show that, since X is generic, $e + mf$ does not contain a positive divisor for $m < [\frac{g}{2}]$; and using the Nakai-Moišezon criterion, that $\phi + mf$ is ample for $m \in \mathbb{Z}^+$ (i.e., $e + mf$ contains $\frac{1}{k} S$ where $k \in \mathbb{Z}^+$ and S is a hyperplane section for some projective embedding). Consequently $e + mf$ contains $\frac{1}{k}\omega$ where ω is the Kähler form induced on X by some projective embedding. Therefore $e + mf$, $m \in \mathbb{Z}^+$, contains a positive definite (1,1) form, while for $m < [\frac{g}{2}]$ it does not contain a positive divisor.

REFERENCES

1. Atiyah, M. F. and F. Hirzebruch, *Analytic cycles on complex manifolds*, Topology 1 (1961), 25-45.

2. Federer, H., *Geometric Measure Theory*, Springer-Verlag, New York, 1969.

3. Grothendieck, A., *Hodge's general conjecture is false for trivial reasons*, Topology 8 (1969), 299-303.

4. Harvey, R. and J. King, *On the structure of positive currents*, Inventiones Math. 15 (1972), 47-52.

5. __________, and B. Shiffman, *A characterization of holomorphic chains*, (to appear in Ann. Math.).

6. Hodge, W. V. D., *The topological invariants of algebraic varieties*, in *Proceedings of the International Congress of Mathematicians*, Vol. I, Harvard (1950), 182-191.

7. King, J., *The currents defined by analytic varieties*, Acta. Math. 127 (1971), 185-220.

8. Kodaira, K., *On Kähler varieties of restricted type*, Ann. Math. 60 (1954), 28-48.

9. __________, and D. C. Spencer, *Groups of complex line bundles over compact Kähler varieties; Divisor class groups on algebraic varieties*, Proc. Nat. Acad. Sci. 39 (1953), 868-877.

10. Lelong, P., *Plurisubharmonic Functions and Positive Differential Forms*, Gordon and Breach, New York, 1969.

11. Maruyama, M., *On Classification of Ruled Surfaces*, Lecture Notes in Mathematics, Kyoto University.

12. Nagota, M., *On self-intersection number of a section on a ruled surface*, Nagoya Math. J. 37 (1970), 191-196.

13. Weil, A., *Variétés Kählériennes*, Hermann, Paris, 1958.

14. Kleiman, S., *Geometry on grassmannians and applications to splitting bundles and smoothing cycles*, Publ. Math. I.H.E.S., 36 (1969), 281-297.

APPLICATIONS OF GEOMETRIC MEASURE THEORY TO VALUE DISTRIBUTION THEORY FOR MEROMORPHIC MAPS

Bernard Shiffman*
Johns Hopkins University

INTRODUCTION

The purpose of this report is to introduce to value distribution theory a new technique--the geometric calculus of currents--that simplifies and clarifies proofs in the Nevanlinna theory of functions of several variables. In so doing, we obtain some new results. In Section 2, we state defect relations for meromorphic maps (Corollary 2.4), generalizing results of Carlson-Griffiths [2] and Griffiths-King [7], and also for certain collections of divisors without normal crossings (Corollary 2.7). The results of Section 2 are proved by current calculus, eliminating the usual argument involving the unintegrated main theorems. (Complete proofs of the results of Section 2 will appear elsewhere.) Sections 3-5 explicate some averaging formulas of Stoll [17], [18], [19]. We show how these averaging formulas are direct consequences of Santaló's "Crofton Formula" in complex projective space [15]. In Section 3, we give a short elementary proof and generalization of the Crofton Formula in $\mathbb{P}^n$ using techniques of currents. In Section 4, Federer's geometric measure theory [5] enters the picture, and the results of Section 3 are recast in this general framework. The order functions of a meromorphic map are interpreted as average counting functions in Section 5. In Section 6, we compute all the defects of an exponential map of two variables using Weyl's computation of the order function of an exponential curve [22] and the averaging formulas. The example in Section 6 illustrates the defect relations for divisors without normal crossings discussed in Section 2.

It is my pleasure to thank Phillip Griffiths and Reese Harvey who

* Research partially supported by N.S.F. Grant GP-27624.

explained to me the concepts of value distribution theory and currents that inspired the results of Section 2.

1. NOTATION AND TERMINOLOGY

For $z = (z_1,\ldots,z_n) \in \mathbb{C}^n$, we write

$$|z| = (|z_1|^2+\ldots+|z_n|^2)^{\frac{1}{2}}.$$

We let B_r denote the ball $\{z \in \mathbb{C}^n : |z| < r\}$ of radius r about the origin. We let

$$\nu = \frac{1}{4\pi} dd^c|z|^2 = \frac{i}{2\pi}\sum_{j=1}^{n} dz_j \wedge d\bar{z}_j$$

be the Euclidean Kähler form on $\mathbb{C}^n$ normalized so that $\int_{B_r} \nu^n = r^{2n}$.

Here, the real differential operator d^c is given by

$$d^c = i(\bar{\partial} - \partial).$$

We also consider the form

$$\tilde{\omega} = \frac{1}{4\pi} dd^c \log |z|^2$$

on $\mathbb{C}^n - \{0\}$. Letting $p: \mathbb{C}^n - \{0\} \to \mathbb{P}^{n-1}$ be the natural projection, we note that $\tilde{\omega} = p^*\omega$, where ω is the Kähler form for the Fubini-Study metric on projective $(n-1)$-space $\mathbb{P}^{n-1}$ normalized so that $\int_{\mathbb{P}^{n-1}} \omega^{n-1} = 1$.

Let M be an oriented C^∞ manifold of real dimension m, and let X be a complex n-dimensional manifold. We let $\mathcal{D}^k(M)$ and $\mathcal{D}^{p,q}(X)$ denote the spaces of compactly supported (complex valued) C^∞ k-forms on M and (p,q)-forms on X respectively with the usual inductive limit topology. We let $A^k(M)$ and $A^{p,q}(X)$ denote the corresponding spaces of compactly supported forms with continuous (instead of C^∞) coefficients, with the C^0 inductive limit topology. We let $\mathcal{D}'^k(M) = \mathcal{D}^{m-k}(M)'$ and $\mathcal{D}'^{p,q}(X) = \mathcal{D}^{n-p,n-q}(X)'$ denote the spaces of *currents* of degree k on M and of type (p,q) on X respectively (see [14], [16]). A current $T \in \mathcal{D}'^{p,q}(X)$ can be expanded on each coordinate neighborhood Ω (with complex coordinates $\{z_i\}$) in the form $\Sigma\, T_{IJ}\, dz^I \wedge d\bar{z}^J$, where I and J run through all multi-indices of length p and length q respectively and the T_{IJ} are elements of $\mathcal{D}'^{0,0}(\Omega)$, which we call *distributions* on Ω. We say that T is *representable by integration* if T extends to a continuous linear functional on $A^{n-p,n-q}(X)$, or equivalently, if the T_{IJ} are given by (regular) Borel measures on Ω. We also let $\mathcal{E}^k(M)$ and $\mathcal{E}^{p,q}(X)$ denote the spaces of C^∞ k-forms on M and (p,q)-forms on X respectively, and we imbed $\mathcal{E}^k(M)$ in $\mathcal{D}'^k(M)$ and $\mathcal{E}^{p,q}(X)$ in $\mathcal{D}'^{p,q}(X)$ in the usual way. If $T \in \mathcal{D}'^k(M)$ and $g: M \longrightarrow M'$ is a C^∞ map (where M' is an oriented n-dimensional manifold) such that $g|\text{supp } T$ is a proper map, then the current $g_*T \in \mathcal{D}'^{k+n-m}(M')$ is given by

$$(g_*T, \phi) = (T \wedge g^*\phi, 1)$$

for $\phi \in \mathcal{D}^{m-k}(M')$.

We now define abstract counting functions that generalize both the usual counting functions and order functions of value distribution theory in several complex variables. For simplicity, we restrict our domain of definition to $\mathbb{C}^n$.

Definition 1.1: Let $\Theta \in \mathcal{D}'^{p,p}(\mathbb{C}^n)$ such that Θ is representable by integration and $d\Theta = 0$, where $1 \le p \le n$. We define the *unintegrated counting function*

$$n(\Theta, r) = (\Theta, r^{2p-2n} \chi_{B_r} \nu^{n-p}) \quad \text{for } r > 0, \tag{1.1}$$

and the (*integrated*) *counting function*

$$N(\Theta, r) = \int_s^r n(\Theta, t)t^{-1}\, dt, \text{ for } r > s \tag{1.2}$$

where s is a fixed arbitrary positive number.

We use the notation χ_E for the characteristic function of a set E. The pairing in (1.1) makes sense since a current which is representable by integration can be paired with a form whose coefficients are bounded Borel measurable functions with compact support. We can further refine our use of the pairing notation: Let $T \in \mathcal{D}'^p(X)$ and $\Phi \in \mathcal{D}'^{2n-p}(X)$. We let $(T, \Phi) = (T \wedge \Phi, 1)$, whenever $T \wedge \Phi$ defines a current in $\mathcal{D}'^{2n}(X)$ with compact support. For example, $T \wedge \Phi$ is defined (by multiplying coefficients) if each point of X has a neighborhood on which either T or Φ has C^∞ coefficients; $T \wedge \Phi$ has compact support if $\text{supp } T \cap \text{supp } \Phi$ is compact.

Let

$$\ell_{sr}(z) = \log^+ \frac{r}{|z|} - \log^+ \frac{s}{|z|},$$

where $\log^+ a = \max(\log a, 0)$. Using Stokes' Theorem and the fact that $\tilde{\omega} = r^{-2}\nu$ on the tangent space of ∂B_r one shows as in Lelong

[12, pp. 260-261] that

$$n(\Theta, r) = (\Theta, \chi_{B_r} \tilde{\omega}^{n-p}) \tag{1.3}$$

if Θ is as in Definition 1.1 and the coefficients of Θ are C^∞ near 0. It then follows from (1.3) by means of Fubini's Theorem that

$$N(\Theta, r) = (\Theta, \ell_{sr} \tilde{\omega}^{n-p}). \tag{1.4}$$

The counting function for an analytic subvariety of $\mathbb{C}^n$ is an important example of Definition 1.1 which occurs often in value distribution theory. We now describe the currents defined by analytic subvarieties. First suppose M is an oriented real j-dimensional C^1 submanifold of X with locally finite volume in X. Then M defines a current $[M] \in \mathcal{D}'^{2n-j}(X)$ given by

$$([M], \phi) = \int_M \phi ,$$

and $[M]$ is representable by integration. Now suppose V is a pure k-dimensional analytic subvariety of X, and let V' denote the set of regular points of V. We then define the current $[V] = [V'] \in \mathcal{D}'^{n-k,n-k}(X)$ and note that $d[V] = 0$. Now let V be a pure k-dimensional analytic subvariety of $\mathbb{C}^n$. Then the unintegrated counting function for V is given by

$$n([V], r) = r^{-2k} \text{ vol } (V \cap B_r),$$

(where vol = volume, normalized so that unit balls have unit volume). If $0 \notin V$, then (1.3) and (1.4) are valid [see (3.16) for the case $0 \in V$].

We shall consider the counting function for divisors with multiplicities, and we are thus led to the concept of a *holomorphic chain* [9], [11]. A current $T \in \mathcal{D}'^{n-k,n-k}(X)$ is said to be a *holomorphic k-chain* on X if T is of the form $\Sigma\, a_j[V_j]$, where each a_j is an integer and the V_j are the irreducible components of a pure k-dimensional analytic set $V = \cup V_j$ in X. Thus, if $T = \Sigma\, a_j[V_j]$ is a holomorphic k-chain, then

$$N(T,r) = \Sigma\, a_j\, N([V_j], r).$$

Holomorphic $(n - 1)$-chains on X are called *divisors* on X. If f is a (not identically zero) meromorphic function on X, we define the *divisor of* f

$$\text{Div } f = \Sigma\, a_j[V_j] - \Sigma\, b_j[W_j]$$

where f has zeroes of multiplicity a_j on V_j and poles of multiplicity

b_j on W_j. We then have the *Poincaré-Lelong Formula* [13] (see also [9], 1.10):

$$\text{Div } f = \frac{1}{2\pi} dd^c \log |f| , \tag{1.5}$$

which is fundamental to the current viewpoint of value distribution theory. (In (1.5), we regard the locally integrable function $\log |f|$ as an element of $\mathcal{D}'^{0}(X)$.)

Let X, Y be complex manifolds. A *meromorphic map* f from X into Y is given by a holomorphic map $f_0: X_0 \to Y$, where X_0 is dense in X, such that the set $G = \text{Cl}_{X\times Y}$ Graph f_0 is an analytic set in $X \times Y$ and the projection $\pi: G \to X$ is a proper map. The set X_0 can be chosen such that $X - X_0$ is an analytic set of codimension at least 2 in X. We write

$$f: X \xrightarrow[m]{} Y$$

to signify that f is a meromorphic map from X into Y. Meromorphic maps $f: X \xrightarrow[m]{} \mathbb{P}^n$ can be represented locally in the form $f = (f_0:\dots:f_n)$, where the f_j are holomorphic functions; this representation can be obtained globally if X is Stein. Let D be a divisor on Y and suppose $f: X \to Y$ is a holomorphic map. If $f(X) \not\subset \text{supp } D$, we can define the pull-back divisor $f^{-1}D$ as follows: If D is given locally as the divisor of g on a domain $\Omega \subset Y$, then $f^{-1}D = \text{Div } g\circ f$ on $f^{-1}(\Omega)$. If, instead, f is meromorphic, then $f^{-1}D$ is the natural extension to X of $f_0^{-1}D$ on X_0. (Alternately, we can let $\tilde{f}: G \to Y$ be the projection, and then define $f^{-1}D = \pi_*(\tilde{f}^{-1}D)$.)

Let L be a holomorphic line bundle on a complex manifold V given by transition functions $\{g_{\alpha\beta}\}$ with respect to an open covering $\{U_\alpha\}$. Let h be a hermitian metric on L; i.e., $h = \{h_\alpha\}$, where $h_\alpha = |g_{\alpha\beta}|^{-2} h_\beta$. We shall consider the *curvature form* η of h defined by

$$\eta = -\frac{1}{4\pi} dd^c \log h_\alpha ,$$

and we note that the 2-form η on V is a representative of the Chern class $c_1(L) \in H^2(V, \mathbb{R})$.

We now let

$$f: \mathbb{C}^n \xrightarrow[m]{} V$$

be a meromorphic map, where V is a compact complex manifold. We are interested in value distribution theory for such maps. (For the defect relations given in Section 2, V shall be an algebraic manifold with

$\dim V \leq n$.) If D is a divisor on V such that $f(\mathbb{C}^n) \not\subset \operatorname{supp} D$, we write

$$N(D,r) = N_f(D,r) = N(f^{-1}D,r) \ , \tag{1.6}$$

which we call the counting function for D.

Suppose A is an analytic set of pure codimension q in V, and suppose $f: \mathbb{C}^n \to V$ is a holomorphic map such that $f^{-1}(A)$ has pure codimension q. One then has a holomorphic chain $f^{-1}[A]$ on $\mathbb{C}^n$ of the form $f^{-1}[A] = \Sigma\, a_j[W_j]$, where the W_j are the irreducible components of $f^{-1}(A)$, and the a_j are positive integers representing multiplicities. These multiplicities can be defined in various ways; see, for example, Section 4 for the case that A is locally a complete intersection. For general A, one can use slicing to define the pull-back of A. (See Definition 5.5.) If T is a holomorphic chain of codimension q on V, one then has a holomorphic chain $f^{-1}\,T \in \mathcal{D}'^{q,q}(\mathbb{C}^n)$ provided that $\operatorname{codim} f^{-1}(\operatorname{supp} T) = q$. If f is meromorphic, then one defines $f^{-1}T = \pi_*(\tilde{f}^{-1}T)$, as in the case of divisors. (Using intersection theory, one also has the defining equation $f^{-1}[A] = \pi_*([G] \cap [\mathbb{C}^n \times A]$.) One then defines, analogously to (1.6), the counting function

$$N(T,r) = N_f(T,r) = N(f^{-1}T,r).$$

Our next examples of counting functions are the order functions of Carlson-Griffiths [2] and Griffiths-King [7] which generalize Stoll's order function [17] which in turn is derived from the Ahlfors-Shimizu characteristic function of classical Nevanlinna theory. Let L be a hermitian line bundle on V with curvature form η. For $1 \leq q \leq \min(n, \dim V)$, we define the current $f^*\,\eta^q \in \mathcal{D}'^{q,q}(\mathbb{C}^n)$ as follows: If f is holomorphic, then $f^*\,\eta^q$ is the usual pull-back of the form η^q (and thus has $\mathcal{C}^\infty$ coefficients). If f is meromorphic, then one shows that the coefficients of the form $f_o^*\,\eta^q$ on X_o (we use the notation in the definition of a meromorphic map, with $X = \mathbb{C}^n$, $Y = V$) are locally integrable on $\mathbb{C}^n$, and thus $f_o^*\,\eta^q$ defines a current (with coefficients in L^1_{loc}) on $\mathbb{C}^n$ that we call $f^*\,\eta^q$. An alternate approach is to let $p_1: \mathbb{C}^n \times V \longrightarrow \mathbb{C}^n$, $p_2: \mathbb{C}^n \times V \longrightarrow V$ denote the projections and to define

$$f^*\,\eta^q = p_{1*}([G] \wedge p_2^{\,*}\eta^q). \tag{1.7}$$

One sees immediately from (1.7) that $f^*\,\eta^q$ is d-closed. We now define the *order functions for* f *relative to* L,

$$T^q(L,r) = T_f^{\,q}(L,r) = N(f^*\,\eta^q,r), \tag{1.8}$$

for $1 \le q \le \min(n, \dim V)$. The order functions $T^q(L,r)$ are defined in [7] (and in [2] for $q = 1$) for *holomorphic maps*. The function $T^q(L,r)$ does not depend modulo $O(1)$ on the choice of hermitian metric for L. We write

$$T(L,r) = T^1(L,r).$$

We let

$$|L| = \{\mathrm{Div}\ \Psi : \Psi \in \Gamma(V, \mathcal{O}(L))\}$$

where $\mathcal{O}(L)$ denotes the sheaf of germs of holomorphic sections of L. We shall show in Corollary 5.4 that under appropriate hypotheses, $T(L,r)$ is the average over all $D \in |L|$ of the counting function $N(D,r)$. In the case studied by Stoll [17] in which $V = \mathbb{P}^m$ and L is the hyperplane section bundle, then $\eta = \omega$ and $T(L,r)$ is the usual order function.

We consider the currents $\sigma_r \in \mathcal{D}'^{2n}(\mathbb{C}^n)$, for $r > 0$, defined by setting (σ_r, ϕ) equal to the average value of ϕ on ∂B_r, for $\phi \in \mathcal{D}^0(\mathbb{C}^n)$. We note that

$$\sigma_r = [\partial B_r] \wedge \frac{1}{4\pi} d^c \log|z|^2 \wedge \tilde{\omega}^{n-1}$$

For $D = \mathrm{Div}\ \Psi \in |L|$, where $\Psi \in \Gamma(V, \mathcal{O}(L))$, we define the proximity term

$$m(D,r) = (\sigma_r, -\log|\Psi \circ f|).$$

which is well-defined, since $\log|\Psi \circ f|$ can easily be shown to be an (almost everywhere defined) integrable function on ∂B_r (for $r > 0$). Here, $|v|$ means the length of $v \in L$. Note that modulo a constant, $m(D,r)$ does not depend on the choice of the section Ψ defining D. One calls $m(D,r)$ the proximity term since it measures the proximity of $f(\partial B_r)$ to D; $m(D,r)$ is large when f maps points of ∂B_r close to D.

2. DEFECT RELATIONS FOR MEROMORPHIC MAPS

In this section we announce new results in value distribution theory for meromorphic maps. We obtain a First Main Theorem and Defect Relations for meromorphic maps, generalizing the results of Carlson-Griffiths and Griffiths-King for holomorphic maps. For simplicity, we state our results for maps defined on $\mathbb{C}^n$. Using techniques of currents, which we briefly indicate here, one obtains a simplification of the proofs in [2] and [7].

We also obtain defect relations for certain families of smooth divisors that do not have normal crossings. Complete proofs and further generalizations of the results of this section will appear elsewhere.

We begin by stating our First Main Theorem.

Proposition 2.1 (F.M.T.): *Let* L *be a hermitian line bundle on a compact complex manifold* V *and let* $f\colon \mathbb{C}^n \xrightarrow[m]{} V$ *be a meromorphic map. Let* $D \in |L|$ *such that* $f(\mathbb{C}^n) \not\subset \operatorname{supp} D$. *Then*

$$N(D,r) + m(D,r) - m(D,s) = T(L,r).$$

It follows that under the hypothesis of Theorem 2.1, we have

$$N(D,r) \leq T(L,r) + O(1). \tag{2.1}$$

We give the formal part of the proof of Theorem 2.1 below. The individual steps of the proof are not difficult to justify. Let Ψ be a section of L whose divisor is D. It follows from the Poincaré-Lelong Formula (1.5) that

$$\frac{1}{2\pi}\, dd^c \log |\Psi| = D - \eta \tag{2.2}$$

(see also [2], p.564).

By "pulling back" equation (2.2) to $\mathbb{C}^n$, we obtain

$$\frac{1}{2\pi}\, dd^c \log |\Psi \circ f| = f^{-1}D - f^*\eta$$

and therefore

$$N(\frac{1}{2\pi}\, dd^c \log |\Psi \circ f|,\ r) = N(D,r) - T(L,r). \tag{2.3}$$

We can assume that f is holomorphic at 0 and $f(0) \notin \operatorname{supp} D$, and thus, recalling (1.4),

$$N(\frac{1}{2\pi}\, dd^c \log |\Psi \circ f|,\ r) = (\frac{1}{2\pi}\, dd^c \log |\Psi \circ f|,\ \ell_{sr}\, \tilde{\omega}^{n-1}).$$

It follows from the identity

$$\frac{1}{2\pi}\, dd^c\, \ell_{sr}\, \tilde{\omega}^{n-1} = \sigma_r - \sigma_s \tag{2.4}$$

that

$$\begin{aligned}(\frac{1}{2\pi}\, dd^c \log |\Psi \circ f|,\ \ell_{sr}\, \tilde{\omega}^{n-1}) &= \\ = (\frac{1}{2\pi}\, dd^c\, \ell_{sr}\, \tilde{\omega}^{n-1},\ \log |\Psi \circ f|) &= (\sigma_r - \sigma_s,\ \log |\Psi \circ f|)\end{aligned} \tag{2.5}$$

The result then follows from (2.3), (2.5), and the definition of $m(D,r)$.

A First Main Theorem for holomorphic maps is given by Carlson and Griffiths ([2],5.7) and for meromorphic maps into $\mathbb{P}^k$ by Stoll [17]. Our proof differs from the usual proofs in that we use (2.5) instead of integrating twice. The proof of our Second Main Theorem (Theorem 2.3) uses the same method, although the details are more technical.

We need some further notation and terminology before we can state our defect relations. Let L be a positive line bundle on a compact manifold V (V must be algebraic by Kodaira's imbedding theorem), and let $f: \mathbb{C}^n \xrightarrow[m]{} V$. For $D \in |L|$, we define the *defect*

$$\delta(D) = \liminf_{r \to \infty} \left[1 - \frac{N(D,r)}{T(L,r)} \right].$$

The defects $\delta(D)$ do not depend on the choice of metric for L. One chooses a metric for L with positive curvature; then $T(L,r) > 0$ for $r > s$, and $T(L,r)$ grows at least as fast as $\log r$. It follows from (2.1) that

$$\delta(D) = \liminf_{r \to \infty} \frac{m(D,r)}{T(L,r)}$$

and thus

$$0 \le \delta(D) \le 1.$$

We say that a collection of smooth analytic hypersurfaces $\{D_1,\ldots D_q\}$ in a complex manifold V has *normal crossings* if for all $x_o \in D = \cup D_j$, there exist a coordinate neighborhood U with coordinates $\{z_1,\ldots,z_n\}$ centered at x_o and an integer k such that

$$\text{Div } z_1 \ldots z_k = [D \cap U] \quad \text{on } U.$$

For example, a collection of hyperplanes in $\mathbb{P}^n$ has normal crossings if and only if the hyperplanes are "in general position."

Definition 2.2: Let L_1, L_2 be line bundles over an algebraic manifold V. We define $[L_1/L_2] \in \mathbb{R} \cup \{+\infty, -\infty\}$ to be the infimum of the set of real numbers t such that $tc_1(L_2) - c_1(L_1) \ge 0$, where the expression $\gamma \ge 0$ means that γ is represented by an everywhere positive-semi-definite $(1,1)$-form on V.

We shall use the standard notation

$$\| P(r)$$

to mean that the statement $P(r)$ is true for all $r \in [s, +\infty) - I$ for some open subset I of $[s, +\infty]$ such that $\int_I d(\log x) < +\infty$.

We now state our *Second Main Theorem.* To simplify notation, if D is a hypersurface, we also let D denote the divisor $[D]$. We let K_V denote the canonical line bundle of $(n,0)$-forms on V.

Theorem 2.3 (S.M.T.): *Let* L *be a positive line bundle on an algebraic manifold* V, *and let* $f: \mathbb{C}^n \xrightarrow[m]{} V$ *be a non-degenerate meromorphic map such that* $\dim V \le n$. *Let* $\{D_1, \ldots, D_q\}$ *be a collection of smooth hypersurfaces in* V *with normal crossings such that* $D_j \in |L|$ *for* $1 \le j \le q$. *Then*

$$\| \sum_{j=1}^{q} m(D_j, r) \le [K_V^* / L]\, T(L,r) - N_1(r) + o(T(L,r))$$

Corollary 2.4 (Defect Relations): *Under the hypotheses of Theorem 2.3, we have*

$$\sum_{j=1}^{q} \delta(D_j) \le [K_V^* / L] - \liminf_{r \to \infty} \frac{N_1(r)}{T(L,r)} .$$

In the above theorem and corollary, the *ramification term* $N_1(r)$ is given by

$$N_1(r) = N(R,r)$$

where R is the divisor of the Jacobian determinant of the map $\hat{f}: \mathbb{C}^n \xrightarrow[m]{} V \times \mathbb{C}^q$ (where $q = n - \dim V$) given by $\hat{f}(z) = (f(z), z_1, \ldots, z_q)$ (if $\dim V = n$, then $\hat{f} = f$).

Similar defect relations are given in Carlson-Griffiths [2] and Griffiths-King [7] for holomorphic maps. If $V = \mathbb{P}^k$ and L is the hyperplane section bundle, then $[K_V^* / L] = k + 1$; defect relations for this case were proved by Stoll [17] for meromorphic maps. (Stoll's result is more general than Corollary 2.4, since Stoll considers $f: \mathbb{C}^n \xrightarrow[m]{} \mathbb{P}^k$, where k can also be greater than n.) If one applies Corollary 2.4 to $F: \mathbb{C}^1 \to \mathbb{P}^1$, one gets Nevanlinna's classical defect relations.

The methods of proof of Theorem 2.3 can be applied to the situation in which the divisors are smooth but do not have normal crossings. A result of M. Green [6] states that any map $f: \mathbb{C}^n \to \mathbb{P}^n$ that omits $n + 2$ *distinct* hyperplanes must be degenerate. However, the sum of the defects over distinct hyperplanes not in general position can be infinite, as we illustrate in Section 6. Corollary 2.7 below gives defect relations for smooth hypersurfaces without normal crossings. This defect relation contains an extra term involving a proximity term for subvarieties of codimension greater than 1, which we now define.

Definition 2.5: Let L be a hermitian line bundle on a compact manifold V, and let $f: \mathbb{C}^n \xrightarrow[m]{} V$. Let $\{D_1,\ldots,D_p\}$ be a collection of smooth hypersurfaces in V with normal crossings such that $[D_j]$ is the divisor of a holomorphic section Ψ_j of L, for $1 \le j \le p$. Let $S = D_1\cap\ldots\cap D_p$. We define the *proximity* to S,

$$m(S,r) = (\sigma_r, -\tfrac{1}{2}\log[|\Psi_1\circ f|^2+\ldots+|\Psi_p\circ f|^2]).$$

We now state our generalized Second Main Theorem in which we allow k too many divisors to intersect along a smooth subvariety S.

Theorem 2.6 (S.M.T.): *Let* L, V, f *be as in Theorem* 2.3. *Let* S *be a smooth subvariety of* V *of pure codimension* p, *and let* $D_1,\ldots,D_q$ *be smooth hypersurfaces in* V *such that:*

i) $\{D_1,\ldots,D_q\}$ *has normal crossings in* $V - S$;

ii) $D_1\cap\ldots\cap D_{p+k} = S$;

iii) $\{D_{j_1},\ldots,D_{j_p}, D_{p+k+1},\ldots,D_q\}$ *has normal crossings in* V

whenever $1 \le j_1 < \ldots < j_p \le p + k$.

Then

$$\Big\| \sum_{j=1}^{q} m(D_j, r) \le [K_V^*/L]T(L,r) + k\, m(S,r) - N_1(r) + o(T(L,r)).$$

Corollary 2.7 (Defect Relations): *Under the hypotheses of Theorem* 2.6, *we have*

$$\sum_{j=1}^{q} \delta(D_j) \le [K_V^*/L] + \limsup_{r\to\infty} \frac{k\, m(S,r) - N_1(r)}{T(L,r)}.$$

The extra term involving $\limsup m(S,r)/T(L,r)$ in Corollary 2.7 does not necessarily vanish, and we give an example in Section 6 in which the maximum defect given by Corollary 2.7 is attained. (Note that if we let $k = 0$ in Theorem 2.6 and Corollary 2.7, we obtain Theorem 2.3 and Corollary 2.4.) Theorem 2.7 is proved by considering the following diagram:

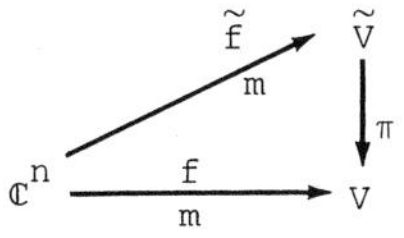

where $\tilde{V}$ is obtained from V by applying the monoidal transformation with center S. One lifts the hypersurfaces D_j to $\tilde{V}$ to obtain a collection of hypersurfaces in $\tilde{V}$ with normal crossings (we also include the hypersurface $\pi^{-1}(S) \subset \tilde{V}$). We then apply the methods of the proof of Theorem 2.3 to $\tilde{f}: \mathbb{C}^n \xrightarrow[m]{} \tilde{V}$. Here it is essential that we consider meromorphic maps since $\tilde{f}$ is (in general) meromorphic when f is holomorphic.

The following Corollary describes a situation in which $m(S,r)$ is essentially as large as $m(D_1,r)$; i.e., the values of f are close to D_1 essentially as often as they are close to S.

Corollary 2.8: *Assume the hypotheses of Theorem* 2.6, *with* $k = 1$. *If*

$$\sum_{j=2}^{q} \delta(D_j) = [K_V^*/L] - \liminf \frac{N_1(r)}{T(L,r)} ,$$

then

$$\| \; m(D_1,r) = m(S,r) + o(T(L,r)).$$

Proof: By definition, $m(S,r) \le m(D_1,r)$ for all r. Let

$$\lambda = [K_V^*/L] - \liminf \frac{N_1(r)}{T(L,r)} .$$

It follows from Theorem 2.6 that

$$\| \sum_{j=1}^{q} \frac{m(D_j,r)}{T(L,r)} \le \lambda + \frac{m(S,r)}{T(L,r)} + o(1) .$$

But

$$\frac{m(D_j,r)}{T(L,r)} \ge \delta(D_j) + o(1),$$

and it therefore follows that

$$\| \; \frac{m(D_1,r)}{T(L,r)} \le \frac{m(S,r)}{T(L,r)} + o(1),$$

which completes the proof.

The phenomenon described in Corollary 2.8 is illustrated in the example given in Section 6.

3. THE CROFTON FORMULA IN COMPLEX PROJECTIVE SPACE

In this section, we apply the Crofton Formula, Theorem 3.1, to obtain an averaging formula for the counting function of an analytic set in $\mathbb{C}^m$ (Theorem 3.6). We prove a new general form of the Crofton Formula in $\mathbb{P}^n$, using elementary techniques involving currents. In the next section we restate the results of this section in the more general framework of geometric measure theory.

We let $G_{m,k}$ denote the Grassmann manifold of complex k-dimensional subspaces of $\mathbb{C}^m$, which we identify with the projective $(k - 1)$-planes in $\mathbb{P}^{m-1}$. The unitary group U_m acts on the left on $G_{m,k}$ and on $\mathbb{P}^{m-1} = G_{m,1}$. (In the following discussion, one can replace U_m by the "projective unitary group" U_m/S^1, which acts effectively on $G_{m,k}$.) We let λ denote Haar measure on U_m normalized such that $\lambda(U_m) = 1$, and we let μ denote the U_m-invariant measure on $G_{m,k}$ normalized such that $\mu(G_{m,k}) = 1$.

We say that a subset A of $\mathbb{P}^n$ is a k-dimensional *immersed analytic space* if there exists a pure k-dimensional analytic space V, a holomorphic map $f: V \to \mathbb{P}^n$ and a relatively compact open subset W of V such that $A = f(W)$. If $A \subset \mathbb{P}^n$ is a k-dimensional immersed analytic space, we let

$$\mathrm{vol}_{2k}(A) = \int_A \omega^k = \int_{W'} f^*\omega^k$$

(where W' equals the set of regular points of W) denote the normalized volume of A (ω is the Kahler form on $\mathbb{P}^n$ defined in Section 1).

Recall that $\mathrm{vol}_{2k}(A) = 1$ for $A \in G_{n+1,k+1}$. If $\dim A = 0$, then $\mathrm{vol}_0(A)$ equals the cardinality of A. Theorems 3.1 and 3.2 below are the Crofton Formulas in $\mathbb{P}^n$ due to L. A. Santaló [15].

Theorem 3.1: *Let* A *be a* k-*dimensional immersed analytic space in* $\mathbb{P}^n$ *Then*

$$\int_{G_{n+1,n+1-q}} \mathrm{vol}_{2k-2q}(A \cap P)\, d\mu(P) = \mathrm{vol}_{2k}(A)$$

for $1 \le q \le k$.

Theorem 3.2: *Let* A *and* B *be immersed analytic spaces in* $\mathbb{P}^n$ *of dimension* j *and* k *respectively such that* $j + k \ge n$. *Then*

$$\int_{U_{n+1}} \mathrm{vol}_{2h}(A \cap gB)\, d\lambda(g) = \mathrm{vol}_{2j}(A)\ \mathrm{vol}_{2k}(B),$$

where $h = j + k - n$.

We observe that Theorem 3.1 is an immediate consequence of Theorem 3.2: Let Q be a fixed projective $(n-q)$-plane in $\mathbb{P}^n$ and let $\pi: U_{n+1} \to G_{n+1,n+1-q}$ be given by $\pi(g) = gQ$. Then $\mu = \pi_* \lambda$, and therefore

$$\int_{G_{n+1,n+1-q}} \mathrm{vol}\,(A \cap P)\, d\mu(P) = \int_{U_{n+1}} \mathrm{vol}\,(A \cap gQ)\, d\lambda(g) = \mathrm{vol}\,(A),$$

where the last equality is a consequence of Theorem 3.2.

If T is a $\mathcal{D}'^{p}(X)$-valued function on a measure space (S,σ), then the statement

$$\int T(s)\,d\sigma(s) = T \in \mathcal{D}'^{p}(X) \tag{3.1}$$

means that

$$\int (T(s),\ \phi)\,d\sigma(s) = (T,\ \phi) \text{ for all } \phi \in \mathcal{D}^{2n-p}(X). \tag{3.2}$$

If T and almost all $T(s)$ are representable by integration, and (3.2) is valid for all $\phi \in A^{2n-p}(X)$, then we say that (3.1) is *valid for* C° *forms.* (If (3.2) is valid, then the integral in (3.1) actually exists in a much stronger sense; see [21].)

The following lemma is the central ingredient in our proof of the Crofton Formula and its generalizations.

Lemma 3.3: *Let* $T \in \mathcal{D}'^{q,q}(\mathbb{P}^n)$. *Then*

$$\int_{U_{n+1}} g_*\, T\, d\lambda(g) = c\omega^q$$

where $c = (T,\omega^{n-q})$. *If* T *is representable by integration, then the above formula is valid for* C° *forms.*

Proof: Let $\phi \in \mathcal{D}^{n-q,n-q}(\mathbb{P}^n)$ $[\phi \in A^{n-q,n-q}(\mathbb{P}^n)$ if T is representable by integration] be arbitrary. Let

$$\alpha = \int_{U_{n+1}} g^*\phi\ d\lambda(g). \tag{3.3}$$

Since the map $g \to g^*\phi$ (with values in either $\mathcal{D}^{n-q,n-q}(\mathbb{P}^n)$ or $A^{n-q,n-q}(\mathbb{P}^n)$) is continuous, the "Riemann sums" for the integral in (3.3) converge to α. We therefore have

$$\int_{U_{n+1}} (g_*T,\phi)\,d\lambda(g) = \int_{U_{n+1}} (T,g^*\phi)\,d\lambda(g) = (T,\alpha) \tag{3.4}$$

The form α is U_{n+1}-invariant; i.e., $g^*\alpha = \alpha$ for all $g \in U_{n+1}$. We now show that any invariant (k,k)-form on $\mathbb{P}^n$ is a multiple of ω^k:

Let α be an invariant (k,k)-form on $\mathbb{P}^n$. Since U_{n+1} is transitive on $\mathbb{P}^n$ and ω^k is invariant, it suffices to show that $\alpha_x = a\,\omega_x^k$ where x is an arbitrary point in $\mathbb{P}^n$. Since the isotropy group of a point in $\mathbb{P}^n$ is U_n, we must show that the space of U_n-invariant elements of $\Lambda^{k,k}(\mathbb{C}^n)$ is 1-dimensional. Let $\{e_1,\ldots,e_n\}$ be the standard basis of $\Lambda^{1,0}(\mathbb{C}^n)$. Let $\beta = \Sigma b_{IJ}\, e^I \wedge \overline{e}^J \in \Lambda^{k,k}(\mathbb{C}^n)$ (where $e^I = e_{i_1} \wedge \ldots \wedge e_{i_k}$, $1 \le i_1 < \ldots < i_k \le n$) be U_n-invariant. Suppose $I \neq J$, and let i be an index in I but not in J. By considering the element $g \in U_n$ that multiplies the i-th coordinate by -1 (and leaves the others fixed), we conclude from $g^*\beta = \beta$ that $b_{IJ} = 0$. Similarly, by considering a permutation of coordinates, we conclude that $b_{II} = b_{JJ}$. Therefore $\beta = b\,\Sigma\, e^I \wedge \overline{e}^I$, which completes the proof of our assertion concerning invariant forms.

Thus $\alpha = a\,\omega^{n-q}$. To compute the constant a, we observe that

$$a = \int_{\mathbb{P}^n} \omega^q \wedge \alpha = \int_{\mathbb{P}^n} \left[\int_{U_{n+1}} \omega^q \wedge g^*\phi \; d\lambda(g)\right],$$

and furthermore, for all $g \in U_{n+1}$

$$\int_{\mathbb{P}^n} \omega^q \wedge g^*\,\phi = \int_{\mathbb{P}^n} g^*(\omega^q \wedge \phi) = \int_{\mathbb{P}^n} \omega^q \wedge \phi .$$

Therefore, by Fubini's theorem

$$a = \int_{\mathbb{P}^n} \omega^q \wedge \phi = (\omega^q,\phi).$$

It then follows from (3.4) and (3.5) that

$$\begin{aligned}\int_{U_{n+1}} (g_*T,\phi)\,d\lambda(g) &= (T, a\,\omega^{n-q}) \\ &= (T,\omega^{n-q})(\omega^q,\phi) = (c\,\omega^q,\phi).\end{aligned}$$

Q.E.D.

Note that the proof of Lemma 3.3 does not use the fact that T is of type (q,q). In fact, Lemma 3.3 is valid for arbitrary T, but one can easily see from the proof that if T is of pure type other than (q,q) (or if T is of odd degree), then $\int g_*T\,d\lambda(g) = 0$.

We shall prove the following generalization of Theorem 3.2.

Theorem 3.4: *Let* A, B *be as in Theorem* 3.2. *Then*

$$\int_{U_{n+1}} d\lambda(g) \int_{A\cap gB} \phi = \mathrm{vol}_{2k}(B) \int_A \omega^{n-k} \wedge \phi$$

for all $\phi \in A^{h,h}(\mathbb{P}^n)$, *where* $h = j + k - n$.

Corollary 3.5: *Let* A *be a* j*-dimensional immersed analytic space in* $\mathbb{P}^n$, *and let* $1 \le q \le j$. *Then*

$$\int_{G_{n+1,n+1-q}} d\mu(P) \int_{A\cap P} \phi = \int_A \omega^q \wedge \phi$$

for all $\phi \in A^{j-q,j-q}(\mathbb{P}^n)$.

Theorem 3.2 follows immediately from Theorem 3.4 by setting $\phi = \omega^h$.

We first explain the basic philosophy behind the proof of Theorem 3.4. The conclusion can be restated using the language of currents as follows:

$$\int [A \cap gB] d\lambda(g) = \mathrm{vol}(B) \cdot [A] \wedge \omega^q. \tag{3.6}$$

By applying Lemma 3.3 with $T = [B]$, we have

$$\int [gB] d\lambda(g) = \mathrm{vol}(B) \cdot \omega^q \tag{3.7}$$

Equation (3.6) will follow from (3.7) if we can justify the following statement:

$$[A] \wedge \int [gB] d\lambda(g) = \int [A \cap gB] d\lambda(g). \tag{3.8}$$

Our proof below is primarily concerned with justifying (3.8). In Section 4, we use the general intersection theory of Federer [5] and Brothers [1] to formulate (3.8) abstractly (see Lemma 4.3) so that Theorem 3.4 can be generalized to normal and flat currents.

Proof of Theorem 3.4: Consider the fibre bundle $\pi\colon U_{n+1} \to \mathbb{P}^n$ where π is given by $\pi(g) = g\,x_0$, where x_0 is a point in $\mathbb{P}^n$. Let $\beta \in \mathcal{D}^q(U_{n+1})$ where $q = n^2 + 1 = \dim \pi^{-1}(x_0)$, such that β gives the volume element along the fibres of π. If M is a submanifold of $\mathbb{P}^n$ of real dimension r and $\Psi \in A^r(\mathbb{P}^n)$, then

$$\int_M \Psi = \int_{\pi^{-1}(M)} \beta \wedge \pi^* \Psi. \tag{3.9}$$

Let A, B be given as in the statement of the theorem, and let $\phi \in A^{h,h}(\mathbb{P}^n)$. We first lift everything to U_{n+1}. Let $\tilde{A} = \pi^{-1}(A)$, $\tilde{B} = \pi^{-1}(B)$, $\alpha = \pi^*\omega$, and $\tilde{\phi} = \beta \wedge \pi^*\phi$. We see from (3.9) that it suffices to verify that

$$\int_{U_{n+1}} ([\tilde{A} \cap g\tilde{B}], \tilde{\phi}) d\lambda(g) = c([\tilde{A}], \alpha^{n-k} \wedge \tilde{\phi}) \tag{3.10}$$

where $c = \mathrm{vol}_{2k}(B)$.

Suppose $\theta \in A^{q+2k}(U_{n+1})$. We can write $\theta = \beta \wedge \pi^*\theta' + \gamma$, where $\theta' \in A^{2k}(\mathbb{P}^n)$ and γ has fewer than q fibre variables. It follows from Lemma 3.3 and (3.9) that

$$\int([g\tilde{B}],\theta)d\lambda(g) = \int([g\tilde{B}], \beta \wedge \pi^*\theta')d\lambda(g) = \int([gB], \theta')d\lambda(g)$$
$$= c\int_{\mathbb{P}^n} \omega^{n-k} \wedge \theta' = c\int_{U_{n+1}} \beta \wedge \alpha^{n-k} \wedge \pi^*\theta'$$

and therefore

$$\int([g\tilde{B}],\theta)d\lambda(g) = c(\alpha^{n-k},\theta). \tag{3.11}$$

Let $p_1, \tau: U_{n+1} \times U_{n+1} \longrightarrow U_{n+1}$ be given by $p_1(a,b) = a$, $\tau(a,b) = ab^{-1}$. Then

$$\tilde{A} \cap g\tilde{B} = p_1[(\tilde{A} \times \tilde{B}) \cap \tau^{-1}(g)]$$

and thus

$$\int([\tilde{A} \cap g\tilde{B}], \tilde{\phi})d\lambda(g) = ([\tilde{A} \times \tilde{B}], \tau^*\nu \wedge p_1^*\tilde{\phi}) \tag{3.12}$$

where ν is the volume form on U_{n+1} (given by $d\lambda$).

Let $\Psi_m \in \mathcal{D}^{2n-2j}(U_{n+1})$ for $m = 1,2,\ldots$ be a sequence of smooth forms in U_{n+1} such that $\Psi_m \to [\tilde{A}]$ as $m \to +\infty$. (For instance, take Ψ_m to be the convolution of $[\tilde{A}]$ with ρ_m, where $\{\rho_m\}$ is a sequence of smooth functions on U_{n+1} converging to the delta function at the identity element of U_{n+1}.) It follows from (3.11) that

$$\int([g\tilde{B}], \Psi_m \wedge \tilde{\phi})d\lambda(g) = c(\alpha^{n-k}, \Psi_m \wedge \tilde{\phi}) = c(\Psi_m, \alpha^{n-k} \wedge \tilde{\phi}). \tag{3.13}$$

Combining (3.13) with the identity

$$\int([g\tilde{B}], \Psi_m \wedge \tilde{\phi})d\lambda(g) = ([U_{n+1} \times \tilde{B}], \tau^*\nu \wedge p_1^*(\Psi_m \wedge \tilde{\phi}))$$
$$= (\Psi_m \times [\tilde{B}], \tau^*\nu \wedge p_1^*\tilde{\phi}),$$

and letting $m \to +\infty$, we obtain

$$c([\tilde{A}], \alpha^{n-k} \wedge \tilde{\phi}) = ([\tilde{A} \times \tilde{B}], \tau^*\nu \wedge p_1^*\tilde{\phi}).$$

Equation (3.10) then follows from (3.12). Q.E.D.

We now give an application of Crofton's Formula to value distribution theory.

Theorem 3.6: *Let* A *be a pure* k-*dimensional analytic set in* $B_R \subset \mathbb{C}^m$. *Then*

$$\int_{G_{m,m-q}} n(A \cap P,r)d\mu(P) = n(A,r)$$

for $q \leq k$, $0 < r < R$. *If* $q = k$, *the multiplicity of* $A \cap P$ *at* 0

must be taken into account in defining $n(A \cap P, r)$.

Proof: We first consider the case $0 \notin A$. Let $A_r = A \cap B_r$. Then, using (1.3)

$$n(A,r) = ([A_r], \tilde{\omega}^k) = (p_*[A_r], \omega^k).$$

The current $p_*[A_r]$ is given by integration over the immersed analytic space $p(A_r)$, possibly with (positive integral) multiplicities. Therefore

$$n(A,r) = \mathrm{vol}_{2k}(p(A_r)),$$

where $p(A_r)$ may have multiplicities. We apply Theorem 3.1 to conclude that

$$n(A,r) = \mathrm{vol}_{2k}(p(A_r)) = \int \mathrm{vol}_{2k}(p(A_r \cap P))d\mu(P) = \int n(A \cap P, r)d\mu(P).$$

We now consider the case $0 \in A$. By [12, pp. 260-261], we have

$$n(A,r) - n(A,\varepsilon) = ([A], \chi_{B_r - B_\varepsilon} \tilde{\omega}^k) = (p_*[A \cap (B_r - B_\varepsilon)], \omega^k).$$

Let

$$n(A,0) = \lim_{\varepsilon \to 0} n(A,\varepsilon);$$

$n(A,0)$ is the *Lelong number* of A at 0. By applying Theorem 3.1 to $p_*[A \cap (B_r - B_\varepsilon)]$ and letting $\varepsilon \to 0$, we then obtain

$$n(A,r) - n(A,0) = \int [n(A \cap P, r) - n(A \cap P, 0)]d\mu(P). \tag{3.14}$$

To complete the proof, we observe that

$$n(A \cap P, 0) = n(A,0) \quad \text{for almost all} \quad P \in G_{m,m-q}. \tag{3.15}$$

The Lelong number $n(A,0)$ is an integer and in fact equals the volume of the algebraic *tangent cone* C_A of A regarded as an algebraic subvariety (with multiplicities) of $\mathbb{P}^{m-1}$ (see [20] and [11], 4.2). If $q < k$, then (3.15) follows from the fact that if V is an *algebraic* set in $\mathbb{P}^{m-1}$, then

$$\mathrm{vol}(V) = \mathrm{vol}(V \cap P)$$

for all $P \in G_{m,m-q}$ such that P intersects V transversally outside of a thin subset of $V \cap P$. (This fact can be verified topologically as follows: The set V is homologous to c copies of a linear subspace, where $c = \mathrm{vol}(V) \in \mathbb{Z}$. Then $V \cap P$ is also homologous to c copies of a linear subspace, and therefore $\mathrm{vol}(V \cap P) = c$ (see [11], 4.1, or [3]). If $q = k$, (3.15) follows from the definition of the multiplicity of $A \cap P$ at 0; see Section 4 of [11]. (One can instead use Theorem 3.1 to show that

$n(A,0) = \int n(A \cap P,0)d\mu(P).)$ Q.E.D.

Corollary 3.7: *Under the conditions of Theorem* 3.6,

$$\int_{G_{m,m-q}} N(A \cap P,r)d\mu(P) = N(A,r).$$

Corollary 3.7 also follows from the results of Section 5; see the remark at the end of that section. (Theorem 3.6 follows from Corollary 3.7 by differentiating with respect to log r.) Corollary 3.7 is implicitly given by Stoll; see Satz 9.2 in [17] and Proposition 5.5 and Section 7 in [18]. Theorem 3.4 and Corollary 3.5 remain valid if A is a real submanifold of $\mathbb{P}^n$. (See the remark following Theorem 4.1.)

4. INTERSECTION OF CURRENTS

In this section we generalize the results of Section 3, using the language of *geometric measure theory*. We assume the reader has familiarity with the concepts in Federer [5]. We also use methods from Brothers [1].

A current T is said to be *locally normal* if both T and dT are representable by integration. The space $F^k_{loc}(X)$ of locally flat currents of degree k on an oriented manifold X is the closure in the "locally flat topology" on $\mathcal{D}'^k(X)$ of the space of locally normal currents of degree k on X ([5] 4.1.12) (see also [9], Section 1.1 or [11], 2.1-2.2). If X is compact, we have $F^k(X) = F^k_{loc}(X)$. (Elements of $F^k(X)$ are called *flat currents*.) If X is a compact complex manifold, we let $F^{p,q}(X)$ be $F^{p+q}(X) \cap \mathcal{D}'^{p,q}(X)$.

An important class of locally flat currents that we shall consider are the currents of the form $[M]$ described in Section 1 and, in particular, currents given by integration over immersed analytic spaces. We now describe the concept of the intersection of locally flat currents, which generalizes the (transverse) intersection of submanifolds. Let Ω be an open subset of $\mathbb{R}^n$ and let $S \in F^p_{loc}(\Omega)$, $T \in F^q_{loc}(\Omega)$ such that $S \times T \in F^{p+q}_{loc}(\Omega \times \Omega)$, and $p + q \le n$. Let $p_1: \Omega \times \Omega \to \Omega$, $\tau: \Omega \times \Omega \to \mathbb{R}^n$ be given by $p_1(x,y) = x$, $\tau(x,y) = x - y$. Federer ([5] 4.3.20) defines the *intersection* $S \cap T \in F^{p+q}_{loc}(\Omega)$ by

$$S \cap T = (-1)^{p(n-q)}\, p_{1*}\langle S \times T, \tau, 0\rangle \tag{4.1}$$

provided the *slice* $\langle S \times T, \tau, 0\rangle$ exists. See Federer ([5] 4.3.1) or [4] for the definition of slicing. (See also [9] Section 1.3.)

If $S \in F^p_{loc}(X)$, $T \in F^q_{loc}(X)$ such that $S \times T \in F^{p+q}_{loc}(X \times X)$ where X

is an oriented C^∞ manifold, then we say that the intersection $S \cap T \in F^{p+q}_{loc}(X)$ exists provided that for all orientation preserving coordinate charts $h: U \to \Omega \subset \mathbb{R}^n$ (where U is open in X), we have

$$h_*(S \cap T|U) = h_*(S|U) \cap h_*(T|U).$$

The product $S \times T$ of two locally flat currents S and T is locally flat if S and T are representable by integration or if either S or T is locally normal.

We note that $T \cap S = (-1)^{pq} S \cap T$, provided $S \cap T$ exists. Furthermore, $S \cap \Psi = S \wedge \Psi$ for $\Psi \in \mathcal{E}^q(X)$ ([5] 4.3.20). However, the main fact on intersections of geometric interest is that if M,N are submanifolds of X that intersect transversally, then $[M] \cap [N] = [M \cap N]$. In fact, if A, B are complex analytic sets of dimensions j,k in an n-dimensional complex manifold X such that $\dim A \cap B = j + k - n$, then $[A] \cap [B]$ exists and is the holomorphic chain given by integration on $A \cap B$ with appropriate intersection multiplicities. This fact is proved in the more general case of real-analytic sets by Hardt [8] and can also be shown using Theorem 3.3.2 and Section 4.1 of King [11]. The intersection of analytic sets is furthermore an associative operation. If $f_1,\dots,f_q$ are holomorphic functions on X, we define

$$D(f_1,\dots,f_q) = \mathrm{Div}\, f_1 \cap \dots \cap \mathrm{Div}\, f_q \tag{4.2}$$

provided that $\dim \mathrm{loc}\,(f_1,\dots,f_q) = n - q$. (One alternately can define $D(f_1 \dots, f_q) = \langle 1, f, 0 \rangle$, where $f = (f_1,\dots,f_q) : X \to \mathbb{C}^q$. Another description of $D(f_1,\dots,f_q)$ which generalizes the Poincaré-Lelong formula is given in [10].) If $h: Y \to X$ is a holomorphic map, we then can define

$$h^{-1} D(f_1,\dots,f_q) = D(f_1 \circ h,\dots,f_q \circ h)$$

provided that $\mathrm{loc}\,(f_1 \circ h,\dots,f_q \circ h)$ is of codimension q. In general, the intersection $S \cap T$ may not exist--for example, when $S = [M]$, $T = [N]$ where M, N, and $M \cap N$ are real submanifolds such that $\mathrm{codim}\ M \cap N <$ co-dim M + codim N. We shall use Brothers's description of intersections in a homogeneous space (such as $\mathbb{P}^n$) in which the intersection "almost always exists" [1]. Let G be an n-dimensional Lie group and let $S \in F^p_{loc}(G)$, $T \in F^q_{loc}(G)$ such that $S \times T \in F^{p+q}_{loc}(G \times G)$. We let τ, $p_1: G \times G \to G$ be given by $\tau(a,b) = ab^{-1}$, $p_1(a,b) = a$. If $g \in G$, we also let g denote the left action of g on G. Then for almost all $g \in G$, $S \cap g_*T$ exists, and

$$S \cap g_* T = (-1)^{p(n-q)} \; p_{1*} \langle S \times T, \tau, g \rangle. \tag{4.3}$$

Now suppose G acts transitively on an oriented homogeneous space X. Let $\pi: G \to X$ be given by $\pi(g) = gx_o$, where $x_o \in X$; $\pi: G \to X$ is a principal fibre bundle. The linear transformation $\pi^*: \mathcal{E}^p(X) \to \mathcal{E}^p(G)$ extends continuously to a linear transformation $\pi^*: \mathcal{D}'^p(X) \to \mathcal{D}'^p(G)$. (If one identifies $\pi^{-1}(U)$ with $U \times \pi^{-1}(X_o)$, for a sufficiently small open set U in X, then $\pi^* T = T \times 1$ for $T \in \mathcal{D}'^p(U)$. The map π^* agrees with the map L considered by Brothers [1].) Note that $\pi^*[M] = [\pi^{-1}(M)]$ for a submanifold M of X. If $S \in F^p_{loc}(X)$, $T \in F^q_{loc}(X)$ such that $S \times T \in F^{p+q}_{loc}(X \times X)$, then for almost all $g \in G$, $S \cap g_* T$ exists and

$$\pi^*(S \cap g_* T) = \pi^* S \cap g_*(\pi^* T) \tag{4.4}$$

where the intersection on the right hand side of (4.4) is given by (4.3). If S and T are representable by integration, then so is $S \cap g_* T$ for almost all g. The following result generalizes Theorem 3.4.

Theorem 4.1: *Let* $S \in F^k(\mathbb{P}^n)$, $T \in F^{q,q}(\mathbb{P}^n)$ *such that* $S \times T \in F^{k+2q}(\mathbb{P}^n \times \mathbb{P}^n)$. *Then*

$$\int_{U_{n+1}} S \cap g_* T \, d\lambda(g) = c \, S \wedge \omega^q$$

where

$$c = (T, \omega^{n-q}).$$

If S *and* T *are representable by integration, then the above formula is valid for* C° *forms.*

Remark: It follows from Theorem 4.1 that Theorem 3.4 is also valid for A a real j-dimensional submanifold of $\mathbb{P}^n$ (and $\phi \in A^{j+2k-2n}(\mathbb{P}^n)$). This fact also follows directly from the proof of Theorem 3.4.

Corollary 4.2: *Let* $S \in F^k(\mathbb{P}^n)$. *Then*

$$\int_{G_{n+1,n+1-q}} S \cap [P] \, d\mu(P) = S \wedge \omega^q$$

Theorem 4.1 is an immediate consequence of Lemma 3.3 and the following lemma:

Lemma 4.3: *Let* X *be an oriented homogeneous space with* n-*dimensional Lie group of transformations* G. *Let* $T \in F^q_{loc}(X)$ *and let*

$$\alpha = \int_G g_* T \, d\lambda(g) \in \mathcal{E}^q(X),$$

where $d\lambda = \nu \in \mathcal{D}^n(G)$.

If $S \in F^p_{loc}(X)$ *such that* $S \times T \in F^{p+q}_{loc}(X \times X)$, *then*

$$\int_G S \cap g_* T \, d\lambda(g) = S \wedge \alpha.$$

If S *and* T *are representable by integration, then the above formula is valid for* C° *forms.*

Proof: (This proof is formally the same as the proof of Theorem 3.4 from Lemma 3.3.) By considering the fibre bundle $\pi: G \to X$ and applying π^*, we conclude that it suffices to consider the case $X = G$ (recall (4.4)). It follows from (4.3) and ([5] 4.3.2(1)) that

$$\begin{aligned} \int_G S \cap g_* T \, d\lambda(g) &= (-1)^{p(n-q)} \, p_{1*} \int_G \langle S \times T, \tau, g \rangle d\lambda(g) \\ &= (-1)^{p(n-q)} \, p_{1*}[(S \times T) \wedge \tau^* \nu]. \end{aligned} \tag{4.5}$$

(Equation (3.12) in the proof of Theorem 3.4 corresponds to (4.5).) As in the proof of Theorem 3.4, let $\Psi_m \in \mathcal{E}^p(G)$, for $m = 1, 2, \ldots$, such that $\Psi_m \to S$ as $m \to +\infty$. Applying (4.5) with $S = \Psi_m$, we conclude that

$$\int_G \Psi_m \cap g_* T \, d\lambda(g) \to \int_G S \cap g_* T \, d\lambda(g)$$

as $m \to +\infty$. However

$$\int_G \Psi_m \cap g_* T \, d\lambda(g) = \Psi_m \wedge \int_G g_* T \, d\lambda(g) = \Psi_m \wedge \alpha \tag{4.6}$$

and the conclusion of the lemma then follows by letting $m \to +\infty$ in (4.6).

Remark: The only property of $\mathbb{P}^n$ that we need in the proof of Theorem 4.1 is that the group $G = (U_{n+1}/S^1)$ of holomorphic isometries of $\mathbb{P}^n$ acts transitively on $\mathbb{P}^n$ and has isotropy group equal to U_n. Suppose X is an n-dimensional Kähler manifold with a group G of holomorphic isometries acting transitively and effectively on X such that the isotropy subgroup $\{g \in G: gx_o = x_o\}$ is the full unitary group U_n. Then one can show that there are only three possibilities: 1) $X = \mathbb{P}^n$; 2) $X = \mathbb{C}^n$; 3) $X = B_1$ (the unit ball in $\mathbb{C}^n$) with the hyperbolic metric that is invariant under the group G of all holomorphic automorphisms of B_1. In each case, G is unimodular. Let η denote the Kähler form on X, and let λ denote Haar measure on G normalized such that $\pi_*(d\lambda) = \eta^n$ (where $\pi: G \to X$ is given by $\pi(g) = g\,x_o$). Using the methods in the proof of Theorem 4.1, one obtains the following result:

If $S \in F^k_{loc}(X)$ and $T \in F^{q,q}(X)$ such that $S \times T \in F^{k+2q}_{loc}(X \times X)$, then

$$\int_G S \cap g_* T \, d\lambda(g) = c\, S \wedge \eta^q \qquad (4.7)$$
$$\text{where } c = (T, \eta^{n-q}).$$

(Theorem 4.1 is (4.7) for the case $X = \mathbb{P}^n$.) One similarly extends Theorem 3.4 to the cases $X = \mathbb{C}^n$ and $X = B_1$.

5. AN AVERAGING FORMULA FOR $T^q(L,r)$

We interpret the order functions $T^q(L,r)$ as average counting functions (Theorem 5.3) by means of the following "pull-back" of the Crofton Formula.

Theorem 5.1: *Let* $\alpha: V \to \mathbb{P}^n$ *be a holomorphic map, where* V *is a complex manifold. Then*

$$\int_{U_{n+1,n+1-q}} \alpha^{-1}[P]\, d\mu(P) = \alpha^* \omega^q$$

for $1 \le q \le n$.

We shall obtain Theorem 5.1 as a special case of a general result involving the *pull-back* of a current (Theorem 5.7).

We shall use the following notation and terminology. If E is an m-dimensional complex vector space, we let $G_{E,k}$ denote the Grassmannian of complex k-dimensional subspaces of E, and we let $\mathbb{P}(E) = G_{E,1}$. If $P \in G_{E,k}$, we let

$$P^* = \{V^* \in E^*: V^*(P) = \{0\}\} \in G_{E^*,m-k}$$

(where E^* denotes the dual of E). A hermitian inner product on E defines a Kähler form ω on $\mathbb{P}(E)$ and a measure μ on $G_{E,k}$ by identifying E with $\mathbb{C}^n$ via an orthonormal basis for E.

Definition 5.2: Let L be a holomorphic line bundle on a compact complex manifold V. We say that L is *semi-ample* if for all $x \in V$, there exists a section $\Psi \in \Gamma(V, \mathcal{O}(L))$ such that $\Psi(x) \neq 0$.

Note that the product of semi-ample bundles and the pull-back of a semi-ample bundle are semi-ample. Ample bundles as well as trivial bundles are semi-ample. Suppose L is a non-trivial semi-ample line bundle on a compact complex manifold V, and let $E = \Gamma(V, \mathcal{O}(L))$. We then have the holomorphic map

$$\alpha : V \to \mathbb{P}(E^*)$$

given by $\alpha = p \circ \tilde{\alpha}$, where $\tilde{\alpha}: V \to E^*-\{0\}$ is given by $(\tilde{\alpha}(x), \Psi) = \Psi(x)$ and $p: E^*-\{0\} \to \mathbb{P}(E^*)$ is the natural projection. Then $L = \alpha^* L_o$, where L_o is the hyperplane section bundle on $\mathbb{P}(E^*)$. For $\Psi \in E$, we let H_Ψ denote the hyperplane $\{\Psi = 0\}$ in $\mathbb{P}(E^*)$, and we note that

$$\text{Div } \Psi = \alpha^{-1} H_\Psi. \tag{5.1}$$

Now suppose $P \in G_{E,q}$, and let $\{\Psi_1,\ldots,\Psi_q\}$ be a basis for P. We write

$$D_P = D(\Psi_1,\ldots,\Psi_q) \in \mathcal{D}'^{q,q}(V), \tag{5.2}$$

provided that $D(\Psi_1,\ldots,\Psi_q)$ exists (recall (4.2)). Alternately, we note that $P^* = H_{\Psi_1} \cap \ldots \cap H_{\Psi_q} \in G_{E^*,m-q}$, and therefore

$$D_P = \alpha^{-1} H_{\Psi_1} \cap \ldots \cap \alpha^{-1} H_{\Psi_q} = \alpha^{-1}(P^*), \tag{5.3}$$

which shows that D_P is independent of the choice of basis for P.

Choose a hermitian inner product on E, and let ω denote the corresponding Kähler form on $\mathbb{P}(E^*)$ and μ denote the invariant measure on $G_{E,q}$. Then the form $\eta = \alpha^*\omega$ is a curvature form for L (η is positive-semi-definite everywhere, so L is "semi-positive").

We have the following result.

Theorem 5.3: *Let* L *be a non-trivial semi-ample line bundle on a compact complex manifold* V, *and let* $f: \mathbb{C}^n \xrightarrow[m]{} V$ *be a meromorphic map. Then*

$$\int_{G_{E,q}} N(D_P, r)\, d\mu(P) = T^q(L,r)$$

where $E = \Gamma(V, \mathcal{O}(L))$ *and* $T^q(L,r)$ *is computed with respect to the curvature form* η (η *and* μ *are given as above).*

Remark. Note that the multiplicity of D_P is identically 1, for almost all P. Thus, we can ignore the multiplicities of D_P in the above equation.

Proof: Let X_o denote the set of points $z \in \mathbb{C}^n$ such that f is holomorphic at z, and let $f_o: X_o \to V$ define f. Choose a basis $\{\Psi_1,\ldots,\Psi_m\}$ of $E = \Gamma(V, \mathcal{O}(L))$. Let $\hat{\alpha}: X_o \to \mathbb{P}^{m-1}$ be given by

$$\hat{\alpha} = (\Psi_1 \circ f : \ldots : \Psi_m \circ f).$$

Use $\{\Psi_1,\ldots,\Psi_m\}$ to identify $\mathbb{P}^{m-1}$ with $\mathbb{P}(E^*)$ in the obvious way. Then $\hat{\alpha} = \alpha \circ f_o$. Let

$$\hat{\eta} = \hat{\alpha}^*\omega = f_o^* \eta$$

(where $\eta = \alpha^*\omega$). We conclude from Theorem 5.1 that

$$\int_{G_{m,m-q}} \hat{\alpha}^{-1}[Q]\, d\mu(Q) = \hat{\eta}^q = f_o^* \eta^q. \tag{5.4}$$

It follows from (5.3) that for $P \in G_{E,q}$ such that D_P is defined, we have

$$f_o^{-1} D_P = f_o^{-1} \alpha^{-1} [P^*] = \alpha^{-1} [P^*]. \tag{5.5}$$

It therefore follows from (5.4) and (5.5) (with $Q = P^*$) that

$$\int_{G_{E,q}} f_o^{-1} D_P\, d\mu(P) = f_o^* \eta^q. \tag{5.6}$$

Let $\rho_k \in \mathcal{D}^\circ(X_o)$ for $k = 1,2,\ldots,$ be a sequence of functions such that $0 \le \rho_k \le \rho_{k+1}$ and $\rho_k \to \chi_{X_o \cap B_r}$ pointwise as $k \to +\infty$. Applying (5.6) to the form $\rho_k \nu^{n-q}$, we obtain

$$\int_{G_{E,q}} (f_o^* D_P, \rho_k \nu^{n-q})\, d\mu(P) = (f_o^* \eta^q, \rho_k \nu^{n-q}). \tag{5.7}$$

Letting $k \to +\infty$ in (5.7) and applying Lebesgue's monotone convergence theorem, we obtain

$$\int_{G_{E,q}} n(D_P, r)\, d\mu(P) = n(f^* \eta^q, r)$$

and therefore

$$\int_{G_{E,q}} N(D_P, r)\, d\mu(P) = N(f^* \eta^q, r) = T^q(L,r).$$

Q.E.D.

The averaging formula in Theorem 5.3 with an added $O(1)$ term is proved (for holomorphic maps) indirectly in Griffiths-King ([7], 5.18) using the First Main Theorem. For meromorphic maps $f\colon \mathbb{C}^n \xrightarrow[m]{} \mathbb{P}^k$ the result is proved by Stoll ([19], Theorem 3.2) also using the First Main Theorem.

We note that the map $\Psi \longrightarrow \operatorname{Div} \Psi$ defines a 1-1 correspondence between $\mathbb{P}(E)$ and $|L|$, and we obtain the following result.

Corollary 5.4: *Let* L, V, f *be as in Theorem 5.3. Then*

$$\int_{|L|} N(D,r)\, d\mu(D) = T(L,r).$$

We now define a general method of pulling back currents, generalizing the pull-backs of holomorphic chains.

Definition 5.5: Let $f\colon Y \to X$ be a differentiable map, where $X \subset \mathbb{R}^n$ and

Y is an oriented C^∞ manifold, and let $T \in F^k_{loc}(X)$. Let $\gamma: X \times Y \to \mathbb{R}^n$, $p_2: X \times Y \to Y$ be given by $\gamma(x,y) = x - f(y)$, $p_2(x,y) = y$. Define $f^{\#}T \in F^k_{loc}(X)$ by

$$f^{\#}T = p_{2*}\langle T \times 1, \gamma, 0\rangle$$

provided the slice $\langle T \times 1, \gamma, 0\rangle$ exists. The current $f^{\#}T$ is called the *pull-back* of T by f. If instead X is an oriented C^∞ manifold, then $f^{\#}T$ is defined using coordinate charts on X.

If $f^{\#}T$ exists, then $f^{\#}(dT) = d(f^{\#}T)$. If $\Psi \in E^k(X)$, then $f^{\#}\Psi = f^*\Psi$ (see Lemma 5.6). If f is a diffeomorphism, then $f^{\#}T = (f^{-1})_*T$. (If $\pi: G \to X$ is as in Section 4, then $\pi^{\#}T = \pi^*T$.) Now suppose $f: Y \to X$ is a holomorphic map, where X and Y are complex manifolds. We claim that $f^{\#}T = f^{-1}T$ for holomorphic chains T such that $f^{-1}T$ exists. It suffices to consider $T = [A]$ where A is an analytic set of codimension q. Suppose $[A] = D(g_1,\ldots,g_q)$ and $X \subset \mathbb{C}^n$. Then

$$\langle [A] \times 1, \gamma, 0\rangle = \langle [A \times Y], \gamma, 0\rangle$$
$$= D(g_1(x),\ldots,g_q(x), x - f(y))$$

and therefore

$$f^{\#}[A] = p_{2*}\, D(g_1(x),\ldots,g_q(x), x - f(y))$$
$$= D(g_1\circ f,\ldots,g_q\circ f) = f^{-1}[A].$$

Since f^{-1} and $f^{\#}$ agree for complete intersections A, they agree for arbitrary A by the continuity of slicing of holomorphic chains (see [8] or [11]).

Lemma 5.6: *Let* $f: Y \to X$ *be a differentiable map of oriented* C^∞ *manifolds* X *and* Y *and let* $T \in F^k_{loc}(X)$. *If* $f^{\#}T$ *exists, then* $T \cap f_*\phi$ *exists and*

$$(f^{\#}T, \phi) = (T \cap f_*\phi, 1)$$

for all $\phi \in \mathcal{D}^{n-k}(Y)$, *where* $n = \dim Y$.

Proof: Assume without loss of generality that X is a domain in $\mathbb{R}^m$. Let $\tilde{f} = id_X \times f: X \times Y \to X \times X$. Then $\gamma = \tau \circ \tilde{f}$ where $\tau(x_1,x_2) = x_1 - x_2$. If $f^{\#}T$ exists, then

$$(f^{\#}T, \phi) = (\langle T \times 1_Y, \gamma, 0\rangle, 1_X \times \phi)$$
$$= (\langle T \times \phi, \gamma, 0\rangle, 1_{X \times Y}).$$

By ([5] 4.3.2 (7)) (see also [9] 1.19),

$$\tilde{f}_* \langle T \times \phi, \gamma, 0\rangle = \langle \tilde{f}_*(T \times \phi), \tau, 0\rangle$$
$$= \langle T \times f_* \phi, \tau, 0\rangle,$$

and hence

$$(f^{\#}T, \phi) = (\tilde{f}_* \langle T \times \phi, \gamma, 0\rangle, 1_{X \times X})$$
$$= (\langle T \times f_* \phi, \tau, 0\rangle, 1_{X \times X})$$
$$= (T \cap f_* \phi, 1).$$

Q.E.D.

Finally, we note that if $T \in F^k_{loc}(G)$, where G is a Lie group, and $f\colon Y \to G$ is differentiable, then for almost all $g \in G$, $f^{\#}(g_*T)$ exists and

$$f^{\#}(g_*T) = (g^{-1} \circ f)^{\#}T = p_{2*}\langle T \times 1, \gamma', g\rangle,$$

where $\gamma'\colon G \times Y \to G$ is given by $\gamma'(a,y) = f(y)a^{-1}$. If X is a homogeneous space with transitive group of transformations G and $f\colon Y \to X$ is a differentiable map, then by lifting f locally to G one can then show that for almost all $g \in G$, the pull-back $f^{\#}(g_*T)$ exists (and equals $(g^{-1} \circ f)^{\#}T$).

Theorem 5.7: *Let* $f\colon Y \to \mathbb{P}^n$ *be a differentiable map, where* Y *is an oriented manifold, and let* $T \in F^{q,q}(\mathbb{P}^n)$. *Then for almost all* $g \in U_{n+1}$, $f^{\#}(g_*T)$ *exists and*

$$\int_{U_{n+1}} f^{\#}(g_*T)\, d\lambda(g) = c\, f^* \omega^q$$

where $c = (T, \omega^{n-q})$.

Proof: Let $\phi \in \mathcal{D}^{m-2q}(Y)$, where $m = \dim Y$. Then by Theorem 4.1 and Lemma 5.6,

$$\int (f^{\#}(g_*T), \phi)\, d\lambda(g) = \int (g_*T \cap f_* \phi, 1)\, d\lambda(g)$$
$$= (c\, f_* \phi \wedge \omega^k, 1) = c(\phi, f^* \omega^k).$$

Q.E.D.

Theorem 5.1 follows immediately from the following corollary to Theorem 5.7.

Corollary 5.8: *Let* $f\colon Y \to \mathbb{P}^n$ *be a differentiable map, where* Y *is an oriented manifold. Then*

$$\int_{G_{n+1,n+1-q}} f^{\#}[P]\, d\mu(P) = f^* \omega^q.$$

Remark: Theorem 5.3 remains valid with $\mathbb{C}^n$ replaced by a Stein manifold as considered by Stoll [19] or more generally with $\mathbb{C}^n$ replaced by a Stein space. Corollary 3.7 is a consequence of this generalization of Theorem 5.3. Let A be a pure k-dimensional analytic set in $\mathbb{C}^m$ and assume for simplicity that $0 \notin A$. Let $f: A \to \mathbb{P}^{m-1}$ be the "residual map;" i.e., $f = p|A$ where $p: \mathbb{C}^m - \{0\} \to \mathbb{P}^{m-1}$ is the usual projection. We observe that

$$N([A],r) = T_f^q(L_o,r) = (f^* \omega^q, \ell_{sr} \tilde{\omega}^{k-q})$$

for $1 \le q \le k$ (where L_o denotes the hyperplane section bundle on $\mathbb{P}^{m-1}$), and Corollary 3.7 then follows from Theorem 5.3 applied to the map f. Corollary 3.7 also follows directly from Theorem 5.1 applied to the map $f: A \to \mathbb{P}^{m-1}$. (See [11] for the definition of a current on a complex space.)

6. AN EXAMPLE

We shall compute the hyperplane-section defects for the holomorphic map $F: \mathbb{C}^2 \to \mathbb{P}^2$ given by

$$F(z_1,z_2) = (1: \exp(z_1): \exp(z_2)). \tag{6.1}$$

We first develop some general machinery that we shall use for computing the order function and counting functions for F.

Let L be a hermitian line bundle on a compact manifold V and let $f: \mathbb{C}^n \to V$ be a holomorphic map. (The results of the following discussion are also valid for meromorphic maps f.) We claim that

$$T_f(L,r) = \int_{G_{n,k}} T_{f|P}(L,r)\, d\mu(P) \tag{6.2}$$

for $1 \le k < n$. The averaging equation (6.2) is verified as follows: For $\Psi \in A^j(\mathbb{C}^n)$ such that Ψ has compact support, we define the current $p_* \Psi \in \mathcal{D}'^{j-2}(\mathbb{P}^{n-1})$ by

$$(p_* \Psi, \phi) = \int_{\mathbb{C}^n-\{0\}} \Psi \wedge p^* \phi = (p^* \phi, \Psi) \tag{6.3}$$

for $\phi \in \mathcal{D}^{2n-j}(\mathbb{P}^{n-1})$ where $p^* \phi \in \mathcal{D}'^{2n-j}(\mathbb{C}^n)$ is defined as in (1.7) (regarding p as a meromorphic map from $\mathbb{C}^n$ into $\mathbb{P}^{n-1}$). One easily observes that in fact $p_* \Psi \in A^{j-2}(\mathbb{P}^{n-1})$. Then

$$T_f(L,r) = (f^*\eta, \ell_{sr}\tilde{\omega}^{n-1}) = (\omega^{n-1}, \alpha)$$

where $\alpha = p_*(\ell_{sr} f^*\eta) \in A^{\circ}(\mathbb{P}^{n-1})$. Similarly, for $P \in G_{n,k}$,

$$T_{f|P}(L,r) = ([P] \wedge \omega^{k-1}, \alpha).$$

It therefore follows from Lemma 3.3 (or from Corollary 3.5) that

$$\begin{aligned} \int T_{f|P}(L,r)\, d\mu(P) &= \int([P], \omega^{k-1} \wedge \alpha)\, d\mu(P) \\ &= (\omega^{n-k}, \omega^{k-1} \wedge \alpha) = T_f(L,r). \end{aligned}$$

Suppose D is an analytic hypersurface in V such that $f(\mathbb{C}^n) \not\subset D$. It follows from Theorem 3.6 that

$$N_f(D,r) = \int_{G_{n,k}} N_{f|P}(D,r)\, d\mu(P) \tag{6.4}$$

for $1 \le k < n$. For $\lambda \in \mathbb{C}^n$ let $f_\lambda\colon \mathbb{C}^1 \to V$ be given by $f_\lambda(t) = f(t\lambda)$ for $t \in \mathbb{C}$. Let ν denote the usual measure on the unit $(2n-1)$-sphere $S = \{z \in \mathbb{C}^n\colon |z| = 1\}$ normalized such that $\nu(S) = 1$. For $k = 1$, equations (6.2) and (6.4) can be written in terms of integration on S as follows:

$$T(L,r) = \int_S T_\lambda(L,r)\, d\nu(\lambda), \tag{6.5}$$

$$N(D,r) = \int_S N_\lambda(D,r)\, d\nu(\lambda), \tag{6.6}$$

where we drop the f from $T_f(L,r)$, $T_{f_\lambda}(L,r)$, etc., to simplify the notation.

Now consider the map $F\colon \mathbb{C}^2 \to \mathbb{P}^2$ defined by (6.1); we then have

$$F_\lambda(t) = (1\colon \exp(\lambda_1 t)\colon \exp(\lambda_2 t)).$$

We let $T_F(r) = T_F(L_o,r)$ and $T_\lambda(r) = T_\lambda(L_o,r)$, where L_o is the hyperplane-section bundle in $\mathbb{P}^2$, denote the classical characteristic functions of F and F_λ respectively. According to H. Weyl ([22], pp.94-98),

$$T_\lambda(r) = \frac{|\lambda_1|+|\lambda_2|+|\lambda_1-\lambda_2|}{2\pi}(r-s) + c(r) \tag{6.7}$$

$$\text{where } |c(r)| \le \tfrac{1}{2}\log 3.$$

(Recall that s is a fixed positive number; see (1.2).) It follows from (6.5) and (6.7) by elementary calculus that

$$T_F(r) = \frac{2+\sqrt{2}}{3\pi} r + O(1) . \tag{6.8}$$

To describe the defects for F, we consider the following projective lines in $\mathbb{P}^2$ given in terms of the homogeneous coordinates $(w_o : w_1 : w_2)$:

$$A^j = \{w_j = 0\}, \text{ for } j = 0, 1, 2;$$

$$B_a{}^o = \{w_2 = aw_1\},\ B_a{}^1 = \{w_2 = aw_o\},\ B_a{}^2 = \{w_1 = aw_o\},$$

$$\text{for } a \in \mathbb{C} - \{0\}.$$

Since $F^{-1}(A^j) = \emptyset$, we have $N(A^j,r) = 0$ and hence

$$\delta(A^j) = 1 \quad \text{for } j = 0, 1, 2. \tag{6.9}$$

One easily computes that

$$N_\lambda(B_a{}^j, r) = \frac{\alpha_j}{\pi}(r - s) + O(1) \quad \text{for } j = 0, 1, 2,$$

where $\alpha_o = |\lambda_1 - \lambda_2|$, $\alpha_1 = |\lambda_2|$, $\alpha_2 = |\lambda_1|$, and the $O(1)$ term is bounded independently of r and λ. It then follows from (6.6) that

$$\left.\begin{aligned} N_F(B_a{}^o, r) &= \frac{2\sqrt{2}}{3\pi} r + O(1)\ , \\ N_F(B_a{}^j, r) &= \frac{2}{3\pi} r + O(1) \quad \text{for } j = 1, 2. \end{aligned}\right. \tag{6.10}$$

Thus

$$\delta(B_a{}^o) = 3 - 2\sqrt{2},\ \delta(B_a{}^1) = \delta(B_a{}^2) = \sqrt{2} - 1 \tag{6.11}$$

for all $a \in \mathbb{C} - \{0\}$.

The defect relations for F are:

$$\Sigma\ \delta(L_j) \le 3 \tag{6.12}$$

where the sum is over any family $\{L_j\}$ of lines in $\mathbb{P}^2$ with normal crossings (no more than two intersect at any point in $\mathbb{P}^2$). The defect relations (6.12) are due to Stoll [17] and also follow from Corollary 2.4. It follows from (6.9) and (6.12) (or from [22] pp. 98-105) that all other lines in $\mathbb{P}^2$ have zero defect. We summarize the defects for F in the illustration at the top of the following page.

Let $D_1 = B_a{}^o$, $D_2 = A^o$, $D_3 = A^1$, $D_4 = A^2$, and let $S = \{(1{:}0{:}0)\}$. Then the hypotheses of Theorem 2.6 (and Corollaries 2.7 and 2.8) are satisfied with $p = 2$, $k = 1$, $q = 4$. Corollary 2.7 gives the defect relation

$$\sum_{j=1}^{4} \delta(D_j) \le 3 + \limsup_{r \to +\infty} \frac{m(S,r)}{T(L,r)}\ . \tag{6.13}$$

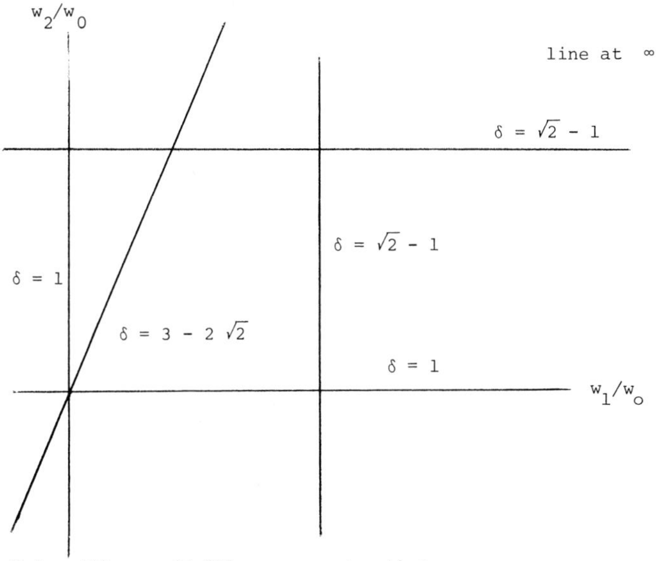

Using [22, pp. 94-98] one computes that

$$m_\lambda(S,r) = \frac{|\lambda_1|+|\lambda_2|-|\lambda_1-\lambda_2|}{2\pi} r + O(1),$$

where the $O(1)$ term is bounded independently of r and λ, and therefore

$$m(S,r) = \frac{2-\sqrt{2}}{3\pi} r + O(1). \tag{6.14}$$

It follows from the First Main Theorem, (6.8), and (6.10) that

$$m(D_1,r) = \frac{2-\sqrt{2}}{3\pi} r + O(1)$$

and therefore

$$m(D_1,r) = m(S,r) + O(1). \tag{6.15}$$

(Compare Corollary 2.8.) Using (6.8) and (6.14), the defect relation (6.13) becomes

$$\sum_{j=1}^{4} \delta(D_j) \leq 6 - 2\sqrt{2}\ .$$

However, in this case

$$\sum_{j=1}^{4} \delta(D_j) = 6 - 2\sqrt{2}\ ,$$

and therefore the maximum sum of the defects given by Corollary 2.7 is attained in this example.

REFERENCES

1. Brothers, J. E., *Integral geometry in homogeneous spaces*, Trans. Am. Math. Soc. 124 (1966), 480-517.

2. Carlson, J., and P. Griffiths, *A defect relation for equidimensional holomorphic mappings between algebraic varieties*, Ann. of Math. 95 (1972), 557-584.

3. Draper, R. N., *Intersection theory in analytic geometry*, Math. Ann. 180 (1969), 175-204.

4. Federer, H., *Some theorems on integral currents*, Trans. Am. Math. Soc. 117 (1965), 43-67.

5. ______, *Geometric Measure Theory*, Springer-Verlag, New York, 1969.

6. Green, M., *Holomorphic maps into* $\mathbb{P}_n$ *omitting hyperplanes*, Trans. Am. Math. Soc. 169 (1972), 89-103.

7. Griffiths, P., and J. King, *Nevanlinna theory and holomorphic mappings between algebraic varieties*, Acta Math. 130 (1973), 145-220.

8. Hardt, R. M., *Slicing and intersection theory for chains associated with real analytic varieties*, Acta Math. 129 (1972), 75-136.

9. Harvey, R., and B. Shiffman, *A characterization of holomorphic chains*, to appear in Ann. of Math.

10. King, J., *A residue formula for complex subvarieties*, Carolina Conference Proceedings, Chapel Hill, N.C., 1970.

11. ______, *The currents defined by analytic varieties*, Acta Math. 127 (1971), 185-220.

12. Lelong, P., *Intégration sur un ensemble analytique complexe*, Bull. Soc. Math. France 85 (1957), 239-262.

13. ______, *Fonctions entières (n variables) et fonctions plurisousharmoniques d'ordre fini dans* $\mathbb{C}^n$, J. Anal. Math. 12 (1964), 365-407.

14. deRham, G., *Variétés differentiables*, Hermann, Paris, 1960.

15. Santaló, L. A., *Integral geometry in Hermitian spaces*, Am. J. Math. 74 (1952), 423-434.

16. Schwartz, L., *Theorié des distributions*, Hermann, Paris, 1966.

17. Stoll, W., *Die beiden Hauptsätze der Wertverteilungstheorie bei Funktionen mehrerer komplexen Veränderlichen* (I), (II), Acta Math. 90 (1953), 1-115, and 92 (1954), 55-169.

18. __________, *The growth of the area of a transcendental analytic set, II*, Math. Ann. 156 (1964), 144-170.

19. __________, *About the value distribution of holomorphic maps into the projective space*, Acta Math. 123 (1969), 83-114.

20. Thie, P., *The Lelong number of a point of a complex analytic set*, Math. Ann. 172 (1967), 269-312.

21. Thomas, G. E. F., *Integration of functions with values in a locally convex Suslin space*, to appear.

22. Weyl, H., *Meromorphic Functions and Analytic Curves*, Princeton Univ. Press, 1943.

HOLOMORPHIC EXTENSION THEOREMS

Peter Kiernan

University of British Columbia

1. INTRODUCTION

The purpose of this expository paper is to discuss some recent extension theorems in several complex variables which generalize the classical big Picard theorem. The generalizations can be separated into two categories. In one category the extension is always a holomorphic map, while in the other category the extension is a meromorphic map. Our main concern will be to discuss the techniques which have been used to obtain holomorphic extension theorems. We will also compare these techniques to the methods used to proving meromorphic extension theorems.

To be more precise, we recall the big Picard theorem. It says that any holomorphic map f from the punctured disk D^* into the Riemann sphere $\mathbb{P}_1(\mathbb{C})$ which omits three points can be extended to a holomorphic map $f: D \to \mathbb{P}_1(\mathbb{C})$. If we let $X = D$, $A = \{0\}$, $M = \mathbb{P}_1(\mathbb{C}) - \{0, 1, \infty\}$ and $Y = \mathbb{P}_1(\mathbb{C})$, then the theorem says that any holomorphic map $f: X - A \to M$ extends to a holomorphic map $f: X \to Y$. In trying to generalize this, we are led to the following basic problem.

Let A be a closed complex subspace of a complex space X and let M be a relatively compact, open set in a complex space Y. Does every holomorphic map $f: X - A \to M$ extend to a holomorphic map $f: X \to Y$?

It is clear that in order to get an affirmative answer we must place fairly strong restrictions on M and Y, and that these restrictions become stronger as we let X and A become more general spaces. One form of our main result says that if (1) X is nonsingular, (2) the singularities of A are normal crossings and (3) $\mathrm{Hol}(D,M)$ is relatively compact in $\mathrm{Hol}(D,Y)$, then every holomorphic map $f: X - A \to M$ extends to a holomorphic map $f: X \to Y$. This says that if we can generalize Montel's theorem on normal families to the spaces M and Y, then we can get a strong generalization of the big Picard theorem for mappings into M.

The essential tool for investigating our extension problem is the Kobayashi pseudo-distance associated to a complex space. In the next section we will give its basic properties and some examples.

2. KOBAYASHI PSEUDO-DISTANCE

The Kobayashi pseudo-distance d_M which is associated to a complex space M is defined as follows. Let p and q be points in M. By a *chain* α from p to q, we mean a sequence of points $P_0, P_1, \ldots, P_k$ in M, points $a_1, \ldots, a_k$ in the unit disk $D = \{z \in \mathbb{C} \mid |z| < 1\}$ and holomorphic maps $f_1, \ldots, f_k$ of D into M with $f_i(0) = p_{i-1}$ and $f_i(a_i) = p_i$. The length $|\alpha|$ of α is defined by

$$|\alpha| = \sum_{i=1}^{k} d(0,a_i) = \sum_{i=1}^{k} \log \frac{1 + |a_i|}{1 - |a_i|}$$

where d is the Poincaré distance on D. It is induced by the metric

$$ds^2 = (1 - |z|^2)^{-2} dzd\bar{z} .$$

We set $d_M(p,q) = \inf_{\alpha \in A} |\alpha|$ where A is the set of all chains from p to q. It is easy to see that d_M is a pseudo-distance. If d_M is a proper distance, then M is called *hyperbolic* and if d_M is a complete distance, then M is called *complete hyperbolic*.

The definition of d_M is analogous to the definition of the distance induced by a Riemannian metric. We can think of a chain from p to q as a piecewise analytic curve. The length of an analytic curve is determined by the Poincaré metric on D and the length of a chain is just the sum of the lengths of the analytic pieces. Then $d_M(p,q)$ is just the infimum of the lengths of curves joining p and q. The pseudo-distance d_M gives a geometric description of the space Hol(D,M) of holomorphic mappings of D into M. The larger the distance is, the smaller Hol(D,M) is.

Examples: (1) $d_D = d$. This follows immediately from Schwarz's lemma which says that if $f:D \to D$ is holomorphic, then $d(f(z_1), f(z_2)) \le d(z_1,z_2)$ for every $z_1, z_2 \in D$.

(2) $d_{\mathbb{C}} \equiv 0$. To see this, let $f_n:D \to \mathbb{C}$ be defined by $f_n(z) = nz$. For $|z'| < n$, we have

$$d_{\mathbb{C}}(z', 0) = d_{\mathbb{C}}(f_n(\tfrac{z'}{n}), f_n(0)) \le d_D(\tfrac{z'}{n}, 0)$$

Since $d_D(\frac{z'}{n}, 0) \to 0$ as $n \to \infty$, it follows that $d_{\mathbb{C}}(z', 0) = 0$.

(3) If M and N are complex spaces and $\pi:M \to N$ is a holomorphic covering map, then M is (complete) hyperbolic iff N is (complete) hyperbolic. Furthermore,

$$d_N(p,q) = \inf \{d_M(\tilde{p},\tilde{q}) \mid \pi(\tilde{p}) = p \text{ and } \pi(\tilde{q}) = q\}.$$

This implies that a Riemann surface is hyperbolic iff its universal covering space is D. In particular, $\mathbb{P}_1(\mathbb{C}) - \{0,1,\infty\}$ is hyperbolic.

(4) A (complete) hermitian manifold M with holomorphic sectional curvature bounded above by a negative constant is (complete) hyperbolic. This follows from Ahlfors generalization of Schwarz's lemma and the fact that the holomorphic sectional curvature of a submanifold is bounded above by the curvature of the manifold.

The most important property of the intrinsic pseudo-distance is that holomorphic mappings are distance decreasing. That is, if $f: M \to N$ is holomorphic, then $d_N(f(p), f(q)) \le d_M(p,q)$ for every $p, q \in M$. This is an immediate consequence of the definition. A simple application of this property is the following generalization of the little Picard theorem.

Proposition. *If* M *is hyperbolic, then any holomorphic map* $f: \mathbb{C} \to M$ *is a constant map.*

Proof: For any $z_1, z_2 \in \mathbb{C}$, we have

$$d_M(f(z_1), f(z_2)) \le d_{\mathbb{C}}(z_1, z_2) = 0.$$

Since M is hyperbolic, $f(z_1) = f(z_2)$. Q.E.D.

The reader is referred to Kobayashi's book [12] for further details concerning hyperbolic manifolds.

3. HOLOMORPHIC EXTENSION THEOREMS

The first extension theorem which used the intrinsic distance was proved in 1968 by Mrs. Kwack [14].

Theorem 1 (Kwack): *Let* A *be a closed complex subspace of a complex manifold* X *and let* M *be a hyperbolic complex space. Then every holomorphic map* $f: X - A \to M$ *extends to a holomorphic map* $f: X \to M$ *if one of the following conditions is satisfied:*

(*i*) M *is compact.*

(*ii*) M *is complete hyperbolic and* codim $A \geq 2$.

This shows that the concept of hyperbolic is useful for attacking our extension problem, but it does not explain the big Picard theorem since $\mathbb{P}_1(C) - \{0,1,\infty\}$ is not compact. It is natural to ask if condition (*i*) in Kwack's theorem can be replaced by the assumption that M is a hyperbolic, relatively compact, open set in a complex space Y. To see that the answer is no, consider the following example.

Example (5): Let $M = \mathbb{C}^2 - \{(z_1,z_2) \mid z_1z_2(z_1 - 1) = 0 \text{ or } z_2 = \exp(1/z_1)\}$. Since M is biholomorphic to $(\mathbb{C} - \{0,1\}) \times (\mathbb{C} - \{0,1\})$, it follows that M is a hyperbolic, relatively compact, open set in $\mathbb{P}_2(\mathbb{C})$. Let $f(z) = (z, \frac{1}{2}\exp(1/z))$ for $z \in D^*$. Clearly f cannot be extended to a holomorphic map defined on D.

In the last example, we could construct a mapping which has no extension precisely because the boundary of M in $\mathbb{P}_2(\mathbb{C})$ is so badly behaved. For example, the boundary is not even an algebraic subvariety of $\mathbb{P}_2(\mathbb{C})$. The next definition gives sufficient restrictions on the boundary in order to imply a strong extension theorem.

Definition. Let M be a relatively compact, open set in a complex space Y. M is said to be *hyperbolically imbedded* in Y if

(*i*) M is hyperbolic

(*ii*) whenever p_n and q_n are sequences in M which converge to distinct boundary points, then $d_M(p_n,q_n)$ does not converge to 0.

The first condition above just says that the distance between any two distinct points of M is greater than zero. Intuitively, the second condition says that the "distance" between any two distinct boundary points is also greater than zero. Since the intrinsic distance on $\mathbb{P}_1(\mathbb{C}) - \{0,1,\infty\}$ is the distance corresponding to an hermitian metric and since the boundary is a discrete set, it is easy to see that this space is hyperbolically imbedded in $\mathbb{P}_1(\mathbb{C})$. Thus Theorem 2 below contains the big Picard theorem as a special case.

Theorem 2. *Let* A *be a closed complex subspace of a complex manifold* X. *If the singularities of* A *are normal crossings and if* M *is hyperbolically imbedded in* Y, *then any holomorphic map* $f: X - A \to M$ *extends to a holomorphic map* $f: X \to Y$.

Kobayashi proved this result for the case where A is nonsingular and the author extended it to the case where A has normal crossings. By normal crossings we mean that $X - A$ is locally biholomorphic to $(D^*)^k \times D^n$. The proof of Theorem 2 depends on a slightly generalized version of a lemma used by Kwack to prove Theorem 1.

Lemma. *Let* M *be hyperbolically imbedded in* Y *and let* $f_k: D^* \to M$ *be a sequence of holomorphic maps. Let* $\{z_k\}$ *and* $\{z_k'\}$ *be sequences in* D^* *converging to* 0 *and such that* $f_k(z_k) \to p \in Y$. *Then*

(*i*) $f_k(z_k') \to p$

(*ii*) *Any holomorphic map* $f: D^* \to M$ *extends to a holomorphic map* $f: D \to Y$

(*iii*) $f_k(0) \to p$.

The proof of the lemma uses the fact that if $C_k = \{z \in D^* \mid |z| = |z_k|\}$, then the diameters of the sets C_k with respect to the distance d_{D^*} are converging to 0 as $k \to \infty$. Since the mappings f_k are distance decreasing, the diameters of the sets $f_k(C_k)$ with respect to the distance d_M are also converging to 0. Since M is hyperbolically imbedded in Y, this implies that $f_k(C_k)$ converges to p as $k \to \infty$. If (a) is false, then we can assume that $f_k(z_k') \to q \neq p$ and for $C_k' = \{z \mid |z| = |z_k'|\}$, it follows that $f_k(C_k')$ converges to q as $k \to \infty$. For k large, we get a contradiction by using a winding number argument due to Grauert and Reckziegel.

In proving Theorem 2, one can assume that $X - A = (D^*)^n \times D^m$. The lemma takes care of the case $X - A = D^*$. The first step is to show that if the theorem is true for $X - A = (D^*)^n$, then it is true for $X - A = (D^*)^n \times D^m$. The second step is to show that the case $X - A = (D^*)^n \times D^m$ implies the case $X - A = (D^*)^{n+1}$. The basic idea in both steps is to restrict the mapping f to lower dimensional slices, extend the restricted mappings and then use the lemma to show that these extensions fit together to give a holomorphic mapping. The proofs of Theorem 2 and the lemma are given in detail in [9].

We now give some examples of hyperbolically imbedded spaces.

Theorem 3 (Borel): *Let $\mathcal{D}$ be a bounded symmetric domain and let Γ be an arithmetically defined torsion free group of automorphisms of $\mathcal{D}$. Let $V = \mathcal{D}/\Gamma$ and let V^* be the Baily-Borel-Satake compactification of* V. *Then* V *is hyperbolically imbedded in* V^* *and every holomorphic map* $f\colon (D^*)^n \to V$ *extends to a holomorphic map* $f\colon D^n \to V^*$.

In [3], Borel uses Theorem 2 in proving Theorem 3. However, it should be pointed out that Borel obtained the result in 1968 which was before Theorem 2 was proved. Let V^{**} be the set V^* with the topology defined by Piateckii-Shapiro. The identity map $i\colon V^* \to V^{**}$ is continuous [1] and therefore is a homeomorphism if V^{**} is Hausdorff. It seems very likely that this fact will follow from a joint work by Borel and Serre which has not been written up yet. At the end of this section, Theorem 3 and another extension theorem will be used to show that V^{**} is Hausdorff. In [13], Kobayashi and Ochiai show that V is hyperbolically imbedded in V^{**}.

Theorem 4: *Let $H_1,\ldots,H_{2n+1}$ be $2n+1$ hyperplanes in general position in $\mathbb{P}_n(\mathbb{C})$ and let $M = \mathbb{P}_n(\mathbb{C}) - H_1\cup\ldots\cup H_{2n+1}$. Then M is hyperbolically imbedded in $\mathbb{P}_n(\mathbb{C})$. If X and A are as in Theorem 2, then every holomorphic map $f\colon X - A \to M$ extends to a holomorphic map $f\colon X \to \mathbb{P}_n(\mathbb{C})$.*

Bloch [2] and Cartan [5] obtained a result which implies that $\mathrm{Hol}(D,M)$ is relatively compact in $\mathrm{Hol}(D,\mathbb{P}_n(\mathbb{C}))$. The next theorem says that this is equivalent to proving the first statement in Theorem 4. The proof of Theorem 5 is contained in [10].

Theorem 5: *Let* M *be an open set in a compact complex manifold* Y. *The following are equivalent:*

(*i*) M *is hyperbolically imbedded in* Y.

(*ii*) Hol(D,M) *is relatively compact in* Hol(D,Y).

(*iii*) *There exists an hermitian metric* h *on* Y *such that for any* $f \in \mathrm{Hol}(D,M)$ *and any* $z_1, z_2 \in D$, *we have*

$$d_h(f(z_1), f(z_2)) \le d_D(z_1, z_2)$$

where d_h *is the distance induced by* h.

We now consider the restrictions on the spaces X and A in Theorem 2. Let $X = \mathbb{C}^2$ and let $A = \{z_1 z_2 (z_1 - z_2) = 0\}$. If $f: X - A \to \mathbb{P}_1(\mathbb{C}) - \{0, 1, \infty\}$ is the natural projection, then it is clear that f does not extend holomorphically to X. Thus the assumption that A has normal crossings cannot be omitted. To see that in general X must be nonsingular, let L be a negative line bundle over a compact Riemann surface M of genus at least 2 and let X be the space obtained from L by blowing down the zero section to a point. If A denotes the point corresponding to the zero section, then the projection $\pi: X - A \to M$ does not extend to a holomorphic map of X into M. There are, however, special cases in which the conclusion of Theorem 2 is valid even though X and A do not satisfy these restrictions. The most interesting one is:

Theorem 6: *Let* V *and* V* *be as in Theorem* 3 *and let* M *be hyperbolically imbedded in* Y. *Then every holomorphic map* $f: V \to M$ *extends to a holomorphic map* $f: V^* \to Y$.

In Theorem 1 of [11], Kobayashi and the author proved that f extends to a continuous map of V** into Y. Since the identity map $i: V^* \to V^{**}$ is continuous, the extended map is a holomorphic map of V* into Y.

Corollary: *The identity map* $i: V^* \to V^{**}$ *is a homeomorphism. Thus* V** *is Hausdorff.*

Proof: Let $j: V \to V$ denote the identity. Since V is hyperbolically imbedded in V* by Theorem 3, the mapping j extends to a continuous mapping $j: V^{**} \to V^*$. Clearly j is the inverse of i. Q.E.D.

4. MEROMORPHIC MAPPINGS

We shall use Remmert's definition of a meromorphic mapping.

Definition: A *meromorphic mapping* f from a complex space X into a complex space Y is a correspondence satisfying:

(*i*) For each point $x \in X$, $f(x)$ is a nonempty compact subset of Y.

(*ii*) The graph $\Gamma_f = \{(x,y) \in X \times Y \mid y \in f(x)\}$ of f is a connected complex subspace of $X \times Y$ with $\dim \Gamma_f = \dim X$.

(*iii*) There exists a dense subset X' of X such that $f(x)$ is a single point for each $x \in X'$.

There have been many extension theorems obtained recently in which the extended mapping is meromorphic. Some of these results are contained in [4], [6], [7] and [8]. In some cases, it is shown that the mapping extends meromorphically and then additional information is used to conclude that the extension is actually holomorphic. For example, both Fujimoto and Green have proved Theorem 4 under the additional assumption that A is nonsingular. The basic method for showing that a mapping extends meromorphically depends upon estimating the growth of the mapping.

To see how this can work, let X and Y be Kähler manifolds and let A be a complex subspace of X. A holomorphic mapping $f: X - A \to Y$ extends meromorphically if and only if the closure $\overline{\Gamma}_f$ of the graph of f is a complex subspace of $X \times Y$. Bishop has shown that this is satisfied if and only if each $x \in X$ has a neighborhood U such that the volume of $\Gamma_f \cap (U \times Y)$ is finite. (If $\dim X = k$, then the volume is computed with respect to the k-dimensional measure determined by any Kähler metric on $X \times Y$.) Thus we can solve the meromorphic extension problem if we can compute the size of the graph of f.

A nice application of this technique is Griffith's proof of part of Kwack's theorem.

Theorem 7: *Let* M *be a compact, hyperbolic, Kähler manifold. If* A *is a closed complex subspace of a complex space* X, *then any holomorphic mapping* $f: X - A \to M$ *extends to a meromorphic mapping of* X *into* M.

Proof: By Hironaka's resolution of singularities, we can reduce the problem to the case where $X - A = (D^*)^n$. For simplicity we assume that

$n = 1$. Let w_0 be the 2-form associated to the invariant hermitian metric on D^* which induces the distance d_{D^*}. Let w_1 be the 2-form associated with a Kähler metric h on M which satisfies condition *(iii)* of Theorem 5. It follows that $f^*(w_1) \leq w_0$. Let U be a relatively compact neighborhood of 0 in D and let w be the Euclidean volume form on D. The volume v of $\Gamma_f \cap (U \times M)$ determined by the Euclidean metric on D and the metric h on M is given by:

$$v = \int_U w + f^*(w_1) \leq 2 \int_U w_0 < \infty.$$

Therefore f is meromorphic. Q.E.D.

It is important to note that the proof outlined above does not yield all of Kwack's theorem. That is, it does not tell you that the extended map is holomorphic if X is nonsingular. This seems to be a basic limitation on any method which estimates the growth of the mapping. To prove that a particular mapping extends holomorphically, one must examine the image space more carefully.

Finally, we note that Theorem 2 implies a meromorphic extension theorem.

Theorem 8: *Let* A *be a closed complex subspace of a complex space* X. *If* M *is hyperbolically imbedded in* Y, *then any holomorphic mapping* $f: X - A \to M$ *extends to a meromorphic mapping of* X *into* Y.

Proof: By Hironaka's resolution of singularities, there exists a complex manifold $\hat{X}$ and a holomorphic mapping ϕ of $\hat{X}$ onto X such that

(i) $\hat{A} = \phi^{-1}(A)$ is a complex subspace of $\hat{X}$ whose singularities are normal crossings, and

(ii) ϕ^{-1} is meromorphic.

Let $\hat{f}: \hat{X} - \hat{A} \to M$ be defined by $\hat{f} = f \circ \phi$. By Theorem 2, $\hat{f}$ extends to a holomorphic map $\hat{f}: X \to Y$. It follows immediately that f is meromorphic. Q.E.D.

Since the singularities of $\hat{A}$ cannot in general be removed, it is important that Theorem 2 allows for normal crossings.

5. REMARKS

In the theorems above, it was assumed that the range space M is hyperbolically imbedded in Y. In order to see the relevancy of this assumption, we consider two examples.

Example (6) : Let $M = (\mathbb{C} - \{0,1\}) \times (\mathbb{C} - \{0,1\})$ and let $M_1 = M$ considered as a subspace of $\mathbb{P}_1(\mathbb{C}) \times \mathbb{P}_1(\mathbb{C})$ and let $M_2 = M$ considered as a subspace of $\mathbb{P}_2(\mathbb{C})$. It is easy to see that M_1 is hyperbolically imbedded in $\mathbb{P}_1(\mathbb{C}) \times \mathbb{P}_1(\mathbb{C})$, but since the identity mapping $i: M_1 \to M_2$ does not extend to a holomorphic mapping of $\mathbb{P}_1(\mathbb{C}) \times \mathbb{P}_1(\mathbb{C})$ into $\mathbb{P}_2(\mathbb{C})$, M_2 is not hyperbolically imbedded in $\mathbb{P}_2(\mathbb{C})$. However, it is easy to show that every holomorphic map $f: D^* \times D^n \to M_2$ extends to a holomorphic map $f: D^{n+1} \to \mathbb{P}_2(\mathbb{C})$ [9]. Thus, the assumption that M is hyperbolically imbedded is not essential in Theorem 2 if we add the assumption that A is nonsingular.

Example (7): Let $M_3 = \mathbb{P}_2(\mathbb{C}) - B$ where $B = \{(z_0,z_1,z_2) \mid z_1 = \pm z_0$ or $z_1 z_2 = \pm z_0^2\}$. Since there is a bimeromorphic map $\phi: \mathbb{P}_1(\mathbb{C}) \times \mathbb{P}_1(\mathbb{C}) \to \mathbb{P}_2(\mathbb{C})$ which maps M_1 biholomorphically onto M_3, Theorem 8 is valid for mappings into $M_3 \subset \mathbb{P}_2(\mathbb{C})$. Since the mapping $f: D^* \times D \to M_3$ given by $f(z,w) = (z,z^2,w)$ does not extend to a holomorphic map $f: D \times D \to \mathbb{P}_2(\mathbb{C})$, M_3 is not hyperbolically imbedded in $\mathbb{P}_2(\mathbb{C})$.

The basic point of the last examples is that hyperbolically imbedded is a much stronger assumption than is necessary in order to prove meromorphic extension theorems. More precisely, Theorem 8 is probably still valid if M is a hyperbolic, Zariski open set in a projective algebraic variety. Intuitively, this conjecture seems plausible for the following reasons. If f is a mapping into M then the fact that M is hyperbolic implies that the graph Γ_f can become large only by approaching the boundary. But since the boundary is algebraic, the mapping cannot grow too fast by following the boundary (compare with Example 5).

REFERENCES

1. Bailey, W. L.,Jr., *Fourier-Jacobi series, algebraic groups and discontinuous subgroups*, Proc. Sympos. Pure Math. Vol. 9, Amer. Math. Soc. 1966, 296-300.

2. Bloch, A., *Sur les systèmes de fonctions holomorphes à variétés linéaires lacunaires*, Ann. de l'Ecole Normale, 43 (1926), 309-362.

3. Borel, A., *Some metric properties of arithmetic quotients of symmetric spaces and an extension theorem,* J. Differential Geometry, 6 (1972), 543-560.

4. Carlson, J., *Some degeneracy theorems for entire functions with values in an algebraic variety,* Trans. A.M.S., 168 (1972), 273-301.

5. Cartan, H., *Sur les systèmes de fonctions holomorphes à variétés linéaires lacunaires et leurs applications,* Ann. de l'Ecole Normale, 45 (1928), 255-346.

6. Fujimoto, H., *Extensions of the big Picard theorem,* to appear in Tohoku Math. J.

7. Green, M. *Some Picard theorems for holomorphic maps to algebraic varieties,* Thesis (1972), Princeton.

8. Griffiths, P. and J. King, *Nevanlinna theory and holomorphic mappings between algebraic varieties,* to appear.

9. Kiernan, P., *Extension of holomorphic maps,* Trans. A.M.S., 172 (1972), 347-355.

10. _________, *Hyperbolically imbedded spaces and the big Picard theorem,* to appear in Math. Annalen.

11. _________, and S. Kobayashi, *Satake compactification and extension of holomorphic mappings,* Inventiones Math., 16 (1972), 237-248.

12. Kobayashi, S., *Hyperbolic Manifolds and Holomorphic Mappings,* Marcel Dekker, New York, 1970.

13. _________, and T. Ochiai, *Satake compactification and the great Picard theorem,* J. Math. Soc. Japan, 23 (1971), 340-350.

14. Kwack, M. H., *Generalization of the big Picard theorem,* Ann. of Math. (2) 90 (1969), 9-22.

EXTENDING HOLOMORPHIC FUNCTIONS WITH
BOUNDED GROWTH FROM CERTAIN GRAPHS

I. Cnop
Vrije Universiteit Brussel

1. INTRODUCTION

A holomorphic functional calculus for elements with unbounded spectra in a class of algebras containing unitary Banach algebras was developed by L. Waelbroeck [9], [10]. The functions which operate are holomorphic on the spectrum and satisfy certain growth conditions. This theory has allowed J.-P. Ferrier [4] to find properties of holomorphic functions with growth restrictions in unbounded domains of holomorphy in the space $\mathbb{C}^n$ of n complex variables. As one more application, we investigate the extension of a holomorphic function with growth restriction on a polynomial graph in a domain of holomorphy, to that domain. In the case of bounded domains of holomorphy, the condition on the submanifold (i.e., to be the graph of polynomials) can be relaxed.

Holomorphic extensions (without growth restrictions) were obtained in the Cartan seminar [2] and, in a constructive way, by E. Bishop [1]. Recently, G. M. Henkin [5] proved that a bounded holomorphic function on a closed submanifold of a (bounded) strictly pseudoconvex domain in $\mathbb{C}^n$ has a bounded extension, if the manifold can be extended through the boundary (see also R. M. Range and Y.-T. Siu [7]).

2. FUNCTIONAL CALCULUS

All algebras encountered here will be commutative. In what follows, δ_o will be the function

$$\delta_o(s) = (1 + |s|^2)^{-\frac{1}{2}} \quad , \ |s|^2 = \sum_1^n |s_i|^2$$

which tends to zero at infinity like $|s|^{-1}$ on the space $\mathbb{C}^n$ of n complex variables (n may vary), and δ will be a bounded nonnegative Lipschitz function on $\mathbb{C}^n$ such that for some positive number N_o, $\delta^{N_o}\delta_o^{-1}$ is bounded.

$\mathfrak{O}(\delta)$ is the commutative unitary algebra of the functions f defined and holomorphic on the set Ω where δ is positive, satisfying $f\delta^N$ is bounded on Ω for some positive number N. A set B is bounded in $\mathfrak{O}(\delta)$ if $f\cdot\delta^N$ is bounded on Ω for some positive number N, uniformly for f in B. We will not discuss topologies on $\mathfrak{O}(\delta)$, but simply remark that it is the increasing union of the Banach spaces of the functions f such that $f\delta^N$ is bounded, N fixed. They are called b-algebras in part of the literature.

The algebra $\mathfrak{O}(\delta)$, and its structure, does not change if we replace δ by an equivalent δ_1: two weight functions are equivalent if each one of them is larger than a positive multiple of a positive power of the other. In fact, everything that follows will be invariant under this equivalence. Therefore, "Lipschitz" will always mean: "equivalent to a function which satisfies a Lipschitz condition with exponent one."

Let now A, A' be commutative unitary algebras with this b-algebra structure (i.e., which are a filtrating union of Banach spaces). A linear mapping $T: A' \to A$ is bounded if it maps bounded sets of A' into bounded sets of A. If A' is a subset of A, we call it a b-subalgebra of A if the canonical injection $i: A' \to A$ is bounded; it is a b-ideal if moreover it is, algebraically, an ideal in A.

The (joint) spectrum of some elements will be replaced by a weight function δ. Let $a_1,\dots,a_n$ be elements in A and α a b-ideal in A. We say that a bounded nonnegative function ϕ on $\mathbb{C}^n$ is spectral for $a_1,\dots,a_n$ in A modulo α, or ϕ belongs to $\Delta(a_1,\dots,a_n; A|\alpha)$, if we can find A-valued functions $u_o, u_1,\dots,u_n$ and an α-valued function v on $\mathbb{C}^n$ satisfying

$$1 = \phi(s)u_o(s) + \textstyle\sum_1^n (a_i - s_i)u_i(s) + v(s)$$

and such that for some positive number N

$$\{u_j(s)\ \delta_o^N(s) \mid 0 \le j \le n,\ s \in \mathbb{C}^n\}$$

is bounded in A and

$$\{v(s)\ \delta_o^N(s) \mid s \in \mathbb{C}^n\}$$

is bounded in α.

If a nonnegative bounded function ϕ belongs to the spectrum, then so does the largest Lipschitz minorant δ of $\min(\phi,\delta_o)$.

If δ is spectral for $a_1,\ldots,a_n$ in A modulo α, the holomorphic functional calculus defines an algebra homomorphism

$$\mathcal{O}(\delta) \to A$$

defined modulo elements in α, sending 1 onto 1 and the coordinate function z_i onto a_i.

3. UNBOUNDED DOMAINS

We consider the situation where δ is a bounded nonnegative Lipschitz function on $\mathbb{C}^n$ with $\delta^{N_o}\delta_o^{-1}$ bounded, and with the property that $-\log\delta$ is plurisubharmonic on the set $\Omega = \{z \in \mathbb{C}^n \mid \delta(z) > 0\}$, which is then a domain of holomorphy. Let p be a k-tuple $(k < n)$ of polynomials $p_1,\ldots,p_k$ in $n - k$ variables. There is a 1-1 correspondence (which we will not write explicitly) between $\mathbb{C}^{n-k}$ and the graph M of these polynomials in $\mathbb{C}^{n-k} \times \mathbb{C}^k$. The restriction $f|_{M\cap\Omega}$ of a function f in $\mathcal{O}(\delta)$ to M can be considered as a function in the $n - k$ first variables, and, as such, belongs to $\mathcal{O}(\delta')$, where δ' is defined on $\mathbb{C}^{n-k}$ by restricting δ to M.

Our main result says that the restriction

$$\mathcal{O}(\delta) \to \mathcal{O}(\delta')$$

is, in fact, surjective:

Theorem 1: *Let δ be a bounded nonnegative Lipschitz function on $\mathbb{C}^n$ with $\delta^{N_o}\delta_o^{-1}$ bounded and $-\log\delta$ plurisubharmonic on the set Ω where δ does not vanish. Let M be the graph of k polynomials $p = (p_1,\ldots,p_k)$ in $n - k$ variables, and δ' be defined on the first $n - k$ variables by restricting δ to M. Then, for each function f' in $\mathcal{O}(\delta')$, we can find a function f in $\mathcal{O}(\delta)$ such that*

$$f|_{M\cap\Omega} = f'$$

and we have the isomorphism:

$$\mathcal{O}(\delta') \simeq \mathcal{O}(\delta)/\alpha$$

where α *is the ideal generated by* $z_{n-k+1} - p_1, \ldots, z_n - p_k$ *in* $\mathcal{O}(\delta)$.

The polynomial growth is used in the following.

Lemma: *There exists a small* $\varepsilon > 0$ *and a positive number* N_1, *such that for all* s' *in* $\mathbb{C}^{n-k}$:

$$\delta_o(p(s')) \geq \varepsilon\, \delta_o^{N_1}(s').$$

(Here, the notation δ_o was used in spaces of different dimension.) Indeed: *we can find* $\varepsilon > 0$ *and* N_1 *large such that*

$$1 + \sum_1^k |p_i(s')|^2 \leq \varepsilon^{-2}(1 + |s'|^2)^{N_1}$$

(and this is the required estimate).

Proof of Theorem 1: The main theorem in the author's thesis (see [3]) says that if $-\log \delta$ is plurisubharmonic, then δ belongs to $\Delta(z_1, \ldots, z_n;\, \mathcal{O}(\delta))$:

$$1 = \sum_1^n (z_i - s_i)\, u_i(s,z) + \delta(s)\, u_o(s,z)$$

$$\text{when } s \in \mathbb{C}^n,\ z \in \Omega,$$

with the right estimates on the coefficients $u_j\ (0 \leq j \leq n)$. Restricting this relation to s in M, we obtain:

$$1 = \sum_1^{n-k} (z_i - s_i)\, u_i'(s',z) + \delta'(s')\, u_o'(s',z)$$
$$+ \sum_1^k (z_{n-k+i} - p_i(s'))\, u'_{n-k+i}(s',z)$$

where s' is the variable in $\mathbb{C}^{n-k}$ and

$$u_j'(s',z) = u_j(s',p(s'),z) \qquad 0 \leq j \leq n.$$

The last sum can be written as

$$\sum_1^k (p_i(z_1,\ldots,z_{n-k}) - p_i(s'))u'_{n-k+i}(s',z)$$

$$+ \sum_1^k (z_{n-k+i} - p_i(z_1,\ldots,z_{n-k}))u'_{n-k+i}(s',z)$$

$$= \sum_{j=1}^{n-k} \sum_{i=1}^{k} (z_j - s_j)p_{ij}(s',z)u'_{n-k+i}(s',z)$$

$$+ \sum_1^k (z_{n-k+i} - p_i(z_1,\ldots,z_{n-k}))u'_{n-k+i}(s',z)$$

where the p_{ij} are polynomials. Finally

$$1 = \sum_1^{n-k} (z_i - s_i)\Big[u'_i(s',z) + \sum_{j=1}^{k} p_{ji}(s',z)u'_{n-k+j}(s',z)\Big]$$

$$+ \delta'(s')u'_o(s',z) + \sum_1^k (z_{n-k+i} - p_i(z_1,\ldots,z_{n-k}))u'_{n-k+i}(s',z).$$

Omitting finite sums and factors which are polynomials, we just have to estimate, for $0 \le j \le n$, s' in $\mathbb{C}^{n-k}$ and z in Ω:

$$|u'_j(s',z)|(\delta_o(s')\ \delta(z))^N = |u_j(s',p(s'),z)|(\delta_o(s')\delta(z))^N$$

$$\le \varepsilon^{-N'}|u_j(s',p(s'),z)|\ \delta_o^{N'}(s')\ \delta_o^{N'}(s')\ \delta_o^{N'}(p(s'))\ \delta^N(z)$$

by the Lemma, if $N = N' + N'N_1$. Using that

$$\delta_o(s')\ \delta_o(p(s')) \le \delta_o(s',\ p(s')),$$

the right hand side does not exceed

$$\sup_{s\in\mathbb{C}^n} \varepsilon^{-N'}|u_j(s,z)|\ \delta_o^{N'}(s)\ \delta^N(z)$$

which is bounded if N and N' are sufficiently large. We have proved that δ' belongs to $\Delta(z_1,\ldots,z_{n-k};\ \mathcal{O}(\delta)|\alpha)$ and the holomorphic functional calculus then defines an extension mapping

$$\mathcal{O}(\delta') \to \mathcal{O}(\delta)$$

which is a bounded homomorphism of unitary algebras, defined modulo α and sending the $z_i (1 \le i \le n - k)$ onto themselves (modulo α).

4. A DECOMPOSITION THEOREM

We did not yet prove that the kernel of the restriction mapping is contained in the ideal α generated by $z_{n-k+1} - p_1,\ldots,z_n - p_k$ in $\mathcal{O}(\delta)$. This will use a decomposition result, which can essentially be found in [10] but which was investigated in [4].

Proposition: *Let* δ *be a bounded nonnegative Lipschitz function on* $\mathbb{C}^n$ *with* $\delta^{N_o}\delta_o^{-1}$ *bounded such that the set* Ω *where* δ *does not vanish is pseudoconvex. If* f *belongs to* $\mathcal{O}(\delta)$ *and vanishes at* s, *then*

$$f(z) = \sum_1^n (z_i - s_i)\, h_i(s,z),$$

where $h_i(s,\cdot)$ *belongs to* $\mathcal{O}(\delta)$, *and remains bounded if* f *varies in a bounded set of* $\mathcal{O}(\delta)$ *and* s *in* Ω.

So if f is zero on $M \cap \Omega$, let

$$s = (z_1,\ldots,z_{n-k}, p(z_1,\ldots,z_{n-k}))$$

be any point of $M \cap \Omega$, and

$$f(z) = \Sigma_1^k (z_{n-k+i} - p_i(z_1,\ldots,z_{n-k}))\, h_i(z)$$

which means that f belongs to α. This ends the proof of Theorem 1.

5. BOUNDED DOMAINS

We now prove a result concerning the extension of holomorphic functions with growth conditions from certain bounded graphs in a (not necessarily bounded) domain of holomorphy:

Theorem 2: *Let* δ *be a bounded nonnegative Lipschitz function on* $\mathbb{C}^n$ *with* $\delta^{N_o}\delta_o^{-1}$ *bounded and* $-\log\delta$ *plurisubharmonic on the set* Ω *where* δ *does not vanish. Let* M *be a bounded subset of* Ω, δ' *be the restriction of* δ *to* M, *and* Ω' *be a pseudoconvex (bounded) set in* $\mathbb{C}^{n-k}$ $(1 \le k < n)$ *such that:*

(*i*) Ω *is contained in* $\Omega' \times \mathbb{C}^{n-k}$;

(*ii*) M *is the graph of a* k*-tuple* $h = (h_1,\ldots,h_k)$ *of (bounded) holomorphic functions on* Ω'. *Then, for each function* f' *in* $\mathcal{O}(\delta')$, *we can find a function* f *in* $\mathcal{O}(\delta)$ *such that*

$$f|_M = f'$$

and we have the isomorphism

$$\mathcal{O}(\delta') \simeq \mathcal{O}(\delta)/\alpha$$

where α *is the ideal generated by* $z_{n-k+1} - h_1,\ldots,z_n - h_k$ *in* $\mathcal{O}(\delta)$.

We remark that we do not impose a condition on the behavior of M near $\partial\Omega$, as in [5].

Proof of Theorem 2: By [3], δ belongs to $\Delta(z_1,\dots,z_n; \mathcal{O}(\delta))$. As in Theorem 1 we are going to restrict this relation to a suitable subset of Ω. Restricting ourselves to s in M gives:

$$1 = \sum_1^{n-k} (z_i - s_i)\, u'_i(s', z) + \delta'(s')\, u'_o(s', z)$$
$$+ \Sigma_1^k (h_i(z') - h_i(s'))u'_{n-k+i}(s', z)$$
$$+ \Sigma_1^k (z_{n-k+i} - h_i(z'))u'_{n-k+i}(s', z)$$

where

$$u'_j(s', z) = u_j(s', h(s'), z) \qquad 0 \le j \le n,$$

if the u_j where the coefficients in the original relation, z belongs to Ω, $s' = (s_1,\dots,s_{n-k})$ and $z' = (z_1,\dots,z_{n-k})$ belong to Ω'. The boundary distance function $\delta_{\Omega'}$, defined by

$$\delta_{\Omega'}(s') = \inf \{|z' - s'| \;\big|\; z' \in \mathbb{C}^{n-k}\backslash\Omega'\}$$

is Lipschitz. Since Ω' is pseudoconvex, we can apply Ferrier's decomposition result (see §4) to the functions h_j $(1 \le j \le k)$, which are bounded and therefore belong to $\mathcal{O}(\delta_{\Omega'})$:

$$h_j(z') - h_j(s') = \Sigma_1^{n-k}(z_i - s_i)h_{ji}(s', z')$$

where all $h_{ji}(s',\cdot)$ belong to $\mathcal{O}(\delta_{\Omega'})$ and are bounded in $\mathcal{O}(\delta_{\Omega'})$ when s' varies in Ω'. So

$$1 = \Sigma_1^{n-k}(z_i - s_i)\left[u'_i(s', z) + \Sigma_1^k h_{ji}(s', z')\cdot u'_{n-k+j}(s', z)\right]$$
$$+ \delta'(s')u'_o(s', z) + \Sigma_1^k(z_{n-k+i} - h_i(z'))u'_{n-k+i}(s', z)$$

To estimate the coefficients, we remark that (up to equivalence):

$$\sup \{\delta(z', z'') \mid z'' \text{ in } \mathbb{C}^k \text{ and } (z', z'') \text{ in } \Omega\}$$
$$\le \delta_{\Omega'}(z')$$

since δ is Lipschitz. Therefore

$$|h_{ji}(s', z')|\cdot \delta^N(z) \qquad \le |h_{ji}(s', z')|\,\delta^N_{\Omega'}(z')$$

and the left hand side will remain bounded, uniformly in $s' \in \Omega'$, if N is sufficiently large. The bound on the u'_j is obvious since

$$u'_j(s', z) = u_j(s', h(s'), z)$$

and $(s', h(s'))$ remains in M which is bounded (this fact replaces the lemma in Theorem 1).

Next, restricting ourselves to points s of the form (s', o), with s' in $\mathbb{C}^{n-k}\backslash\Omega'$, we obtain

$$1 = \Sigma_1^{n-k}(z_i - s_i)u'_i(s', z) + \Sigma_1^k h_i(z')u'_{n-k+i}(s', z)$$

$$+ \Sigma_1^k(z_i - h_i(z'))u'_{n-k+i}(s', z)$$

since $\delta(s',o) = 0$ when s' does not belong to Ω'. Here again,

$$u'_j(s', z) = u_j(s', o, z) \qquad 0 \le j \le n,$$

and z belongs to Ω. Once more the bound on the u'_j is obvious since

$$\delta_o(s', o) = \delta_o(s')$$

and moreover, since the h_i are bounded, the term

$$\Sigma_1^k h_i(z')u'_{n-k+i}(s', z)$$

has the right growth to be a coefficient, say of the function which is identically 1 on $\mathbb{C}^{n-k}\backslash\Omega'$.

So far, we have proved that the bounded nonnegative functions ϕ on $\mathbb{C}^{n-k}$, defined by

$$\phi(s') = \delta'(s') \quad \text{if } s' \text{ is in } \Omega' \text{ and}$$

$$\phi(s') = 1 \quad \text{if } s' \text{ is not in } \Omega'$$

belongs to $\Delta(z_1,\ldots,z_{n-k};\ \mathcal{O}(\delta)|\alpha)$. So δ'', the largest Lipschitz minorant of $\min(\phi,\delta_o)$ also belongs to $\Delta(z_1,\ldots,z_{n-k};\ \mathcal{O}(\delta)|\alpha)$. The holomorphic functional calculus defines a bounded algebra homomorphism

$$\mathcal{O}(\delta'') \to \mathcal{O}(\delta),$$

which is defined modulo α, sends 1 onto 1 and z_i onto z_i $(1 \le i \le n - k)$.

On the other hand, we have a canonical mapping

$$\mathcal{O}(\delta') \to \mathcal{O}(\delta'')$$

defined by extending each f' in $\mathcal{O}(\delta')$ by the function identically zero outside $\bar{\Omega}'$. Here we used that δ'' is zero on the boundary $\partial\Omega'$ of Ω', and therefore functions in $\mathcal{O}(\delta'')$ are not defined on $\partial\Omega'$. The composition of the two mappings gives the extension

$$\mathcal{O}(\delta') \to \mathcal{O}(\delta).$$

The same argument as in §4 will prove that also in this case, a function f in $\mathcal{O}(\delta)$ which vanishes on M belongs to α.

6. REMARKS

Combining these theorems with approximation results for algebras $\mathcal{O}(\delta)$ ([4], [8]), or with extension results [6] gives approximation or extension results from certain lower-dimensional sets.

REFERENCES

1. Bishop, E., *Some global problems in the theory of functions of several complex variables*, Amer. J. Math. 83(1961), 479-498.

2. Cartan, H., *Séminaire* E.N.S., 1951-52.

3. Cnop, I., *Spectral study of holomorphic functions with bounded growth*, Ann. Inst. Fourier (Grenoble) 22(1972), 293-309.

4. Ferrier, J.-P., *Spectral Theory and Complex Analysis*, North-Holland Publishing Company, Amsterdam, 1973.

5. Henkin, G. M., *Continuation of bounded holomorphic functions from submanifolds in general position to strictly pseudoconvex domains*, Math. U.S.S.R. Izvestija 6(1972), 536-563.

6. Pflug, P., *Eigenschaften der Fortsetzungen von in speziellen Gebieten holomorphen polynomialen Funktionen in die Holomorphiehülle.* Thesis, Göttingen, 1972.

7. Range, R. M. and Y.-T. Siu, *Uniform extimates for the $\bar{\partial}$-equation on domains with piecewise smooth strictly pseudoconvex boundaries*, Dept. Math., Yale University, 1972.

8. Sibony, N., *Approximation polynomiale pondérée sur un domaine d'holomorphie de* $\mathbb{C}^n$. C. R. Acad. Sci. Paris A, 276(1973), 249-252.

9. Waelbroeck, L., *Etude spectrale des algèbres complètes*, Acad. Roy. Belgique, Mém. Cl. Sci., 31, 7, 1960.

10. __________, *Lectures in spectral theory*, Dept. Math., Yale University, 1963.

THE COMPLEMENT OF THE DUAL OF A PLANE CURVE AND SOME NEW HYPERBOLIC MANIFOLDS

Mark L. Green
University of California, Berkeley

1. INTRODUCTION

In recent years there have developed numerous and mysterious connections between the study of holomorphic mappings of complex manifolds and the differential-geometric properties of the manifolds in question, notably the existence of a hermitian metric with some sort of negative curvature. These questions are made all the more tantalizing by the existence of the Kobayashi pseudo-distance, an intrinsic invariant of a complex manifold M which measures the ways in which the unit disc can be mapped holomorphically to M. The manifolds imposing the strongest conditions on maps into them are those for which this pseudo-distance is actually a distance, named *hyperbolic* manifolds by Kobayashi.

For manifolds which we suspect of being hyperbolic, since on an intuitive level the Kobayashi distance is somehow akin to a metric of negative curvature, it is tempting to try to construct from scratch a negatively curved metric which will allow us to prove the manifold is hyperbolic. In a somewhat different situation, this was the approach applied so successfully by Carlson and Griffiths [2] to obtain defect relations for equidimensional holomorphic maps. In this paper, we will find two classes of hyperbolic manifolds by techniques along these lines (in actual fact, the proofs will be fairly elementary and will make only the slightest call on facts about curvature). The first will be the complements of the duals of algebraic plane curves of genus ≥ 2 with only ordinary singularities or of nonsingular curves of genus 1. The second will be hypersurfaces in $\mathbb{P}_3$ highly branched over the duals of algebraic plane curves like those just mentioned. The first classes were first investigated by Griffiths, who obtained many of the results given in Section 3 by algebro-geometric methods using the period mapping.

Acknowledgments: I am grateful to Phillip Griffiths, S.-S. Chern, James Carlson, and H. Wu for much help and encouragement. I am indebted to the mathematics faculty and visitors at Tulane for their generosity and many stimulating discussions, especially Mike Cowen and Al Vitter.

2. HYPERBOLIC MANIFOLDS

About the strongest statement one can make about a complex manifold from the standpoint of holomorphic mappings is that it is hyperbolic or complete hyperbolic. These concepts were introduced by Kobayashi and their implications are discussed in his book [3]. This section will briefly recall the definitions and facts we will need.

On the unit disc Δ there is the natural distance of non-Euclidean geometry, the *Poincaré distance* d_Δ associated to the metric $\frac{4\,dz\,d\bar{z}}{(1-|z|^2)^2}$ of constant negative curvature -1. This classical distance is invariant under the group of conformal automorphisms of Δ, and is characterized by this property.

The *Kobayashi intrinsic pseudo-distance* d_M on a complex manifold M is defined as follows: let p,q be points of M, and $p_1,p_2,\ldots,p_n$ points of M with $p_1 = p$, $p_n = q$. Assume we are given holomorphic maps $f_i: \Delta \to M$ with $f_i(a_i) = p_i$, $f_i(b_i) = p_{i+1}$ for $i = 1$ to $n-1$. Call such a set-up *a chain from* p *to* q and let $\rho(p,q) = \sum_{i=1}^{n-1} d_\Delta(a_i, b_i)$. Then $d_M(p,q)$ is defined to be the infimum of $\rho(p,q)$ over all chains from p to q.

It is easy to verify that

$$d_M(p,r) \le d_M(p,q) + d_M(q,r)$$
$$d_M(p,q) = d_M(q,p)\ .$$

So this is a pseudo-distance. It may not happen that $d_M(p,q) = 0$ implies $p = q$. For example, $d_{\mathbb{C}}(p,q) = 0$ for all $p,q \in \mathbb{C}$.

Definition: A manifold M is *hyperbolic* if $d_M(p,q) \neq 0$ if $p \neq q$.

Thus d_M makes a hyperbolic manifold M into a metric space. M is said to be *complete hyperbolic* if M is complete with this metric, i.e., every Cauchy sequence in M converges to a point of M.

What relates the Kobayashi pseudo-distance to the study of holomorphic maps is a distance-decreasing property:

Proposition: *Let* $f: M \to N$ *be a holomorphic map. Then*

$$d_M(p,q) \geq d_N(f(p), f(q)) .$$

Since $d_{\mathbb{C}}(p,q) = 0$ for all p,q, this implies a Picard theorem for maps to hyperbolic manifolds:

Proposition: *Let* $f: \mathbb{C} \to N$ *be a holomorphic map. If* N *is hyperbolic then* f *is constant.*

There is a strong relation between hyperbolicity and negative curvature which is not yet completely understood. We will use only the simplest aspect of this relationship, the *Ahlfors-Schwarz Lemma.*

Let $G = g(z)dzd\bar{z}$ be a *pseudo-hermitian* metric, on the unit disc Δ, that is, $g \geq 0$ and g may vanish at an isolated set of points. The curvature $K(G)$ is $-\frac{1}{g}\frac{\partial^2}{\partial z \partial \bar{z}}$ log g. It is easily verified that:

Lemma: *If* $K(G_i) \leq -1$ *for* $i = 1,\ldots n$, *then* $K(G_1+\ldots+G_n) \leq -\frac{1}{n}$.

We will need this fact later on. More important is:

Ahlfors-Schwarz Lemma: If G *is a pseudo-hermitian metric on the unit disc* Δ *satisfying* $K(G) \leq -k < 0$, *and* d_Δ *is the Poincaré metric on the disc, then*

$$d_\Delta \geq kd_G$$

where d_G *is the distance associated to the metric* G.

From the definition of the Kobayashi pseudo-distance and the Ahlfors-Schwarz Lemma, we have:

Proposition: *Let* M *be a complex manifold,* Ω *a pseudo-hermitian metric (it may degenerate on a proper analytic subvariety) which has curvature* $\leq -k < 0$ *when restricted to the holomorphic image of any disc. Then*

$$d_M \geq kd_\Omega$$

where d_M *is the Kobayashi pseudo-distance and* d_Ω *is the pseudo-distance associated to* Ω.

Remark: The condition hypothesized for Ω is equivalent to Ω having all *holomorphic sectional curvatures* $\leq -k$. In the interest of maintaining an elementary level, we will avoid this concept.

From the theory of Riemann surfaces, we will need that any compact Riemann surface of genus ≥ 2 or a Riemann surface of genus 1 with a point deleted has a complete hermitian metric of constant negative curvature -1. Such Riemann surfaces have the disc as universal cover, and the metric in question is the Poincaré metric on the disc pushed down.

The Kobayashi pseudo-distance behaves nicely with respect to covering spaces:

Proposition: *Let* $\tilde{M} \xrightarrow{\pi} M$ *be a holomorphic covering map with* M *and* $\tilde{M}$ *complex manifolds. If* $\pi(\tilde{p}) = p$, *then*

$$d_M(p,q) = \inf_{\pi(\tilde{q})=q} d_{\tilde{M}}(\tilde{p},\tilde{q}) .$$

3. THE COMPLEMENT OF THE DUAL OF AN ALGEBRAIC PLANE CURVE

Let D be a non-singular algebraic curve of degree k in $\mathbb{P}_2$. The *dual* D^* of the curve D is a curve in the dual projective plane $\mathbb{P}_2^*$ of lines in $\mathbb{P}_2$. D^* consists of those lines which are tangent to D. The degree of D^* is $k(k-1)$.

Theorem: $\mathbb{P}_2^* - D^*$ *is complete hyperbolic if degree* $D \geq 3$.

Remark: If degree $D = 2$, then D^* is a conic and $d_{\mathbb{P}_2^*-D^*} = 0$. Later in this section, we will discuss the case where D has singularities.

Proof: We first assume $k \geq 4$, so D is a hyperbolic Riemann surface. Let ω be the hermitian metric of constant negative curvature -1 on D. Any point in $\mathbb{P}_2^* - D^*$ is by duality a line in $\mathbb{P}_2$ intersecting D in $k = \deg D$ distinct points. Locally, these roots are holomorphic functions $a_1(z),\ldots a_k(z)$ on a neighborhood of a point in $\mathbb{P}_2^* - D^*$. Locally, we define a hermitian metric Ω on $\mathbb{P}_2^* - D^*$ by

$$\Omega = a_1^*(\omega)+\ldots+a_k^*(\omega)$$

Since this involves the roots $a_1,\ldots a_k$ symmetrically, it is independent of the neighborhood picked. Thus Ω is globally defined.

Another way of looking at what has just been done is this--let $D^{(k)}$ be the k-fold symmetric product of D with itself consisting of k-tuples of points of D without respect to order. This can be given the structure of a complex manifold by employing the elementary symmetric functions.

Let Σ be the subvariety of $D^{(k)}$ consisting of k-tuples with a repeated entry. We then have a holomorphic map

$$\mathbb{P}_2^* - D^* \xrightarrow{F} D^{(k)} - \Sigma$$

taking a line in $\mathbb{P}_2$ to the k-tuple of its intersections with D. The map F is an embedding, essentially because two points determine a line.

If D^k is the ordinary k-fold Cartesian product of D and $\tilde{\Sigma}$ the set of k-tuples with a repeated entry, we have the order-oblivious map

$$\pi\colon D^k - \tilde{\Sigma} \to D^{(k)} - \Sigma$$

which is a holomorphic covering. The product metric on D^k induced by ω drops to $D^{(k)} - \Sigma$ by π. The pullback of this metric by F is the metric Ω defined earlier. Because F is an embedding, Ω is a metric (otherwise it might vanish or be degenerate at some points).

By the estimate on the curvature of a sum of metrics mentioned in Section 2, we get that the curvature of Ω restricted to any holomorphic image of a disc has curvature $\le -1/k$. (What's true, though we won't need this, is that Ω has holomorphic sectional curvatures $\le -1/k$.) Thus by the proposition stated in Section 2, we have

$$d_{\mathbb{P}_2^*-D^*} \ge 1/k\ \Omega$$

where $d_{\mathbb{P}_2^*-D^*}$ is the Kobayashi pseudo-distance and d_Ω is the distance associated to the hermitian metric Ω. This implies $\mathbb{P}_2^* - D^*$ is hyperbolic.

It is possible to show $\mathbb{P}_2^* - D^*$ is hyperbolic using only the distance-decreasing property of the Kobayashi pseudo-distance. This method is best for proving $\mathbb{P}_2^* - D^*$ is complete hyperbolic.

We have the diagram

$$\begin{array}{ccc} & & D^k - \tilde{\Sigma} \\ & & \big\downarrow \pi \\ \mathbb{P}_2^* - D^* & \xrightarrow{F} & D^{(k)} - \Sigma \end{array}$$

Since F is an embedding and π a covering map, we conclude from the distance-decreasing property and the proposition quoted in Section 2 about covering spaces that $\mathbb{P}_2^* - D^*$ is complete hyperbolic if $D^k - \tilde{\Sigma}$ is. We know D^k is hyperbolic (using the product of the metric of constant negative curvature on D), so $D^k - \tilde{\Sigma}$ is. We can do better. $\tilde{\Sigma}$ is reducible, and

$$\tilde{\Sigma} = \bigcup_{1\le i<j\le k} \tilde{\Sigma}_{ij}$$

where $\tilde{\Sigma}_{ij}$ is the set of the k-tuples of points whose i'th and j'th entries are equal. By the distance-decreasing property, $D^k - \Sigma$ is complete hyperbolic if $D^k - \tilde{\Sigma}_{ij}$ is for all $i \neq j$. The latter spaces are all equivalent, and we have a natural biholomorphic map

$$D^k - \tilde{\Sigma}_{ij} \to D^{k-2} \times (D^2 - S)$$

where S is the diagonal in D^2. It suffices to prove $D^2 - S$ is complete hyperbolic.

The unit disc Δ is the universal cover of D, and we have the covering

$$\rho: \Delta^2 - \tilde{S} \to D^2 - S$$

where

$$\tilde{S} = \bigcup_{\sigma\in\Gamma} \tilde{S}_\sigma$$

where Γ is the group of covering transformations of $\Delta \to D$ and

$$\tilde{S}_\sigma = \{(z_1,z_2) \in \Delta^2 | z_1 = \sigma(z_2)\}.$$

We will be done if we can show $\Delta^2 - \tilde{S}_\sigma$ is complete hyperbolic for all $\sigma \in \Gamma$. These spaces are all biholomorphic to $\Delta^2 - \tilde{S}_1$, where $1: \Delta \to \Delta$ is the identity, by the map

$$\Delta^2 - \tilde{S}_\sigma \to \Delta^2 - \tilde{S}_1 \quad \text{defined by} \quad (z_1,z_2) \mapsto (z_1,\sigma(z_2))$$

Now there is a map

$$\Delta^2 - \tilde{S}_1 \to \Delta - \{0\} \quad \text{defined by} \quad (z_1,z_2) \mapsto \frac{z_1-z_2}{2}$$

It thus suffices to know that $\Delta - \{0\}$ is complete hyperbolic, and this is true classically. So $\mathbb{P}_2^* - D^*$ is complete hyperbolic if degree $D \ge 4$.

For D a non-singular curve of degree 3, a variation of the preceding argument is necessary. Since D is an elliptic curve, it is a group under the following classical operation (see for example Walker [4]): Let 0 be a flex (point of inflexion) of D. If P,Q are two points of D, let R be the third intersection of the line PQ with D. Then define P + Q to be the third point of the line RO on D.

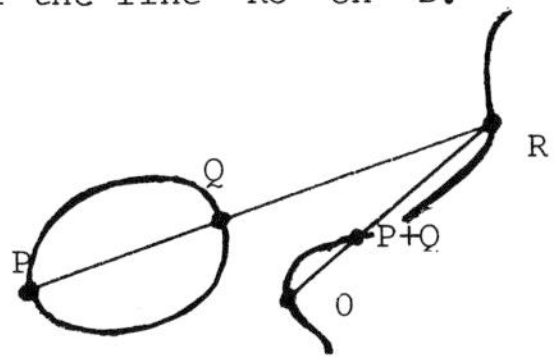

If P and Q or R and 0 coincide, we use tangents to replace the lines PQ or R0. Under this addition, with identity element 0, we easily see R = - (P + Q).

Thus, if a line intersects D in a_1, a_2, a_3 we have $a_3 = -(a_1 + a_2)$. We define a metric on $\mathbb{P}_2^* - D^*$ as follows: let ω be the metric of constant negative curvature -1 on $D - \{0\}$. Then let

$$\Omega = (a_2 - a_1)^*(\omega) + (a_1 - a_3)^*(\omega) + (a_2 - a_3)^*(\omega)$$

This is really symmetric in a_1, a_2, a_3 since $(a_i - a_j)^*(\omega) = (a_j - a_i)^*(\omega)$ because ω is preserved by the map $x \mapsto -x$. So Ω is globally defined on $\mathbb{P}_2^* - D^*$. Since $a_3 = -(a_1 + a_2)$, we can also write

$$\Omega = (a_2 - a_1)^*(\omega) + (a_2 + 2a_1)^*(\omega) + (a_1 + 2a_2)^*(\omega).$$

Since the two points a_1 and a_2 determine a line in $\mathbb{P}_2^*$, it is easily verified that Ω is a metric. It has the same curvature property as the metric defined earlier in this section. Hence

$$d_{\mathbb{P}_2^* - D^*} \geq \frac{1}{3} d_\Omega .$$

So $\mathbb{P}_2^* - D^*$ is hyperbolic in this case as well. Further, using the fact ω is complete (it blows up near 0), we easily conclude Ω is a complete metric; hence $\mathbb{P}_2^* - D^*$ is complete hyperbolic for deg $D^* = 3$.

This completes the proof of the theorem. If D has no singularities worse than ordinary multiple points, (all tangents are distinct), and the genus of D is ≥ 2, the same argument to prove hyperbolicity goes through. The only change is to notice that even though $a_1, \ldots a_k$ are no longer always distinct, they are still each holomorphic functions (intuitively, when a line goes through an ordinary multiple point without becoming tangent, we can keep track of which intersection is which and don't get branching).

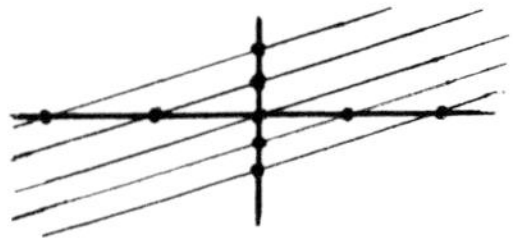

If D has worse singularities, it is necessary to remove all lines through these singularities from $\mathbb{P}_2^*$. These will give a line removed from $\mathbb{P}_2^*$ for every bad singular point. Since $D^{**} = D$ and genus(D^*) = genus(D), we can dualize this theorem to state:

Theorem: *Let* D *be an algebraic plane curve of genus* ≥ 2. *Then* $\mathbb{P}_2 - \{D$ *union all inflexional tangents to* $D\}$ *is hyperbolic.*

4. A CLASS OF COMPACT HYPERBOLIC SURFACES

It is a reasonable conjecture that a generic hypersurface of sufficiently high degree in $\mathbb{P}_n$ is hyperbolic. It is a sign of how great a road lies before us in value-distribution theory that virtually no hyperbolic hypersurfaces in $\mathbb{P}_n$, other than plane curves, are known. Our examples will be a class of hyperbolic hypersurfaces in $\mathbb{P}_3$, singular with codimension-two singularities. Any hypersurface in $\mathbb{P}_n$, $n > 2$, is simply connected by a theorem of Lefschetz.

Let D be a non-singular curve in $\mathbb{P}_2$, D^* its dual curve of tangent lines of D in the dual projective plane $\mathbb{P}_2^*$. For a generic D, the singularities of D^* are no worse than ordinary double points or cusps, corresponding to double tangents and inflexional tangents of D. We assume D to be such a curve with degree $D = k \geq 4$.

We construct a surface M branched over D^* by a method of James Carlson's thesis [1]. Let D^* be given in $\mathbb{P}_2^*$ by the vanishing of a homogeneous polynomial $P^*(w_0,w_1,w_2)$, so degree $P^* = k(k-1)$. Let $M \subset \mathbb{P}_3$ be given by

$$M = \{w_3^{k(k-1)} = P^*(w_0,w_1,w_2)\}.$$

Theorem. M *is hyperbolic.*

Proof. Let $\pi: M \to \mathbb{P}_2^*$ be the projection $(w_0,\ldots w_3) \mapsto (w_0,\ldots w_2)$. On $\mathbb{P}_2^* - D^*$, we have the metric Ω constructed in the last section. Thus $\tilde{\Omega} = \pi^*(\Omega)$ is a hermitian metric on $M - \pi^{-1}(D^*)$ and has negative curvature properties. If we can show that the map π branches over D^* enough to override the tendency of Ω to blow up near D^*, so that $\tilde{\Omega}$ is defined over all of M, it will then by virtue of its curvature properties lie below a multiple of the Kobayashi pseudo-distance on M, the argument being the same as that of the last section.

We first investigate the behavior of Ω on a neighborhood of a simple point p^* of D^*. Pick a patch U for D for the point of tangency p of the tangent p^* and pick a neighborhood U^* of p^* in $\mathbb{P}_2^*$ so that there are exactly two intersections of any line $\ell^* \in U^*$ with D in the patch U (this may be accomplished by picking the patch U not to include any intersections of p^* with D other than p.)

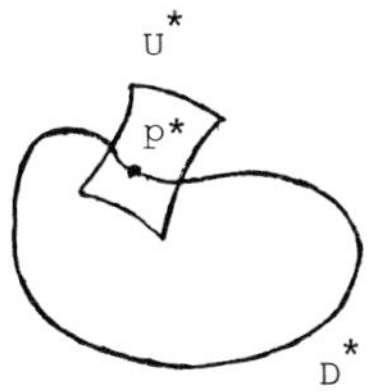

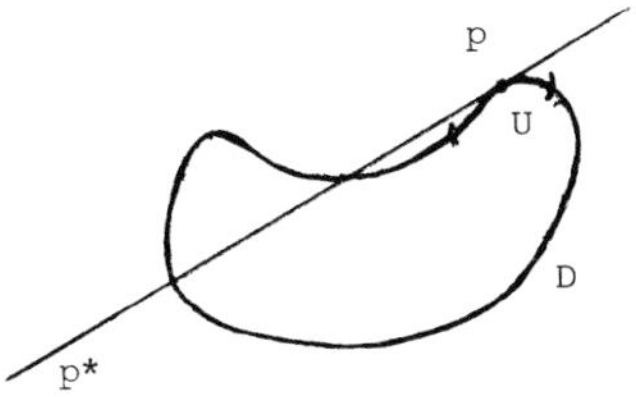

Let $a_1(\ell^*)$, $a_2(\ell^*)$ be these two intersections. Since we are in a single patch, we can add and multiply, so let

$$w = \frac{a_1(\ell^*) + a_2(\ell^*)}{2}$$

$$z = w^2 - a_1(\ell^*)a_2(\ell^*)$$

These are holomorphic functions on U^* as they depend only on symmetric functions of the intersecting a_1, a_2. In fact, z, w are coordinates on U^* and we have $D^* = \{z = 0\}$.

At this point, we recall how Ω was defined. If ω is the metric of constant negative curvature -1 on D, and if

$$\mathbb{P}_2^* - D^* \xrightarrow{(a_1, \ldots a_k)} D^{(k)} - \Sigma$$

where we let $a_1, \ldots a_k$ be the k-distinct intersections of a line in $\mathbb{P}_2^* - D^*$ with D, then

$$\Omega = a_1^*(\omega) + \ldots + a_k^*(\omega)$$

On U^*, labelling the intersections $a_1,\dots a_k$ so that a_1,a_2 are the intersections in the patch U, since $a_3,\dots a_k$ are holomorphic functions on U^* (because all points of $D^* \cap U^*$ are simple points of D^* if we take our neighborhood U^* of the simple point p^* small enough), we only get singularities from $a_1^*(\omega) + a_2^*(\omega)$. We can write a_1 and a_2 in terms of z,w as

$$a_1(z,w) = w + \sqrt{z}$$
$$a_2(z,w) = w - \sqrt{z}$$

If the expression for the metric ω in terms of the local coordinate v on U is $g(v)dv\,d\bar{v}$ then

$$\Omega = g(w + \sqrt{z})(dw + \frac{1}{2\sqrt{z}}dz)(d\bar{w} + \frac{1}{2\sqrt{\bar{z}}}d\bar{z})$$
$$+ g(w - \sqrt{z})(dw - \frac{1}{2\sqrt{z}}dz)(d\bar{w} - \frac{1}{2\sqrt{\bar{z}}}d\bar{z})$$

Now, if y_1,y_2,y_3 are affine coordinates on $\mathbb{P}_3$ associated to homogeneous coordinates $w_0,\dots,w_3$ by $y_i = \frac{w_i}{w_0}$, and $\pi: M \to \mathbb{P}_2^*$ is as before, we have

$$\pi^*(dz) = k(k-1)y_3^{k(k-1)-1}dy_3$$

and hence

$$\pi^*(\frac{dz}{\sqrt{z}}) = k(k-1)y_3^{\frac{1}{2}k(k-1)-1}dy_3$$

Since $k \geq 4$, we see that $\pi^*(\frac{dz}{\sqrt{z}}) \to 0$ as we approach $\pi^{-1}(p^*)$. So

$$\tilde{\Omega} = \pi^*(\Omega) \to 2\pi^*(g)\ \pi^*(dw\,d\bar{w})$$
$$+ \pi^*a_3^*(\omega) + \dots + \pi^*a_k^*(\omega).$$

as we approach $\pi^{-1}(p^*)$.

We conclude:

(1) We can extend $\tilde{\Omega}$ across π^{-1}(simple points of D^*).

(2) $\tilde{\Omega}$ may be zero at $\pi^{-1}(D^*)$ in the normal direction to $\pi^{-1}(D^*)$, but at simple points of D^* it does not vanish in directions tangential to D^*.

We next consider the singularities of D^*, which by our assumptions on D are no worse than ordinary double points and cusps. Ordinary double points of D^* correspond to lines which are tangent to D at two distinct points p_1,p_2, as shown in the diagram on the next page.

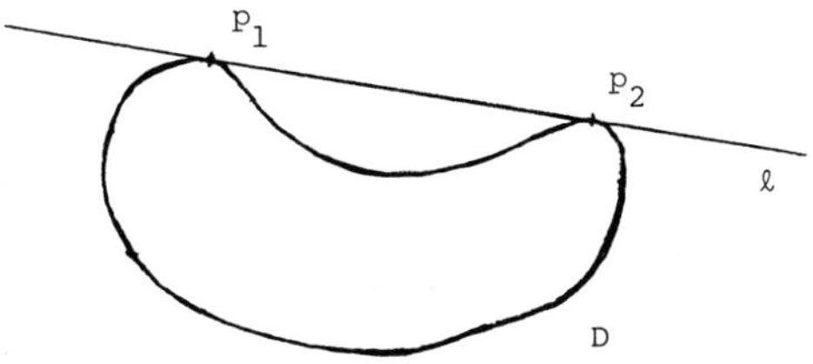

Near each point of tangency, we set up coordinates as before using a_1, a_2, the two intersections near p_1 and a_3, a_4 the intersections near p_2. The singularities of Ω are no worse than in the case just discussed, and $\tilde{\Omega}$ extends.

Cusps of D^* correspond to inflexional tangents of D. These will require a somewhat lengthy argument. If $Q \in D$ is a point of inflexion, $Q^* \in D^*$ the corresponding inflexional tangent, we can find a coordinate neighborhood U of Q for D and a neighborhood U^* of Q^* in $\mathbb{P}_2^*$ so that for any $\ell^* \in U^*$ there are exactly three intersections a_1, a_2, a_3 of ℓ^* with D in U. We further arrange that D^* have no other singularities in U^*. Then a_1, a_2, a_3 are solutions of a cubic

$$x^3 + u_1 x^2 + u_2 x + u_3 = 0$$

where u_1, u_2, u_3 are holomorphic on U^*. Then $a_1 + \frac{u_1}{3}$, $a_2 + \frac{u_1}{3}$, $a_3 + \frac{u_1}{3}$ satisfy an equation

$$x^3 + px + q = 0$$

where p and q are holomorphic on U^*. We now invoke the formula for the solution of a cubic equation. The discriminant of a cubic is

$$\delta = -4p^3 - 27q^2$$

So $\delta(\ell^*) = 0 \iff \ell^* \in D^*$. Let s be defined by

$$s^3 = -\frac{27}{2} q + \frac{3}{2} \sqrt{-3\delta}$$

Then

$$a_i = -\frac{u_1}{3} + \frac{1}{3} (\alpha_i s - \frac{3p}{\alpha_i s}) \qquad i = 1,2,3$$

where α_1, α_2, α_3 are the cube roots of 1.

Let $\omega = g(v)dv\bar{dv}$ be the metric of constant negative curvature -1 on D expressed in terms of the local coordinate on U. Then

$$\Omega = a_1^*(\omega)+\ldots+a_k(\omega)$$
$$= a_1^*(\omega) + a_2^*(\omega) + a_3^*(\omega) + \text{a term that is smooth on } U^*.$$

Now

$$a_i^*(dv) = \frac{1}{3}du_1 + \frac{\alpha_i}{3}ds - \frac{3}{\alpha_i}d(\frac{p}{s})$$

Pulling back by $\pi: M \to \mathbb{P}_2^*$, we need to show that $d(s\circ\pi)$ and $d(\frac{p\circ\pi}{s\circ\pi})$ don't blow up as we approach $\pi^{-1}(Q^*)$. As π ramifies over D^* to order $k(k-1)$, and $\delta(\ell^*) = 0 \iff \ell^* \in D^*$, we have that $\delta\pi$ vanishes to order at least $k(k-1)$ as we approach $\pi^{-1}(D^*)$. As $k \geq 4$, $\delta\circ\pi$ vanishes to at least order 12. From the formula for δ, we must have $p\circ\pi$ vanish to order at least 2 and $q\circ\pi$ vanish to order at least 3 as we approach $\pi^{-1}(D^*)$.

From the expression for s, we conclude $(s\circ\pi)^3$ vanishes to order at least 3 as we approach $\pi^{-1}(Q^*)$, as both q and $\sqrt{\delta}$ vanish at least that much. This implies that $d(s\circ\pi)$ doesn't blow up as we approach $\pi^{-1}(Q^*)$.

A formula that follows from those already given is

$$27p^3 = s^3(s^3 + 27q)$$

Thus, as we approach $\pi^{-1}(Q^*)$, since $(s\circ\pi)^3 + 27(q\circ\pi)$ vanishes to order at least 3, so $(p\circ\pi)$ vanishes to order at least one greater than that of $s\circ\pi$. Thus, $d(\frac{p\circ\pi}{s\circ\pi})$ doesn't blow up as we approach $\pi^{-1}(Q^*)$. Thus $\pi^*a_i^*(\omega)$ is continuous at $\pi^{-1}(Q^*)$. Hence $\tilde{\Omega}$ is well-behaved at the cusps of D^*.

From the foregoing, we have that $\tilde{\Omega}$ is continuous on all of M, is a metric on $M - \pi^{-1}(D^*)$, vanishes on $\pi^{-1}(D^*)$, if at all, only in some directions not tangent to $\pi^{-1}(D^*)$. In particular, $d_{\tilde{\Omega}}(Q_1, Q_2) > 0$ if Q_1 and Q_2 are distinct points of M. Since $\frac{1}{k}\tilde{\Omega}$ lies below the Kobayashi pseudo-distance on M by its curvature properties, we get that M is hyperbolic.

References

1. Carlson, James, *Some degeneracy theorems for entire functions with values in an algebraic variety*, Trans. Amer. Math. Soc. 168(1972), 273-301.

2. __________, and Phillip Griffiths, *A defect relation for equidimensional holomorphic mappings between algebraic varieties*, Ann. of Math. 95, No. 3(May 1972), 557-584.

3. Kobayashi, Shoshichi, *Hyperbolic Manifolds and Holomorphic Mappings*, Marcel Dekker, Inc., New York, 1970.

4. Walker, Robert, *Algebraic Curves*, Dover, New York, 1950.

A REMARK ON THE TRANSCENDENTAL BEZOUT PROBLEM

James A. Carlson

Brandeis University

INTRODUCTION

In algebraic geometry one defines the degree of an algebraic variety $W \subset \mathbb{P}^n$ as the number of points in the intersection of W with a generic linear space L of complementary dimension: $\dim W + \dim L = n$. If W_1 and W_2 have complementary dimensions and degrees d_1, d_2 respectively, then the number of points in the intersection $W_1 \cdot W_2$ (counted with multiplicity) is $d_1 d_2$. This is *Bezout's Theorem*. Note that W_1 and W_2 are required to be in *general position* in the sense that $W_1 \cdot W_2$ is discrete. More generally, if $W_1, W_2 \subset \mathbb{P}^n$ do not have complementary dimension, but intersect in a variety of the right dimension, namely $\dim W_1 + \dim W_2 - n$, then $\text{degree}\,(W_1 \cdot W_2) = d_1 d_2$.

To ask whether an analogue of Bezout's Theorem holds for noncompact varieties $W \subset \mathbb{C}^n$, we make the following definitions:

$$d^c = \frac{i}{4\pi}(\bar{\partial} - \partial)$$

$$\omega = \text{standard Kähler form on } \mathbb{P}^n \quad *$$

$$\psi_k = (dd^c \log\|z\|^2)^k \quad \text{where} \quad \|z\|^2 = \Sigma\, |z_i|^2$$

$$\mathbb{C}^n[r] = \{z \mid \|z\| < r\}$$

$$W[r] = W \cap \mathbb{C}^n[r] \quad \text{where} \quad W \subset \mathbb{C}^n \text{ is an analytic set}$$

---

* Thus, if H is the hyperplane at infinity and $z_1, \ldots, z_n$ are the usual coordinates on $\mathbb{C}^n = \mathbb{P}^n - H$, we have $\omega = dd^c \log(1 + \|z\|^2)$.

$$n(W,r) = \int_{W[r]} \psi_p \qquad \text{(volume of } W[r]) \text{ where } \dim_{\mathbb{C}} W = p.$$

$$N(W,r) = \int_0^r n(W,t)\,\frac{dt}{t}$$

$$\operatorname{ord}(W) = \overline{\lim_{r\to\infty}} \frac{\log N(W,r)}{\log r}$$

Now Wirtinger's Theorem says that when W is an algebraic variety

$$\int_W \omega^p = \deg(W)$$

where $p = \dim_{\mathbb{C}} W$. Thus the degree of W, a topological quantity, may be thought of as volume. In this sense, the functions $n(W,r)$ and $N(W,r)$; which measure the growth of an analytic set, are the correct generalization to analytic sets of the degree of a projective variety. In fact, if $W \subset \mathbb{C}^n \subset \mathbb{P}^n$ is algebraic, $N(W,r)$ is asymptotic to $\deg(W)\log r$: $\lim_{r\to\infty} \frac{N(W,r)}{\log r} = \deg(W)$.

The Bezout problem [2] in analytic geometry is to show that if $\operatorname{ord}(W_i) < \infty$, then $\operatorname{ord}(W_1 \cdot W_2)$ can be calculated from $\operatorname{ord}(W_i)$. For example, if A is a linear space, one could hope for an estimate like

$$N(W\cdot A,r) \le C'(N(W,C''r) + 1)$$

for suitable constants $C', C'' > 1$.

In this strong form, the Bezout problem has a negative answer, as is shown by the beautiful example of Cornalba and Shiffman [1]: They construct an analytic curve C in $\mathbb{C}^3$ which has order zero, but for which the intersection with a certain hyperplane has infinite order. Nevertheless, as we shall see in Section 1, there are many hyperplanes $A \subset \mathbb{C}^3$ for which $C \cdot A$ has order zero. Thus it makes sense to ask the weaker question: for which A can we estimate $N(W\cdot A,r)$ by $N(W,r)$? To state the answer, let $\mathbb{P}G_{q,n}$ be the Grassmann manifold of linear spaces of codimension q in $\mathbb{P}^n$, and let μ be the standard invariant measure on $\mathbb{P}G_{q,n}$. In Section 2 we show that for almost all $A \in \mathbb{P}G_{q,n}$, a Bezout estimate holds. In particular, if W has finite order, then for almost all A, $W \cdot A$ has finite order. One should regard this as an analogue of the algebro-geometric situation, where the Bezout Theorem holds only when the varieties are in suitably general position. In Section 3, we give an example of an analytic set with irregular growth properties, and in Section 4 we discuss the order of $W \cdot F^{-1}(A)$, where $f: W \to \mathbb{C}^m$ is a holomorphic map of finite order and $A \subset \mathbb{C}^m$ is an affine subspace (i.e. a translate of a linear subspace).

1. THE CORNALBA-SHIFFMAN EXAMPLE

The Cornalba-Shiffman curve $\mathcal{C}$ (see [1]) is the union of irreducible components $\mathcal{C}_k \subset \mathbb{C}^3$ given by

$$\mathcal{C}_k = \begin{cases} z = 3^{-c_k^2} y^{c_k} \\ x = 2^k \end{cases}$$

where c_k is an increasing sequence of natural numbers. Then $\operatorname{ord}(\mathcal{C}) = 0$ because of the damping factor $3^{-c_k^2}$, but $\operatorname{ord}(\mathcal{C}) \cap \{z = 0\}) = \infty$ if $\{c_k\}$ increases rapidly enough, for example, if $c_k \geq e^k$.

Now consider $\mathcal{C} \cdot L_a = \mathcal{C} \cap \{z = a\}$. We will show that $\mathcal{C} \cdot L_a$ has order zero for $a \neq 0$ and $c_k = e^k$. Define $\lambda(k)$ = number of k such that $\mathcal{C}_k \cdot L_a \subset \mathbb{C}^3[r]$. Now $\mathcal{C}_k \cdot L_a = \{(2^k, 3^{c_k} a^{1/c_k}, a)\}$, so we find that $\lambda(k) \leq \#\{k \mid 3^{e^k} \leq Ar\}$ for suitable $A > 0$, hence $\lambda(k) \leq \log\log Ar$. Then

$$n(\mathcal{C} \cdot L_a, r) \leq \sum_{1}^{\log\log Ar} e^k \leq (\log\log Ar)(\log Ar) .$$

One sees easily that $\overline{\lim} \dfrac{\log N(V,r)}{\log r} = \overline{\lim} \dfrac{\log n(V,r)}{\log r}$, so that $\mathcal{C} \cdot L_a$ has order zero if $a \neq 0$. Thus the "bad" intersection $\mathcal{C} \cdot L_0$ can be perturbed by an abritrarily small amount to a "good" intersection $\mathcal{C} \cdot L_\varepsilon$. The same method shows that in general $\mathcal{C} \cdot L_a$, $a \neq 0$, has order zero.

2. A BEZOUT ESTIMATE

Our Bezout estimate will follow easily from the Crofton formula given in [3], which should be viewed as an average Bezout estimate. Given a complex submanifold $W \subset \mathbb{C}^N$, and a map $f: W \to \mathbb{P}^m$, we define an order function by

$$T_q(r) = \int_0^r \frac{dt}{t} \int_{W[t]} \omega_f^q \wedge \psi_{n-q} \qquad (2.1)$$

where $n = \dim_{\mathbb{C}} W$. If $\mathbb{P}G_{q,m}$ is the Grassmannian of linear spaces of codimension q and μ is the standard normalized invariant measure on $\mathbb{P}G_{q,m}$, the Crofton formula is*

$$\int_{\mathbb{P}G_{q,m}} N(W \cdot f^{-1}(A), r) d\mu_A = T_q(r) \qquad (2.2)$$

If $W \subset \mathbb{C}^N$ does not pass through the origin (which we assume henceforth), then $\pi\colon \mathbb{C}^N - \{0\} \to \mathbb{P}^{N-1}$ defines a holomorphic map $\pi\colon W \to \mathbb{P}^{N-1}$. Now $\pi^*\omega = \psi$, so (2.1) gives

$$\int_{G_{q,N}} N(W \cdot B, r) d\mu_B = N(W,r) \qquad (2.3)$$

where now $G_{q,N}$ is the Grassmannian of linear subspaces of codimension q in $\mathbb{C}^N$. The main result is:

Theorem A: *Fix* $\alpha > 0$. *Then for any* $\varepsilon > 0$, *there is a measurable set* $F_\varepsilon \subset G_{q,N}$ *and an* $r_\varepsilon > 0$ *such that*

$$N(W \cdot B, r) \le (r+2)^{1+\alpha} N(W, r+2)$$

for all $B \notin F_\varepsilon$ *and all* $r \ge r_\varepsilon$.

Proof: Set $E_k = \{B \mid N(W \cdot B, k) \ge k^{1+\alpha} N(W,k)\}$. Then E_k is closed, and

$$\mu(E_k) k^{1+\alpha} N(W,k) \le \int_{E_k} N(W \cdot B, k) d\mu_B \le N(W,k),$$

hence $\mu(E_k) \le k^{-(1+\alpha)}$. Define $F_k = \bigcup_{j \ge k} E_j$. Then $\mu(F_k) \le \sum_{j=k}^{\infty} j^{-(1+\alpha)}$. Since $\sum_1^\infty j^{-(1+\alpha)} < \infty$, $\mu(F_k) < \varepsilon$ for k large enough. However, on the complement of F_k, we have

$$N(W \cdot B, j) < j^{1+\alpha} N(W,j) \qquad j \ge k.$$

Now

$$N(W \cdot B, r) \le N(W \cdot B, [r]+1) \le ([r]+1)^{1+\alpha} N(W, [r]+1) \le (r+2)^{1+\alpha} N(W, r+2),$$

where $[r]$ = greatest integer in r. Q.E.D.

---

* The exhaustion function $\tau(z) = \log\|z\|^2$ on $W \subset \mathbb{C}^N$ is not special, i.e. $(dd^c\tau)^n \ne 0$, where $\dim W = n$. This does not affect the validity of (2.2).

Corollary 1: *Consider the estimate*

$$\text{(B.E.)} \qquad N(W\cdot B,r) \le (r+2)^{1+\alpha}\, N(W,r+2), \qquad r \ge r_B.$$

The set of B *for which this estimate fails is of measure zero in* $G_{q,N}$.

Corollary 2: *Given* $B_o \in G_{q,N}$, *there is a holomorphic family of linear spaces* $\{B_t\}_{|t|<2\varepsilon}$ *such that* $W\cdot B_\varepsilon$ *satisfies* (B.E.)

Thus every intersection $W\cdot B_o$ is holomorphically homotopic (by an arbitrarily small perturbation B_ε) to a "good" intersection in the sense that a Bezout estimate holds. Finally, we have

Corollary 3: $\operatorname{ord}(W\cdot B) \le \operatorname{ord}(W) + 1 \quad \text{a.e.}(B)$
where a.e.(B) *means that the inequality holds outside a set of measure zero in* $G_{q,N}$.

The proof follows immediately from Corollary 2 and the easily proved

Lemma: *If* g *and* h *are positive increasing functions such that*

$$g(r) \le (Ar+B)^{\beta}\, h(Ar+B)$$

for $\beta, A, B > 0$, *then* $\operatorname{ord}(g) \le \operatorname{ord}(h) + \beta$.

With additional assumptions on the growth of W, we can sharpen the inequality of Corollary 3. To this end, let us redefine E_k by

$$E_k = \{B \mid N(W\cdot B,k) \ge N^{\beta}(W,k)\}$$

where $\beta > 1$. Then the Crofton Formula (2.3) gives

$$\mu(E_k) \le \frac{1}{N^{\beta-1}(W,k)}$$

We temporarily assume that W is *regular of order* ρ: that is, $N(W,r) \ge cr^{\rho}$ for $r \ge r_o$. Then $\mu(F_k) \le c\sum_{j=k}^{\infty} r^{\rho(1-\beta)}$. Convergence is assured if $\rho(\beta-1) > 1$, that is if $\beta > 1 + \frac{1}{\rho}$. Proceeding as before, we see that given $\delta > 0$, there is an ε so that

$$\begin{cases} N(W\cdot B,r) \le [N(W,r+2)]^{1+\frac{1}{\rho}+\delta} \\ \text{where} \quad B \notin F_\varepsilon \quad \text{and} \quad \mu(F_\varepsilon) < \varepsilon \\ \text{where} \quad r \ge r_\varepsilon . \end{cases}$$

Thus we have

Proposition: *If* W *is regular of order* ρ, *then* $\operatorname{ord}(W \cdot B) \le \operatorname{ord}(W) + \frac{1}{\rho}$ a.e.(B).

To remove the assumption of regularity we appeal to the following

Lemma: *For all* n, $0 \le n \le N$, *and for all* $p \ge 0$, *there is an analytic set* W *of pure dimension* n *in* $\mathbb{C}^N$ *which is regular of order* ρ.

Proof: Let W be the union of linear spaces of dimension n, where we position $[j^{\rho-1}]$ linear spaces tangent to the sphere of radius j. A simple computation verifies the lemma. Q.E.D.

Now let W, W' be analytic sets of dimension n and order ρ, where W' is regular of order ρ. Then $W \cup W'$ is regular of order ρ, and so we have

$$\begin{cases} N(W \cdot B,r) \le N((W \cup W') \cdot B,r) \le [N(W \cup W',r+2)]^{1+\frac{1}{\rho}+\delta} \\ r \ge r_B , \quad \text{a.e.}(B). \end{cases}$$

Since W has order $\rho \iff$ all $\varepsilon > 0$, $N(W,r) \le r^{\rho+\varepsilon}$ if $r \ge r_\varepsilon$, we conclude

Theorem B: $$\operatorname{ord}(W \cdot B) \le \operatorname{ord}(W) + \frac{1}{\operatorname{ord}(W)} \quad \text{a.e.}(B)$$

Note that by defining π to be projection relative to lines through $a \notin W$ rather than lines through the origin, that the above results apply to almost all affine subspaces of $\mathbb{C}^N$.

3. AN EXAMPLE OF IRREGULAR GROWTH

The above detour in the proof seems necessary, since there are analytic sets of finite order $\rho \ne 0$ which are not regular of order ρ' for any $\rho' > 0$. To construct an example, let $E_1(z) = e^z$, and set $E_a(z) = \exp E_{a-1}(z)$. A divisor in $\mathbb{C}$ is given by specifying its multiplicity $\nu(z)$ at each point. We define divisors ν_a as follows:

(i) if $|z| \ne E_a(n)$, $\nu_a(z) = 0$.

(ii) on the circle about the origin of radius $E_a(n)$,

position $[E_a(n) - E_a(n-1)]$ points, where $[x]$ denotes the greatest integer less than x.

Then $n_a(r) = \sum_{|s|\le r} \nu_a(s)$, and we see that $n_a(E_a(n)) = E_a(n) + \text{constant}$. From this it follows easily that ν_a has order one. Now set $r_n = E_a(n)$. If ν_a is regular of order $\varepsilon > 0$, then we have $n_a(r) \ge r^\varepsilon$ for $r \ge r_o$, hence $n_a(r_{n-1}) = \lim_{r\uparrow r_n} n_a(r) \ge \lim_{r\uparrow r_n} r^\varepsilon = r_n^\varepsilon$. We examine $n_a(r)$ for $a = 1,2,3$:

(i) ν_1 regular $\iff e^{n-1} \ge (e^n)^\varepsilon$, $n \ge n_o$, which holds for $n \ge 2$, $\varepsilon \le 1/2$.

(ii) ν_2 regular $\iff E_2(n-1) \ge (E_2(n))^\varepsilon \iff n-1 \ge \log \varepsilon + n$, which holds for $n \ge 0$ if $\varepsilon = e^{-1}$.

(iii) ν_3 regular $\iff E_3(n-1) \ge (E_3(n))^\varepsilon \iff e^{n-1} \ge \log \varepsilon + e^n$, which fails for any choice of ε and n_o.

Hence ν_3 is not regular of order ρ for any $\rho > 0$.

4. A BEZOUT ESTIMATE FOR COMPLETE INTERSECTIONS

Let $W \subset \mathbb{C}^N$ be an analytic set of pure dimension n. Let $f\colon W \to \mathbb{C}^m$ be a holomorphic map, and define the order of f by

$$\operatorname{ord}(f) = \overline{\lim} \frac{\log M_f(r)}{\log r}$$

where $M_f(r) = \max\{1 + \|f\|^2 \mid \|z\| \le r\}$. Then we have

Theorem A': *Let* $W \subset \mathbb{C}^N$ *be a complex submanifold of pure dimension* $n < N$ *such that* $0 \notin W$. *Let* $f\colon W \to \mathbb{C}^m \subset \mathbb{P}^m$ *be holomorphic, and such that for almost all* $A \in \mathbb{P}G_{q,m}$, $f^{-1}(A)$ *has pure codimension* q. *Then for almost all* A *we have*

(i) $N(W \cdot f^{-1}(A), r) \le C_\theta (r+2)^{1+\alpha} [\log M_f(\theta(r+2))]^q\, N(W, \theta(r+2))$

where $\theta > 1$, $C_\theta > 0$, *and* $r \ge r_A$.

(ii) $\operatorname{ord}(W \cdot f^{-1}(A)) \le q \operatorname{ord}(f) + \operatorname{ord}(W) + 1$

Proof: The second statement follows immediately from the first. To prove the first, we define

$$E_k = \{A \mid N(W \cdot f^{-1}(A), k) \ge k^{1+\alpha}\, T_q(k)\}$$

where $T_q(r)$ is given by (2.1). Proceeding exactly as in the proof of Theorem A, we obtain

$$N(W \cdot f^{-1}(A),r) \leq (r+2)^{1+\alpha} \; T_q(r+2), \quad r \geq r_A \quad \text{a.e.}(A).$$

To complete the proof, we need the following:

Proposition: *Let* $W \subset \mathbb{C}^N$ *be a complex submanifold of pure dimension* n *such that* $0 \notin W$. *Let* $f: W \to \mathbb{C}^m$ *be holomorphic. Then* $T_q(r) \leq C_\theta (\log M_f(\theta r))^q \, N(W,\theta r)$ *where* $\theta > 1$ *and* C_θ *is a suitable constant.*

Proof:

$$\begin{aligned}
T_q(r) &= \int_0^r \frac{dt}{t} \int_{W[t]} \omega_f^q \wedge \psi_{n-q} \\
&= \int_0^r \frac{dt}{t} \int_{\partial W[t]} d^c \log(1+\|f\|^2) \wedge \omega_f^{q-1} \wedge \psi_{n-q} && \text{(Stokes' Theorem)} \\
&= \int_{W[r]} d \log\|z\|^2 \wedge d^c \log(1+\|f\|^2) \wedge \omega_f^{q-1} \wedge \psi_{n-q} && \text{(Fubini's Theorem)} \\
&= \int_{W[r]} d \log(1+\|f\|^2) \wedge d^c \log\|z\|^2 \wedge \omega_f^{q-1} \wedge \psi_{n-q} && \text{(by type considerations)} \\
&= \int_{\partial W[r]} \log(1+\|f\|^2)\, \omega_f^{q-1} \wedge d^c \log\|z\|^2 \wedge \psi_{n-q} \\
&\quad - \int_{W[r]} \log(1+\|f\|^2)\, \omega_f^{q-1} \wedge \psi_{n-q+1} && \text{(Stokes' Theorem)}
\end{aligned}$$

Since $\log(1+\|f\|^2) \geq 0$, we have

$$T_q(r) \leq \log M_f(r) \int_{\partial W[r]} \omega_f^{q-1} \wedge d^c \log\|z\|^2 \wedge \psi_{n-q} ,$$

hence

$$T_q(r) \leq \log M_f(r) \int_{W[r]} \omega_f^{q-1} \wedge \psi_{n-q+1} . \tag{4.1}$$

Now

$$\int_r^{\theta r} \frac{dt}{t} \int_{W[t]} \omega_f^q \wedge \psi_{n-q} \leq T_q(\theta r) ,$$

where $\theta > 1$, hence

$$\int_{W[r]} \omega_f^q \wedge \psi_{n-q} \leq \frac{1}{\log \theta} \; T_q(\theta r) . \tag{4.2}$$

Comparing (4.1) and (4.2), obtain

$$T_q(r) \leq \frac{1}{\log \theta} \; \log M_f(r)\, T_{q-1}(\theta r) .$$

Repeating the argument q times yields

$$T_q(r) \leq \left(\frac{1}{\log \theta}\right)^q \left(\log M_f(\theta r)\right)^q \, T_o(\theta^q r) .$$

But $T_o(r) = N(W,r)$, so we are done. Q.E.D.

If we let $W = \mathbb{C}^N$, then the argument is the same up to the last step. Then

$$T_1(r) = \int_{\partial\mathbb{C}^N[r]} \log(1 + \|f\|^2)\, d^c \log\|z\|^2 \wedge \psi_{N-1}$$

since $\psi_N = 0$.
Hence

$$T_1(r) \leq \log M_f(r) \int_{\partial\mathbb{C}^N[r]} d^c \log\|z\|^2 \wedge \psi_{N-1} .$$

But the latter integral is just 1, so we obtain $T_q(r) \leq C_\theta(\log M_f(\theta r))^q$, and thus

Corollary: *Let* $f\colon \mathbb{C}^N \to \mathbb{C}^m$ *be holomorphic. Suppose that for almost all* $A \in \mathbb{P}G_{q,m}$, $f^{-1}(A)$ *has pure codimension* q. *Then for almost all* A,

(i) $N(f^{-1}(A),r) \leq C_\theta(r+2)^{1+\alpha} \;(\log M_f(\theta(r+2)))^q,\ r \geq r_A.$

(ii) $\operatorname{ord}(f^{-1}(A)) \leq q \operatorname{ord}(f) + 1$

Using the same techniques as in the proof of Theorem B, we find

Theorem B': *Let* $W \subset \mathbb{C}^N$ *be a complex submanifold of pure dimension* w. *Suppose that* $f\colon W \to \mathbb{C}^m$ *is a holomorphic map such that for almost all* $A \in \mathbb{P}G_{q,m}$, $f^{-1}(A)$ *has pure codimension* q. *Then for almost all* A *we have*

$$\operatorname{ord}(W \cdot f^{-1}(A)) \leq q \operatorname{ord}(f) + \operatorname{ord}(W) + \frac{1}{\operatorname{ord}(W)} .$$

5. CONCLUSION

(a) We should point out that several variants of the almost everywhere Bezout Theorem given above are possible. First, if $\overline{V}$ is an m-dimensional projective variety and $D \subset \overline{V}$ is an ample divisor, we may replace $\mathbb{C}^m$ by $V = \overline{V} - D$. If L is a line bundle on $\overline{V}$, we replace $\mathbb{P}G_{q,m}$ by the Grassmannian $G = \{\sigma_1 \wedge \ldots \wedge \sigma_q | \sigma_i \in \Gamma(\overline{V},L)\}$. Thus G parametrizes subvarieties of $\overline{V}$ which are generically complete intersections of codimension q. Then, using affine coordinates ζ_i on V given by ratios of sections of L, we define $\omega = dd^c \log(1 + \Sigma|\zeta_i|^2)$ and

$M_f(r) = \max(1 + \Sigma|\zeta_i \circ f|^2)$. Since the Crofton formula of Griffiths-King [3] holds for this situation, we obtain as corollaries, Theorems A' and B'.

Second, we should note that Stoll's average Bezout Theorem [4] yields as an immediate corollary an almost everywhere estimate. Let $F: \mathbb{C}^N \to \mathbb{C}^q$ be holomorphic and let $W \subset \mathbb{C}^N$ be a complex submanifold. Define $F_t(z) = F(tz)$. Then, if $W \cdot F_t^{-1}(0)$ is a pure codimension q intersection for all $t \in [0,1]$, Stoll shows that

$$\int_0^1 N(W \cdot F_t^{-1}(0), r)\,dt \le C_\theta (\log M_F(\theta r))^q N(V, \theta r)$$

From this it follows that

$$\begin{cases} N(W \cdot F_t^{-1}(0), r) \le C_\theta (r+2)^{1+\alpha} [\log M_f(\theta(r+2))]^q N(W, \theta(r+2)) \\ \text{a.e.}(t), \quad \text{and where} \quad r \ge r_t . \end{cases}$$

(b) Let G be a family of codimension q subvarieties A of a fixed projective or affine variety V. Let W be a Stein manifold, and $f: W \to V$ a holomorphic map such that $f^{-1}(A)$ has codimension q whenever $A \in G' \subset G$. Finally, let us define the *Bezout set* of f, W, G to be the subset $\mathcal{B}$ of G' such that an estimate of the form

$$N(W \cdot f^{-1}(A), r) \le C_\theta (r+b)^{1+\alpha} [T_1(\theta r + b)]^q N(W, \theta r + b)$$

holds for suitable constants α, θ, C_θ and b, and where $r \ge r_A$. We know that in certain situations $\mathcal{B}$ is nonempty, in fact of largest possible measure. Moreover, since G is of finite measure in Theorems A' and B', we know that $\mathcal{B}$ is *dense*. Thus any intersection $W \cdot f^{-1}(A_0)$ has an arbitrarily small perturbation to a "good intersection" $W \cdot f^{-1}(A_1)$ where $A_1 \in \mathcal{B}$. However, we do not know whether $W \cdot f^{-1}(A_1)$ is *stable*, i.e. whether $\mathcal{B}$ has nonempty interior. A more difficult question is to describe $G - \mathcal{B}$ in an *a priori* fashion. For example, is $G - \mathcal{B}$ an analytic set?

In conclusion, what is missing is a precise notion of general position which guarantees a Bezout estimate in analytic geometry.

REFERENCES

1. Cornalba, M. and B. Shiffman, *A counterexample to the "Transcendental Bezout Problem"*, Ann. of Math. 96 (1972), 402-406.

2. Griffiths, P., *Two theorems on extensions of holomorphic mappings*,

Invent. Math. 14 (1971), 27-62.

3. __________, and J. King, *Nevanlinna Theory and holomorphic mappings between algebraic varieties*, to appear in Acta Math.

4. Stoll, W., *A Bezout estimate for complete intersections*, Ann. of Math. 96 (1972), 361-401.

A THEOREM IN COMPLEX GEOMETRIC FUNCTION THEORY

R. E. Greene and H. Wu

University of California,
Los Angeles and Berkeley

This paper gives an expository account of one theorem in complex geometric function theory assuming only an elementary knowledge of functions of several complex variables. By complex geometric function theory, we mean the study of the relationship between geometric invariants (e.g., curvature, second fundamental form) and complex analytic properties (e.g., existence of holomorphic functions, nonexistence of certain holomorphic mappings). One aspect of it deals with the existence and properties of holomorphic functions (or more generally holomorphic p-forms) on noncompact Kähler manifolds. In a series of papers [2]-[6], we have begun such a study by combining various geometric techniques with recent advances in the theory of complex functions of several variables. Such a study invariably presupposes a certain amount of familiarity with Riemannian geometry so that when a detailed account of [2]-[4] is written up, the audience would perhaps be somewhat limited. We therefore take this opportunity to outline the proof of a representative theorem with the hope that we have succeeded in conveying the essential flavor of the subject without encumbering the reader with excessive technical details. Specifically, we shall sketch a proof of:

Main Theorem: *If* M *is a noncompact complete Kähler manifold with positive curvature, then* M *is a Stein manifold.*

A nearly self-contained account of the differential geometry involved in this theorem will be given in the following. Because we want the exposi-

Research partially supported by the National Science Foundation. The second author holds an Alfred P. Sloan Foundation Fellowship.

tion to be completely elementary on the geometric side, we have conscientiously avoided achieving maximal generality in our presentation. In particular, we should point out that the Main Theorem is actually a special case of each of the following two results:

(α) *A noncompact complete Kähler manifold with nonnegative sectional curvature and positive Ricci curvature whose canonical bundle is trivial is a Stein manifold.*

(β) *A complete noncompact Hermitian manifold whose Hermitian metric is, outside a compact set, Kählerian and of positive sectional curvature is obtained from a Stein space by blowing up a finite number of points.*

The proof we outline here of the Main Theorem is, with appropriate modifications, sufficient to prove (α), but the proof of (β) requires more powerful techniques, geometrically as well as function theoretically. (α) has been announced, as Theorem 3 in [4], while (β) is of more recent vintage.

We note that our desire to make this exposition elementary has given rise to occasional imprecise (though correct) statements. In such case, we hope that no confusion will arise.

The Main Theorem is an immediate consequence of three theorems of independent interest. For their statements, let us first recall two key facts from geometry that we need (see §1 for the precise definition of the terms involved):

(A) (Gromoll-Meyer [7]): *If* M *is a noncompact complete Riemannian manifold of positive sectional curvature, then* M *is diffeomorphic to euclidean space.*

(B) (Cheeger-Gromoll [1]): *If* M *is a noncompact complete Riemannian manifold of nonnegative sectional curvature. Then there exists a continuous function* $\tau: M \to [0,\infty)$ *such that*

(i) τ *is* convex, *in the sense that restricted to each geodesic,* τ *is a convex function of one variable.*

(ii) *For* $c \in \mathbb{R}, M_c \equiv \{p \in M: \tau(p) \le c\}$ *is compact.*

The function τ in (B) is not unique, but we will fix one such in all subsequent discussions. As an example of such a τ, let $M = \mathbb{R}^n$. Then $\tau(x_1,\ldots,x_n) = (\Sigma_i x_i^2)^{1/2}$ is one.

Now the three theorems alluded to above are as follows:

Theorem 1: *With* M *as in the Main Theorem and* τ *as in* (B) *let* A(M) *and* $A(M_c)$ *be the rings of holomorphic functions on* M *and* M_c *respectively. Then the restrictions of elements of* A(M) *to* M_c *are dense in* $A(M_c)$ *in the sup norm for all* $c \in \mathbb{R}$.

Theorem 2: *With* M *as above, then there exists a* C^∞ *strictly plurisubharmonic function on* M.

Theorem 3: *With* M *as above, if* M *possesses a* C^∞ *strictly plurisubharmonic function and if* $A(M)|_{M_c}$ *is dense in* $A(M_c)$ *for all* $c \in \mathbb{R}$, *then* M *is Stein.*

The proofs of Theorems 1 and 2 depend on an L^2 estimate on 1-forms, while all three theorems make use of an approximation theorem for convex functions ([5], Theorem 2). We note explicitly that the analytic ideas of the proofs of Theorems 1 and 2 are borrowed from Hörmander [8] and [9]. Particularly relevant are the proofs of Lemma 4.3.1 and Theorem 4.2.2 in [8].

The structure of this paper is as follows. In §1, we give a short course in differential geometry; this section both sets up the notation and summarizes all the geometric facts we need. §2 is the heart of the paper. Here we show how to combine geometric techniques with the L^2 theory to draw analytic conclusions. The lemmas developed in §2 are applied to §3 to prove the above three theorems. There is an appendix which deals with a technical matter concerning the Hilbert Space adjoint of a differential operator.

1. SUMMARY OF RIEMANNIAN GEOMETRY

In this section, we summarize the minimal amount of Riemannian geometry needed for our purpose.

Let M be a real differential manifold. (Throughout this paper, everything is C^∞ unless explicitly denied.) A *connection* D on M is characterized by the following properties: Let X be a vector field and A a tensor field, then

i) $D_X A$ is a tensor field of the same type as A. I.e., if A is a vector field, so is $D_X A$, and if A is a p-form, so is $D_X A$, etc.

ii) If f is a function, $D_X f = Xf$.

iii) $D_{fX} A = f(D_X A)$

iv) If Y is another vector field,

$$D_{X+Y} A = D_X A + D_Y A$$

v) If B is another tensor field, then

$$D_X (A \otimes B) = (D_X A) \otimes B + A \otimes (D_X B).$$

In particular, if μ, ν are differential forms, then

$$D_X (\mu \wedge \nu) = (D_X \mu) \wedge \nu + \mu \wedge (D_X \nu).$$

Let g be a Riemannian metric on M. The *Levi-Civita connection* of g is the unique connection on M satisfying the following two conditions:

vi) $D_X g \equiv 0$

vii) $D_X Y - D_Y X = [X,Y]$.

In the following, whenever we deal with a Riemannian metric g, it is understood that the Levi-Civita connection of g is the only connection we shall use. The most important tensor associated with D is the *curvature tensor* R_{XY}, where X,Y are vector fields on M. By definition, if A is a tensor field, $R_{XY} A$ is a tensor field of the same type as A given by:

$$R_{XY} A = - D_X D_Y A + D_Y D_X A + D_{[X,Y]} A.$$

R_{XY} has the following properties:

a) If f is a function, $R_{XY} f \equiv 0$.

b) If A,B are tensor fields,

$$R_{XY} (A \otimes B) = (R_{XY} A) \otimes B + A \otimes (R_{XY} B).$$

c) For $p \in M$, $(R_{XY} A)(p)$ depends only on $X(p)$, $Y(p)$ and $A(p)$. In particular, given $x,y,z \in M_p$ (tangent space to M at p), it makes sense to write $R_{xy} z$, which is a vector in M_p.

We can now define the *curvature function* K, which assigns to every 2-plane P of a tangent space a real number. Let $\{e_1, e_2\}$ be an orthonormal basis of P, then

$$K(P) \equiv g(R_{e_1 e_2} e_1, e_2).$$

It can be verified that this definition is independent of the choice of $\{e_1, e_2\}$ and that K is a C^∞ function (defined on the Grassmann bundle of 2-planes). $K(P)$ is called the *sectional curvature* of P, or more simply, just the *curvature* of P. By definition, the Riemannian manifold has positive curvature if and only if $K > 0$, has nonnegative curvature if and only if $K \geq 0$, etc. When $\dim M = 2$, then the tangent space itself is the only 2-plane at each point and K therefore becomes a C^∞ function defined on M. It is an instructive computation to verify that, in this case, K coincides with the classical notion of Gaussian curvature.

We next introduce some standard operators on the Riemannian manifold M. From now on, M will be assumed to be oriented and of dimension $2n$. Locally, an orthonormal basis of vector fields $X_1, \ldots, X_{2n}$ will always be assumed to be coherent with the orientation, i.e., $(X_1 \wedge \ldots \wedge X_{2n})(p)$ gives the positive orientation at each p in the domain of definition of $\{X_i\}$. If $\omega^1, \ldots, \omega^{2n}$ are the dual 1-forms of $\{X_i\}$ in the sense that $\omega^i(X_j) = \delta^i_j$, we can extend the metric g to be an inner-product for forms at each point: locally, if

$$\alpha = \sum_{i_1 < \ldots < i_p} \alpha_{i_1 \ldots i_p} \omega^{i_1} \wedge \ldots \wedge \omega^{i_p}, \quad \text{and}$$

$$\beta = \sum_{i_1 < \ldots < i_p} \beta_{i_1 \ldots i_p} \omega^{i_1} \wedge \ldots \wedge \omega^{i_p}, \quad \text{then}$$

$$g(\alpha, \beta) \equiv \sum_{i_1 < \ldots < i_p} \alpha_{i_1 \ldots i_p} \beta_{i_1 \ldots i_p}.$$

This definition is independent of the choice of $X_1, \ldots, X_{2n}$ (by just linear algebra). In particular, $\{\omega^1, \ldots, \omega^{2n}\}$ are orthonormal 1-forms in this neighborhood. Let us denote the p-forms on M by $\Lambda^p(M)$. Define $*: \Lambda^p(M) \to \Lambda^{2n-p}(M)$ by

$$*(\omega^1 \wedge \ldots \wedge \omega^p) \equiv \omega^{p+1} \wedge \ldots \wedge \omega^{2n},$$

and extend by linearity. If d denotes the exterior differential operator, then the co-differential $\delta: \Lambda^p(M) \to \Lambda^{p-1}(M)$ is defined to be $\delta = -*d*$. When the dimension of M is odd, there is a complicated rule for the sign in the definition of δ. Since we shall be concerned only with complex manifolds, which are necessarily even-dimensional, this sign complication does not occur. The Laplacian $\Delta = \Lambda^p(M) \to \Lambda^p(M)$ is defined by $\Delta = \delta d + d\delta$. If $M = \mathbb{R}^{2n}$, then it is not difficult to see that

$$\Delta\left(\sum_{i_1\ldots i_p} f_{i_1\ldots i_p}\, dx^{i_1}\wedge\ldots\wedge dx^{i_p}\right) =$$

$$=\sum_{i_1\ldots i_p}\left(-\sum_j \frac{\partial^2}{(\partial x^j)^2} f_{i_1\ldots i_p}\right)dx^{i_1}\wedge\ldots\wedge dx^{i_p}.$$

With a bit of tedious work, one can prove that for $\mu \in \Lambda^p(M)$:

$$d\mu = \sum_i \omega^i \wedge (D_{X_i}\mu),$$

$$\delta\mu = -\sum_i i(X_i)(D_{X_i}\mu),$$

where $i(X): \Lambda^p(M) \to \Lambda^{p-1}(M)$ for any vector field X is the inner derivation defined by:

$$(i(X)\mu)(Y_1,\ldots,Y_{p-1}) = \mu(X,Y_1,\ldots,Y_{p-1}).$$

Using these formulas for d and δ, one can compute Δ explicitly in terms of D. We see that Δ involves a two-fold covariant differentiation (i.e., $D_X D_Y$) and since so does the curvature tensor, it is not surprising to find that the curvature tensor appears in the explicit expression for Δ. As a psychological preparation for what is to come, we shall give an example (which will not be used again): If $\phi = \sum_i \phi_i \omega^i$ is a 1-form, let W be the unit vector field proportional to $\sum_i \phi_i X_i$ at each point. Then

$$g(\Delta\phi,\phi) = -g(\sum_i D^2_{X_i X_i}\phi,\phi) + g(\phi,\phi)\sum_i g(R_{X_i W}X_i, W),$$

where $D^2_{X_i X_i}$ is a second covariant derivative whose precise definition will be given later.

Still staying with a real differentiable manifold of real dimension $2n$, we are going to *complexify* formally the full tensor algebra. Since we shall only be interested in vector fields and differential forms, we shall describe these explicitly. A *complex vector field* Z on M is a formal sum $Z \equiv X + \sqrt{-1}\,Y$, where X,Y are vector fields on M. A *complex p-form* μ on M is a formal sum $\mu \equiv \alpha + \sqrt{-1}\,\beta$, where $\alpha,\beta \in \Lambda^p(M)$. We shall denote the complex p-forms on M by $\tilde{\Lambda}^p(M)$, and from now on everything is assumed to be complex unless explicitly stated to the contrary. We extend the domain of g to complex vector fields and forms by forcing it to be *complex bilinear* in both variables. Note that $g(Z,Z)$ may be identically zero even if Z is nowhere zero. For example, if $\{X_1,\ldots,X_{2n}\}$

is a local orthonormal basis of real vector fields, and if $Z \equiv X_1 + \sqrt{-1}\, X_2$, then

$$g(Z,Z) = g(X_1, X_1) + 2\sqrt{-1}\, g(X_1, X_2) - g(X_2, X_2)$$
$$= 1 + 0 - 1 = 0.$$

On the other hand, $g(Z,\bar{Z}) > 0$ at each point where Z is non-zero. The domain of definition of the various operators $D,*,d,\delta$ are again extended to complex vector fields and forms by forcing it to be complex linear. Thus, if $Z = X + \sqrt{-1}\, Y$ and $\mu = \alpha + \sqrt{-1}\, \beta$, then

$$D_Z\mu \equiv D_X\alpha + \sqrt{-1}\,(D_Y\alpha + D_X\beta) - D_Y\beta,$$
$$d\mu \equiv d\alpha + \sqrt{-1}\, d\beta, \quad \text{etc.}$$

Denoting by $\tilde{\Lambda}^p_o(M)$ the (differentiable) p-forms of compact support. We define

$$\langle\phi,\psi\rangle_\lambda \equiv \int_M g(\phi,\bar{\psi})e^{-\lambda}$$

for $\phi,\psi \in \Lambda^p_o(M)$, where $\bar{\psi}$ denotes the conjugate of ψ (defined in the obvious manner) and λ is any real-valued continuous function on M. One easily checks that this defines an inner product on $\tilde{\Lambda}^p_o$, and the completion of $\tilde{\Lambda}^p_o$ relative to $\langle\ ,\ \rangle_\lambda$ will be denoted by $L^2_p(\lambda)$. This is a Hilbert space. Obviously, given $\mu \in \tilde{\Lambda}^p$, we can choose a λ growing rapidly enough at ∞ so that $\mu \in L^2_p(\lambda)$.

Since d is defined on $\tilde{\Lambda}^p_o$, $d\colon L^2_p(\lambda) \to L^2_{p+1}(\lambda)$ is a densely defined linear operator. The Hilbert space adjoint of d will be denoted by $\delta_\lambda\colon L^2_{p+1}(\lambda) \to L^2_p(\lambda)$. *Assuming* λ *to be differentiable*, it is easy to see that δ_λ is itself a differential operator:

$$\delta_\lambda\phi = e^\lambda\delta(e^{-\lambda}\phi).$$

Of course if λ is merely continuous, δ_λ is in general not a differential operator.

Now we let M be a complex manifold of complex dimension n. Forgetting the complex structure of M for a moment, we have a real manifold of real dimension $2n$; we are therefore in a position to transfer to this setting all the above apparatus. To be specific, let M be given a *Kähler metric* G. By this we mean G assigns to each point $p \in M$ an Hermitian inner product $G(p)$ on the tangent space M_p (which is a complex vector space because M is complex), such that the following additional property holds: Let $z^1,\ldots,z^n$ be local coordinates in M, then we can write

$$G = \sum_{i,j} G_{ij}\, dz^i \otimes d\bar{z}^j,$$

where $\{G_{ij}(p)\}$ is a positive definite Hermitian matrix for each p. The associated 2-form

$$\Omega = \sqrt{-1} \sum_{i,j} G_{ij}\, dz^i \wedge d\bar{z}^j$$

is called the *Kähler form* of G. One can easily show that Ω so defined is independent of the coordinates $z^1,\ldots,z^n$. If Ω is closed, i.e., $d\Omega = 0$, we say G is a Kähler metric.

Let g be the real part of G. Then g is a Riemannian metric, so that relative to this g on M we can do Riemannian geometry. In particular, to say G has positive curvature means precisely that g has positive curvature. Moreover, we have as before the operators D, d, δ_λ etc. Now since M is complex, $\tilde{\Lambda}^a$ splits into a direct sum, $\sum_{p+q=a} \tilde{\Lambda}^{p,q}$, where $\tilde{\Lambda}^{p,q}$ denotes the complex forms of type (p,q). As a special case, the 1-forms split into a direct sum of (1,0)-forms and (0,1)-forms. A vector field is of type (1,0) if and only if it is annihilated by all (0,1)-forms. Elementary algebra shows that locally there exist (1,0)-forms $\{\omega^1,\ldots,\omega^n\}$ such that:

i) $\{\omega^1,\ldots,\omega^n,\bar{\omega}^1,\ldots,\bar{\omega}^n\}$ is locally a basis of 1-forms.

ii) $g(\omega^i,\omega^j) = g(\bar{\omega}^i,\bar{\omega}^j) = 0,$ while $g(\omega^i,\bar{\omega}^j) = \delta^{ij}$.

We shall call such $\{\omega^1,\ldots,\omega^n\}$ an *orthonormal basis* of (1,0)-forms. Dually, let $\{X_1,\ldots,X_n\}$ be vector fields of type (1,0) such that $\omega^i(X_j) = \delta^i_j$. Then

i)' $\{X_1,\ldots,X_n, \bar{X}_1,\ldots,\bar{X}_n\}$ is locally a basis of vector fields.

ii)' $g(X_i, X_j) = g(\bar{X}_i, \bar{X}_j) = 0$ while $g(X_i, \bar{X}_j) = \delta_{ij}$.

We shall call $\{X_1,\ldots,X_n\}$ the *dual orthonormal vector fields*.

Recall that $d = d' + d''$, where d' is of type (1,0) and d'' of type (0,1), i.e., $d''\colon \tilde{\Lambda}^{p,q}(M) \to \tilde{\Lambda}^{p,q+1}(M)$. Similarly $\delta = \delta' + \delta''$, where $\delta' = \tilde{\Lambda}^{p,q}(M) \to \tilde{\Lambda}^{p-1,q}(M)$ and $\delta''\colon \tilde{\Lambda}^{p,q}(M) \to \tilde{\Lambda}^{p,q-1}(M)$. Another bit of tedious computation gives by way of previous expression of d and δ:

$$d''\phi = \sum_i \bar{\omega}^i \wedge D_{\bar{X}_i}\phi,$$

$$\delta''\phi = -\sum_i i(\bar{X}_i) D_{X_i}\phi,$$

for orthonormal $\{\omega^i\}$ and $\{X_i\}$ as above. Again $d''\colon L^2_{(p,q)}(\lambda) \to L^2_{(p,q+1)}(\lambda)$ is a densely defined operator. *If* λ *is differentiable*, the Hilbert space adjoint $\delta''_\lambda : L^2_{(p,q+1)}(\lambda) \to L^2_{(p,q)}(\lambda)$ is easily computed to be:

$$\delta''_\lambda \phi = e^\lambda \delta''(e^{-\lambda}\phi)$$

for any differentiable ϕ. Thus if λ is differentiable, δ''_λ is a differentiable operator. In our applications, our λ will be a convex function given by Cheeger-Gromoll [1] (see (B) of the introduction), which will unfortunately be only continuous but not even C^1 in general. However, every convex function is Lipschitz continuous on every compact set, and using this fact, a more intricate argument given in the Appendix shows that δ''_λ acting on $L^2_{(p,q+1)}(\lambda)$ is still a first order differential operator, though only with locally bounded measurable coefficients. In fact,

$$\delta''_\lambda \phi = e^\lambda \delta''(e^{-\lambda}\phi)$$

continues to hold for all differentiable ϕ.

We define the *weighted Laplacian* $\square_\lambda$ to be $\square_\lambda = d''\delta''_\lambda + \delta_\lambda d''$. If $\lambda = 0$, we write $\square$ in place of $\square_0$.

2. AN L^2 ESTIMATE FOR (0,1)-FORMS

The L^2-estimate we are after is just an explicit formula for $\langle \square_\lambda \phi, \phi \rangle_\lambda$. For our need, it suffices to do this for a (0,1)-form ϕ.

We shall need fairly elaborate notation. First given any tensor A, the second order covariant derivative of A is defined by:

$$D^2_{XY}A \equiv D_X D_Y A - D_{D_X Y} A ,$$

where X,Y are arbitrary vector fields. Next, given any real-valued differentiable function λ, the *Levi form* L_λ of λ is by definition:

$$L_\lambda = 4 \sum_{i,j} \frac{\partial^2 \lambda}{\partial z^i \partial \bar{z}^j} dz^i \otimes d\bar{z}^j .$$

This is easily seen to be independent of coordinate systems. Using the orthonormal 1-forms $\{\omega^1,\ldots,\omega^n\}$, we may write:

$$L_\lambda = \sum_{i,j} \lambda_{ij}\, \omega^i \otimes \bar{\omega}^j .$$

$\{\lambda_{ij}\}$ is then an Hermitian matrix of C^∞ functions. Its eigenvalues

$\lambda_1, \ldots, \lambda_n$ are continuous functions globally defined on M. In particular $\sum_i \lambda_i$ is a well-defined function on M. One last bit of notation is this: Let the $(0,1)$ form ϕ be written as $\sum_i \phi_i \overline{\omega}^i$. We shall be concerned with the vector field $\sum_i \phi_i \overline{X}_i$. Write $X_i = \frac{1}{\sqrt{2}}(a_i + \sqrt{-1}\, b_i)$, where a_i, b_i are unit real vector fields. Let P_{ij}, Q_{ij} be the functions which at each point $p \in M$ assign the 2-planes

$$P_{ij}(p) = \text{span}\{b_i(p), a_j(p)\},$$

$$Q_{ij}(p) = \text{span}\{a_i(p), a_j(p)\}.$$

$K(P_{ij})$ and $K(Q_{ij})$ are the functions in this neighborhood which assign at each p the curvatures of the 2-planes $P_{ij}(p)$ and $Q_{ij}(p)$ respectively.

The basic formula is this: If $\phi \in \tilde{\Lambda}^{0,1}(M)$,

$$\begin{aligned} g(\Box_\lambda \phi, \overline{\phi}) &= -g(\textstyle\sum_i D^2_{X_i \overline{X}_i} \phi, \phi) + (\sum_i \lambda_i)|\phi|^2 \\ &\quad + \textstyle\sum_j |\phi_j|^2 (\sum_i K(P_{ij}) + K(Q_{ij})), \end{aligned}$$

where we have written $|\phi|^2$ for $g(\phi, \overline{\phi})$. Now integrate both sides relative to the weight factor $e^{-\lambda}$. For this purpose, we assume $\phi \in \tilde{\Lambda}_0^{0,1}(M)$ (i.e., compact support). The left side becomes

$$\begin{aligned} \int_M g(\Box_\lambda \phi, \overline{\phi}) e^{-\lambda} &= \langle \Box_\lambda, \phi \rangle_\lambda \\ &= \langle \delta''_\lambda d''\phi + d''\delta''_\lambda \phi, \phi \rangle \\ &= \langle d''\phi, d''\phi \rangle + \langle \delta''_\lambda \phi, \delta''_\lambda \phi \rangle \\ &\equiv \|d''\phi\|^2_\lambda + \|\delta''_\lambda \phi\|^2_\lambda . \end{aligned}$$

For the right side, we can use Stokes' Theorem to transform the first term. Without giving details, we arrive at:

$$\begin{aligned} -\int_M g(\textstyle\sum_i D^2_{X_i \overline{X}_i} \phi, \overline{\phi}) e^{-\lambda} &= \int_M g(\sum_i D_{X_i}\phi, \overline{\sum_i D_{X_i}\phi}) e^{-\lambda} \\ &= \|\textstyle\sum_i D_{X_i}\phi\|^2 \geq 0. \end{aligned}$$

To simplify the third term, we introduce a continuous function $c: M \to \mathbb{R}$, where $c(p) = \min K(P)$ as P varies over all 2-planes in M_p. We shall call c the *lower bound of curvature.* Thus

$$\int_M \sum_j |\phi_j|^2 \left(\sum_i K(P_{ij}) + K(Q_{ij})\right)e^{-\lambda}$$

$$\geq \int_M (2n-1)c \sum_j |\phi_j|^2 e^{-\lambda} = \int_M (2n-1)c\, |\phi|^2 e^{-\lambda}.$$

Hence we obtain the basic inequality:

$$(*) \quad \|d''\phi\|_\lambda^2 + \|\delta''_\lambda \phi\|_\lambda^2 \geq \int_M (2n-1)c + \sum_i \lambda_i)\, |\phi|^2 e^{-\lambda}.$$

We specifically recall the assumptions that went into this:

1) M is a Kähler manifold.
2) $\phi \in \tilde{\Lambda}_o^{0,1}(M)$.
3) λ is C^∞.
4) $\{\lambda_i\}$ are the eigenvalues of the Levi form L_λ.
5) c is the lower bound of curvature.

We shall tie this up with what we are trying to do. Since we assume positive curvature, $c > 0$. The main objective is to find a λ so that $(2n-1)c + \sum_i \lambda_i$ remains positive. As noted above, our applications will require λ to be a convex function which is merely continuous but not necessarily C^1. So we must translate the inequality $(*)$ into a statement concerning a continuous convex function λ. For this purpose, the numerical factor $(2n-1)$ causes minor though irritating problems if n is allowed to assume the value 1. Fortunately, when $n = 1$, M is a Riemann surface and hence a Stein manifold regardless of any curvature assumptions; our Main Theorem is devoid of content in this case. We shall therefore tacitly assume $n > 1$ in the following. Our first main observation is:

Lemma 1: *Let* M *be a Kähler manifold with positive curvature and let* λ *be a convex function. If* c *is the lower bound of curvature, then for all* $\phi \in \tilde{\Lambda}_o^{0,1}(M)$:

$$(**) \quad \|d''\phi\|_\lambda^2 + \|\delta''_\lambda \phi\|_\lambda^2 \geq \int_M c|\phi|^2 e^{-\lambda}.$$

For the proof of this lemma we need the approximation theorem of [5] already alluded to in the introduction. If λ is a convex function on M, then over any compact set, it must be Lipschitz continuous. This is a standard result in euclidean space convexity theory and the proof extends to the general case of Riemannian manifolds. Now let K be a given compact subset of M and let γ_K be the Lipschitz constant of the convex function λ over K. The *approximation theorem* of [5] (Theorem 2) is: Given $-\delta < 0$, there exists a family of C^∞ functions λ_ε, $0 < \varepsilon \le \varepsilon_0$ (where ε_0 is a positive constant depending on δ), defined on an open neighborhood U of K such that:

(i_A) $\lambda_\varepsilon \downarrow \lambda$ uniformly on K as $\varepsilon \to 0$.

(ii_A) The minimal eigenvalue of L_{λ_ε}, for all $0 < \varepsilon \le \varepsilon_0$, exceeds $-\delta$ over K.

(iii_A) There exists a constant σ such that for all $0 < \varepsilon \le \varepsilon_0$, $|d\lambda_\varepsilon| \le \sigma$, and σ can be chosen to be as close to γ_K as one pleases.

The proof of this theorem in case $M = \mathbb{C}^n$ is quite easy; one convolutes λ with a C_0^∞ function symmetric about the origin to obtain the desired λ_ε. In general, however, the convolution process has to be replaced by something more sophisticated and the proof of (ii_A) is nontrivial.

The proof of Lemma 1 is now easy. Let us fix one $\phi \in \tilde{\Lambda}_0^{0,1}(M)$ and let K = support ϕ. For each λ_ε as given by the approximation theorem, (*) says (regard λ_ε as having been arbitrarily extended to a C^∞ function on M):

$$\|d''\phi\|_{\lambda_\varepsilon}^2 + \|\delta''_{\lambda_\varepsilon}\phi\|_{\lambda_\varepsilon}^2 \ge \int_M \{(2n-1)c - \textstyle\sum_i (\lambda_\varepsilon)_i\}|\phi|^2 e^{-\lambda_\varepsilon}.$$

Let $\delta = \min_K \frac{1}{n}(2n-2)c$ (recall $n > 1$ so that $(2n-2) > 0$), then (ii_A) says that on K, $\{(2n-1)c + \sum_i (\lambda_\varepsilon)_i\} \ge c$. Letting $\varepsilon \to 0$, (i_A) tells us that

$$\|d''\phi\|_\lambda^2 + \|\delta''_\lambda\phi\|_\lambda^2 \ge \int_M c|\phi|^2 e^{-\lambda}$$

as desired.

There is a sleight-of-hand in the above reasoning. While it is clear that $\|d''\phi\|_{\lambda_\varepsilon}^2 \to \|d''\phi\|_\lambda^2$ and $\int_M c|\phi|^2 e^{-\lambda_\varepsilon} \to \int_M c|\phi|^2 e^{-\lambda}$ as $\varepsilon \to 0$, it is not so clear that $\|\delta_{\lambda_\varepsilon}\phi\|_{\lambda_\varepsilon}^2 \to \|\delta_\lambda\phi\|_\lambda^2$ as $\varepsilon \to 0$. To understand this, assume λ is itself C^∞. Then $\delta''_{\lambda_\varepsilon}\phi = e^{\lambda_\varepsilon}\delta''(e^{-\lambda_\varepsilon}\phi)$ and $\delta''_\lambda\phi = e^\lambda\delta''(e^{-\lambda}\phi)$, so that the convergence of $\|\delta''_{\lambda_\varepsilon}\phi\|_{\lambda_\varepsilon}^2$ to $\|\delta''_\lambda\phi\|_\lambda^2$ roughly speaking involves the convergence of the first partials of λ_ε to those of λ uniformly on K. Now if λ is not C^∞ (as in our situation), this is clearly a serious matter. Fortunately, using (iii_A), a more careful analysis gives

$$\varliminf_{\varepsilon\to 0} \|\delta_{\lambda_\varepsilon} \phi\|_{\lambda_\varepsilon} \leq \|\delta''_\lambda \phi\|_\lambda ,$$

which, so far as the *inequality* (**) is concerned, is quite adequate.

To put (**) to use, we must now do some simple functional analysis on the Hilbert space $L^2_{(0,1)}(\lambda)$. We shall assume from now on that λ is a convex function. Recall, $d'': L^2_{(0,1)}(\lambda) \to L^2_{(0,2)}(\lambda)$ and $\delta''_\lambda = L^2_{(0,1)}(\lambda) \to L^2_{(0,0)}(\lambda)$ are densely defined differential operators. (Again we refer to the Appendix.) We can extend their domains of definition using the concept of current, namely, $\phi \in L^2_{(0,1)}(\lambda)$ is in the domain $\mathcal{D}_{d''}$ of d'' if and only if $d''\phi$ in the sense of currents (or distributions) is in $L^2_{(0,2)}(\lambda)$. Likewise for the domain $\mathcal{D}_{\delta''_\lambda}$ of δ''_λ. With this extension, d'' and δ''_λ become closed operators in the sense that the graphs of d'' and δ''_λ are now closed subsets of $L^2_{(0,1)}(\lambda) \oplus L^2_{(0,2)}(\lambda)$ and $L^2_{(0,1)}(\lambda) \oplus L^2_{(0,0)}(\lambda)$, respectively. Note that functional analysts would call these the "weak extensions", and the next lemma, due to Hörmander [8, p. 80] and Andreotti-Vesentini [11, p. 22], says essentially that the strong and weak extensions are identical.

Lemma 2: *If* M *is a complete Kähler manifold, then* $\tilde{\Lambda}^{0,1}_0(M)$ *is dense in* $\mathcal{D}_{\delta''_\lambda} \cap \mathcal{D}_{d''}$ *in the graph norm:*

$$\phi \longrightarrow \|\phi\|^2_\lambda + \|d''\phi\|^2_\lambda + \|\delta''_\lambda \phi\|^2_\lambda .$$

In other words, given any $\phi \in \mathcal{D}_{\delta''_\lambda} \cap \mathcal{D}_{d''}$, there exists a sequence of $C^\infty(0,1)$ forms $\{\phi_i\}$ with compact support such that

$$\left.\begin{aligned} \phi_i &\to \phi \\ \delta''_\lambda \phi_i &\to \delta''_\lambda \phi \\ d''\phi_i &\to d''\phi \end{aligned}\right\} \text{ simultaneously in the Hilbert space norms.}$$

One may recall that the Kähler manifold M is *complete* if the associated Riemannian metric g (the real part of the Kähler metric G) is a complete metric, i.e., M equipped with the distance function induced by g is a complete metric space. The proof of this lemma uses the classical mollifier technique of Friedrichs plus an elementary argument using the completeness of M. In any case, using Lemmas 1 and 2, we deduce immediately:

Lemma 3: *Let* M *be a complete noncompact Kähler manifold of positive curvature. Then there exists a positive continuous function* $c: M \to (0,\infty)$ *such that for any convex function* λ *on* M,

$$\|d''\phi\|_\lambda^2 + \|\delta_\lambda''\phi\|_\lambda^2 \geq \int_M c|\phi|^2 e^{-\lambda}$$

for all $\phi \in \mathcal{D}_{\delta_\lambda''} \cap \mathcal{D}_{d''} \subseteq L^2_{(0,1)}(\lambda)$.

Here the noncompactness of M is invoked for the first time. Of course it is immaterial whether M is noncompact or not insofar as the validity of the lemma is concerned, but if M is compact, the only convex functions are constants (by an elementary argument). The lemma is then not very interesting since it could have been proved much more simply by more elementary methods.

As a first application, we prove:

Lemma 4: *Let* M *be as in Lemma 3. If* ϕ *is a* C^∞ (0,1)*-form such that* $d''\phi = 0$, *then there exists a* C^∞ *function* f *such that* $d''f = \phi$.

Proof: To avoid hopeless confusion, it is necessary to set up some notation. We shall write:

$$T \text{ for the operator } d'': L^2_{(0,0)}(\lambda) \to L^2_{(0,1)}(\lambda),$$

$$S \text{ for the operator } d'': L^2_{(0,1)}(\lambda) \to L^2_{(0,2)}(\lambda),$$

and T_λ^*, S_λ^* for their respective Hilbert space adjoints. Thus Lemma 3 in this notation becomes:

$$\|S\phi\|_\lambda^2 + \|T_\lambda^*\phi\|_\lambda^2 \geq \int_M c|\phi|^2 e^{-\lambda}$$

for all $\phi \in \mathcal{D}_{T_\lambda^*} \cap \mathcal{D}_S$. Furthermore, the lemma says: if $S\phi = 0$ and ϕ is C^∞, then $\phi = Tf$ for some $C^\infty f$.

The formal proof now begins. Assuming for the moment that ϕ is locally an L^2 function, we will find an f which is also locally in L^2 such that $Tf = \phi$. We first choose a λ so that $\phi \in L^2_{(0,1)}(\lambda)$. By the Cheeger-Gromall result quoted in (B) of the introduction (see [1]), there is a convex function τ on M which goes uniformly to ∞ at the ideal boundary of M. Let χ be a C^∞ function, $\chi: [0,\infty) \to [0,\infty)$, such that $\chi > 0$, $\chi' > 0$ and $\chi'' > 0$. Then elementary calculus shows that $\chi \circ \tau$ is still convex and goes to ∞ uniformly at the ideal boundary of M as fast as we wish provided χ grows fast enough. Pick such a χ and let $\lambda = \chi \circ \tau$. Then for this λ, $\phi \in L^2_{(0,1)}(\lambda)$. For the same reason, we may also assume

$$\int_M |\phi|^2 \frac{e^{-\lambda}}{c} < \infty.$$

Let us call this finite positive constant A^2. We claim that for all $\psi \in \mathcal{D}_{T^*_\lambda}$,

(#) $\quad |\langle\phi,\psi\rangle_\lambda| \le A\|T^*_\lambda \psi\|_\lambda$.

Indeed, if $\psi \perp \ker S$ (ker = kernel), then the assumption that $\phi \in \ker S$ implies that the left-side is zero and there is nothing to prove. On the other hand, suppose $\psi \in \ker S$. Then $\psi \in \ker S \cap \mathcal{D}_{T^*_\lambda} \subseteq \mathcal{D}_S \cap \mathcal{D}_{T^*_\lambda}$, and Lemma 3 implies that for all $\psi \in \ker S \cap \mathcal{D}_{T^*_\lambda}$,

(##) $\quad \int_M c|\psi|^2 e^{-\lambda} \le \|T^*_\lambda \psi\|^2_\lambda$.

Thus, $|\langle\phi,\psi\rangle_\lambda|^2 \le \int_M |\phi|^2 \frac{e^{-\lambda}}{c} \cdot \int_M c|\psi|^2 e^{-\lambda} \le A^2\|T^*_\lambda\psi\|^2_\lambda$, which is exactly (#). Since $L^2_{(0,1)}(\lambda) = \ker S + (\ker S)^\perp$ as an orthogonal direct sum (note that $\ker S$ is closed in $L^2_{(0,1)}(\lambda)$ since S is closed), (#) for a general $\psi \in \mathcal{D}_{T^*_\lambda}$ follows.

Now, define a linear functional $\alpha\colon \overline{R}_{T^*_\lambda} \to \mathbb{C}$, where $R_{T^*_\lambda} \subseteq L^2_{(0,0)}(\lambda)$ is the range of T^*_λ, by $\alpha(T^*_\lambda \psi) = \langle\phi,\psi\rangle_\lambda$. Because (##) implies that T^*_λ restricted to $\ker S \cap \mathcal{D}_{T^*_\lambda}$ is injective, α is well-defined if we assume $\psi \in \ker S$. But we can always make this assumption because if $\psi \in (\ker S)^\perp$, then $(\ker S)^\perp \subseteq R_T^\perp = \ker T^*_\lambda$ which implies that $\psi \in \ker T^*_\lambda$; hence $T^*_\lambda \psi = 0$.

By (#), α is a bounded linear functional so that by the Riesz representation theorem, there exists an $f \in \overline{R}_{T^*_\lambda}$ which represents α. Thus

$$\langle\phi,\psi\rangle_\lambda = \alpha(T^*_\lambda \psi) = \langle f, T^*_\lambda \psi\rangle_\lambda$$

for all $\psi \in \ker S$. By definition of an adjoint operator, since both ϕ and Tf belong to $\ker S$, the above equation is exactly the statement that $Tf = \phi$.

Now we assume ϕ is C^∞ and proceed to prove that f is likewise. Since $d''f \equiv Tf = \phi$, $\square f = \delta''d''f + d''\delta''f = \delta''\phi$ ($\delta''f = 0$ since f is a function) and $\delta''\phi$ is a C^∞ function. However, using the formulas given in §1, one sees immediately that $\Box f = -\sum_i D^2_{X_i\overline{X}_i} f$, where $\{X_i\}$ is an orthonormal basis of vector fields of type $(1,0)$. Thus $\Box f = \delta''\phi$ is a second order elliptic equation. By the regularity theorem for elliptic operators, $\delta''\phi$ is C^∞ implies that f is C^∞. Q.E.D.

3. PROOF OF THE MAIN THEOREM

We shall give detailed proofs of Theorems 1 - 3 in this section using the Lemmas of §2. Let us first deal with Theorem 2, i.e., we must show that on M there exists a C^∞ strictly plurisubharmonic function. Let G be the Kähler metric as usual. Locally, $G = \sum_{i,j} G_{ij}\, dz^i \otimes d\overline{z}^j$. The associated Kähler form is

$$\Omega = \sqrt{-1} \sum_{ij} G_{ij}\, dz^i \wedge d\overline{z}^j .$$

One can check that $\Omega = \overline{\Omega}$ so that Ω is a closed *real* 2-form. Since the Gromoll-Meyer result [7] (see (A) of the introduction) implies that $H^2(M,\mathbb{R}) = 0$, there exists a *real* 1-form ω such that $d\omega = \Omega$ (de Rham's Theorem). Since ω is real, we can write $\omega = \phi + \overline{\phi}$, where ϕ is a (1,0)-form and hence $\overline{\phi}$ is of type $(0,1)$ (e.g., $dx = \frac{1}{2}dz + \overline{\frac{1}{2}dz}$). Thus

$$\Omega = (d' + d'')(\phi + \overline{\phi})$$

$$= \underbrace{d'\phi}_{(2,0)} + \underbrace{(d'\overline{\phi} + d''\phi)}_{(1,1)} + \underbrace{d''\overline{\phi}}_{(0,2)} .$$

Since Ω is of type $(1,1)$, we conclude that $d''\overline{\phi} = 0$ and $d'\overline{\phi} + d''\phi = \Omega$. By Lemma 4, there exists a C^∞ function f such that $\overline{\phi} = d''f$. Hence $\Omega = d'd''f + d''d'\overline{f} = d'd''(f - \overline{f})$. Define the real-valued C^∞ function h by $\sqrt{-1}\, h = f - \overline{f}$; then we have $\Omega = \sqrt{-1}\, d'd''f$. This says

$$\frac{\partial^2 h}{\partial z^i \partial \overline{z}^j} = G_{ij} ,$$

and since $\{G_{ij}\}$ is positive definite, the Levi form L_h of h must be positive definite too. Equivalently, h is a C^∞ strictly plurisubharmonic function. (Note that the above reasoning is the standard one used to demonstrate that every Kähler metric is locally given by a potential.)

We next prove Theorem 1. Recall that this asserts, if τ is a convex function as given by Cheeger-Gromoll [1] ((B) of the introduction), and if $M_b = \{p: \tau(p) \le b\}$, then $A(M)$ is dense in $A(M_b)$ by restriction for all $b \in \mathbb{R}$. Here, $A(M_b)$ is the set of holomorphic functions $f: U \to \mathbb{C}$ where U is some open set containing M_b. Using the notation of the proof of Lemma 4, we recall the inequality (##) above: if $\psi \in \ker S \cap \mathcal{D}_{T^*} \subseteq L^2_{(0,1)}(\lambda)$, then

$$(\#\#) \quad \int_M c|\psi|^2 e^{-\lambda} \le \|T^*_\lambda \psi\|^2_\lambda .$$

Here, $\lambda = \chi \circ \tau$, where χ is any C^∞ increasing convex function of one variable. Note that one could equivalently define M_b by $\{p \in M: \lambda(p) \le \chi(b)\}$.

Denoting the Hilbert space of L^2 functions on M_b by $L^2(M_b)$, we have the inclusions $A(M) \subseteq A(M_b) \subseteq L^2(M_b)$. The induced topology on $A(M)$ and $A(M_b)$ from $L^2(M_b)$ is equivalent to the sup norm topology, as is well-known. Thus taking closure relative to $L^2(M_b)$, we ask whether $\overline{A(M)} \subseteq \overline{A(M_b)}$ is equality. Suppose not; then there exists $v \in \overline{A(M_b)}$ such that $v \ne 0$ as an element of $L^2(M_b)$ and $v \in A(M)^\perp$. Extend v to be a function on M by defining it to be zero on $M \setminus M_b$. Thus for all $u \in A(M)$,

$$\int_M v\bar{u} = \int_{M_b} v\bar{u} = 0 .$$

Equivalently, $\int_M (ve^\lambda)\,\bar{u}\,e^{-\lambda} = 0$ for all $u \in A(M)^\perp$. Thus $ve^\lambda \in (\ker T)^\perp = \overline{R}_{T^*_\lambda}$. If $ve^\lambda \in R_{T^*_\lambda}$, then there exists $\psi \in \mathcal{D}_{T^*_\lambda} \cap \ker S$ such that $T^*_\lambda \psi = ve^\lambda$ (see the proof of Lemma 4) and

$$\int_M c|\psi|^2 e^{-\lambda} \le \|ve^\lambda\|^2_\lambda ,$$

by (##). However, we only know in general that $ve^\lambda \in \overline{R}_{T^*_\lambda}$, so we can only assert the existence of a sequence v_i such that $v_i \to ve^\lambda$ in norm, and $v_i = T^*_\lambda \psi_i$ for $\psi_i \in \mathcal{D}_{T^*_\lambda} \cap \ker S$ and

$$\int_M c|\psi_i|^2 e^{-\lambda} \le \|v_i\|^2_\lambda \to \|ve^\lambda\|^2_\lambda .$$

Thus $\{\psi_i\}$ is a bounded set in $L^2_{(0,1)}(\lambda - \log c)$, and hence a subsequence of it (to be denoted also by $\{\psi_i\}$) converges *weakly* to some $\psi \in L^2_{(0,1)}(\lambda - \log c)$. Denoting the norm of $L^2_{(0,1)}(\lambda - \log c)$ simply by $\|\ \|$, we know in general that $\overline{\lim_i}\, \|\psi_i\| \ge \|\psi\| = \int_M c|\psi|^2 e^{-\lambda}$. Thus

$$\int_M c|\psi|^2 e^{-\lambda} \le \|ve^{\lambda}\|_\lambda^2 .$$

Since $\psi_i \to \psi$ weakly, $\psi_i \to \psi$ as distributions, so that (T^* being a differential operator on locally L^2 forms) $T^*_\lambda \psi_i \to T^*_\lambda \psi$. But also $T^*_\lambda \psi_i = v_i \to ve^{\lambda}$, so we know $T^*_\lambda \psi = ve^{\lambda}$. In particular, this means $\psi \in \mathcal{D}_{T^*_\lambda}$ because $ve^{\lambda} \in L^2_{(0,1)}(\lambda)$. Thus $T^*_\lambda \psi = e^{\lambda}\delta''(e^{-\lambda}\psi)$ (see Appendix) and we have

$$\delta''(e^{-\lambda}\psi) = v .$$

Let us rewrite the previous inequality as

$$\int_M c|\psi e^{-\lambda}|^2 e^{\lambda} \le \int_M |v|^2 e^{\lambda}$$

and we recall that $\lambda = \chi \circ \tau$ for any C^∞ increasing convex function χ. We now introduce a sequence of such χ: Let χ_ν be a sequence of C^∞ increasing convex function of a real variable such that

$$\chi_\nu(t) = \chi_{\nu+1}(t) \quad \text{for} \quad t \le b,$$

$$\chi_\nu(t) \uparrow \infty \quad \text{for} \quad t > b.$$

Let $\lambda_\nu = \chi_\nu \circ \tau$, and replace λ above by λ_ν. We have then $\delta''(e^{-\lambda_\nu}\psi) = v$ and

$$\int_M c|\psi e^{-\lambda_\nu}|^2 e^{\lambda_\nu} \le \int_M |v|^2 e^{\lambda_\nu}$$

$$= \int_{M_b} |v|^2 e^{\lambda_\nu} = \int_{M_b} |v|^2 e^{\lambda_1} \equiv a,$$

where the last equality is because $\lambda_\nu = \lambda_{\nu+1}$ on support v which lies in M_b. Let $\xi_\nu = \psi e^{-\lambda_\nu}$, then $\delta''\xi_\nu = v$ and $\int_M c|\xi_\nu|^2 e^{\lambda_\nu} \le a$. This implies $\int_M c|\xi_\nu|^2 e^{\lambda_1} \le a$. Hence $\{\xi_\nu\} \subseteq$ closed ball of radius $\sqrt{a}$ in $L^2_{(0,1)}(-\lambda_1 - \log c)$, so that we may assume $\xi_\nu \to \xi$ weakly for some ξ in this closed ball. In particular,

$$\int_M c|\xi|^2 e^{\lambda_1} \le a .$$

Since also $\xi_\nu \to \xi$ as distributions, $\delta''\xi = v$.

We claim: support $\xi \subseteq M_b$.

Before proving this claim, let us show how this is used to prove the theorem. We shall show that relative to the ordinary L^2 norm $\langle\ ,\ \rangle$ of $L^2(M_b)$, $v \perp A(M_b)$. Since $v \in \overline{A(M_b)}$, this shows $v \equiv 0$ and thereby contradicts our initial assumption on v. So let $u \in A(M_b)$. u is holomorphic in a neighborhood U of M_b. Therefore by altering u on $U\backslash M_b$ if necessary, we may assume that u is a C_o^∞ function on M which is holomorphic on M_b. Hence, using $v|_{M\backslash M_b} \equiv 0$ we have:

$$\langle u,v\rangle = \int_{M_b} v\bar{u} = \int_M v\bar{u} = \langle v,\bar{u}\rangle_o$$
$$= \langle \delta''\xi, u\rangle_o = \langle \xi, d''u\rangle_o$$
$$= \int_M g(\xi, \overline{d''u}) = 0$$

because support $\xi \subseteq M_b$ and support $d''u \subseteq M\backslash M_b$.

It remains to prove the claim. Let us first show that for $t > b$,

$$\int_{M\backslash M_t} c|\xi_\nu|^2 e^{\lambda_1} < \varepsilon$$

for all large ν, where ε is any positive constant. For since $\lambda_\nu \uparrow \infty$ on $M\backslash M_t$, we may choose ν so large that $e^{\lambda_\nu - \lambda_1} \geq a/\varepsilon$ on $M\backslash M_t$. Then

$$\frac{a}{\varepsilon} \int_{M\backslash M_t} c|\xi_\nu|^2 e^{\lambda_1} \leq \int_{M\backslash M_t} c|\xi_\nu|^2 e^{\lambda_\nu} \leq a,$$

implying the asserted inequality above. Thus as $\nu \to \infty$,

$$\int_{M\backslash M_t} c|\xi_\nu|^2 e^{\lambda_1} \to 0.$$

Now
$$\left|\int_{M\backslash M_t} cg(\xi,\bar{\xi}_\nu)\, e^{\lambda_1}\right| \leq \left|\int_{M\backslash M_t} c|\xi|\ |\xi_\nu|\, e^{\lambda_1}\right|$$
$$\leq \left(\int_{M\backslash M_t} c|\xi|^2 e^{\lambda_1}\right)^{1/2} \cdot \left(\int_{M\backslash M_t} c|\xi_\nu|^2 e^{\lambda_1}\right)^{1/2}$$
$$\leq \sqrt{a}\left(\int_{M\backslash M_t} c|\xi_\nu|^2 e^{\lambda_1}\right)^{1/2} \longrightarrow 0 \quad \text{as} \quad \nu \to \infty.$$

On the other hand, if χ_B stands for the characteristic function of the set B,

$$\int_{M\backslash M_t} cg(\xi,\bar{\xi}_\nu) e^{\lambda_1} = \langle \xi, \xi_\nu \chi_{M\backslash M_t}\rangle_{-\lambda_1 - \log c}$$
$$\longrightarrow \langle \xi, \xi\chi_{M\backslash M_t}\rangle_{-\lambda_1 - \log c} = \int_{M\backslash M_t} c|\xi|^2 e^{\lambda_1}.$$

Hence $\int_{M\backslash M_t} c|\xi|^2 e^{\lambda_1} = 0$, implying $\xi|_{M\backslash M_t} \equiv 0$ for all $t > b$.

Thus support $\xi \subseteq M_b$. Q.E.D.

We may now give the simple proof of Theorem 3. Given the fact that $A(M)$ is dense in $A(M_a)$ ($M_a \equiv \{p \in M: \tau(p) \le a\}$) for every $a \in \mathbb{R}$, and the existence of a C^∞ strictly plurisubharmonic function on M, we have to show that M is a Stein manifold. Thus we are asked to prove:

1) $A(M)$ gives local coordinates at every $p \in M$.

2) M is holomorphically convex.

The proofs of those two facts depend on a simple observation: If $a < b$, then in between M_a and M_b, we can interpose a Stein manifold M_o, i.e., $M_a \subset\subset M_o \subset\subset \overset{o}{M}_b$ (= interior of M_b). To prove this, first note that the C^∞ strictly plurisubharmonic function h on M may be assumed to be everywhere positive. Indeed, e^h is a positive C^∞ strictly plurisubharmonic function. Furthermore, multiplying h with a small positive constant if necessary, we may assume $0 < h < c$, where $c < b - a$, on the compact set M_b. Let $M' = \{p \in M: (\tau + h)(p) \le b\}$. Then $M_a \subset\subset \overset{o}{M}{}' \subset\subset \overset{o}{M}_b$. Let the minimal eigenvalue of the Levi form L_h of h achieve the minimum $2\delta > 0$ in M_b. On the compact set M_b, let $\{\tau_\varepsilon\}$ be the family of C^∞ functions approximating τ as given by the approximation theorem of [5] quoted in §2. Let us say ε is so small that $|\tau_\varepsilon - \tau| < \delta'$ on M_b, where δ' is a small positive number to be determined later, and let us assume the eigenvalues of L_{τ_ε} all exceed $-\delta$ on M_b. Let $h_\varepsilon = \tau_\varepsilon + h$. Then h_ε is C^∞ and strictly plurisubharmonic on M_b since the minimal eigenvalue of L_{h_ε} is at least $\delta > 0$ on M_b. Moreover, we will choose δ' so small that the open set $M_o \equiv \{p \in M: h_\varepsilon(p) < b\}$ satisfies $M_a \subset\subset M_o \subset\subset M_b$. By Grauert's solution of the Levi problem, M_o is a Stein manifold.

The proofs of 1) and 2) are now easy. Given $p \in M$, let a be so large that $p \in \overset{o}{M}_a$. Let M_o be a Stein manifold such that $M_a \subset\subset \overset{o}{M}_o \subset\subset \overset{o}{M}_{2a}$. There are $f_1,\ldots,f_n \in A(M_o)$ which give local coordinates at p. The functions $f_1|_{M_a},\ldots,f_n|_{M_a}$ can be uniformly approximated in M_a by elements of $A(M)$, say $g_1,\ldots,g_n$. Since then the partial derivatives of $\{g_i\}$ also uniformly approximate those of $\{f_i|_{M_a}\}$ in M_a, the Jacobian of $\{g_i\}$ is nonzero at p. Thus $\{g_i\}$ gives local coordinates at p, proving 1). To prove 2), it suffices to prove that if $p \notin M_a$, there exists an $f \in A(M)$ such that $|f(p)| > \max_{M_a} |f|$. Suppose $p \in M_b$. Let M_o be a Stein manifold such that $M_b \subset\subset M_o \subset\subset \overset{o}{M}_{2b}$. Hence there exists $g \in A(M_o)$ such that $|g(p)| > \max_{M_a} |g|$. Let $f \in A(M)$ approximate $g|_{M_b}$ sufficiently closely on M_b, then also $|f(p)| > \max_{M_a} |f|$. Q.E.D.

Final Remarks: R. Narasimhan in [10] proved that if a complex manifold has a C^2 strictly plurisubharmonic function and a continuous plurisubharmonic function τ such that each $M_a \equiv \{p \in M: \tau(p) \leq a\}$ is compact, then M is a Stein manifold. It follows quite readily from our approximation theorem of §2 (see §4 of [5]) that the convex function τ above is actually plurisubharmonic. Hence Theorem 2 and Narasimhan's Theorem immediately yield our Main Theorem, thereby bypassing (and at the same time proving as corollaries) Theorems 1 and 3. The reasons we did not follow this procedure in our exposition but instead gave independent proofs of Theorems 1 and 3 are the following:

a) The argument used to prove the Runge type theorem (Theorem 1) is valid when "positive curvature" is replaced by "positive Ricci curvature and nonnegative sectional curvature" (see [4], Theorem 7). Under this weaker hypothesis, Theorem 1 is inaccessible by other methods.

b) Narasimhan's Theorem is proved in [10] with maximal generality in the context of Stein spaces and is very long. Invoking it would be something of an overkill.

c) With the geometric methods at our disposal, the proof of Theorem 3 is very simple.

APPENDIX

We wish to give a detailed discussion of δ''_λ and $\mathcal{D}_{\delta''_\lambda}$ in the case λ is merely Lipschitz continuous on every compact subset of M. Of course we have in mind a convex λ. From a classical measure theory, we know that such a λ possesses first order derivatives that are measurable and locally bounded. Hence if $\phi \in \tilde{\Lambda}_o^{p,q}$, $\delta''(e^{-\lambda}\phi) \in L^2$, and consequently also $e^{\lambda}\delta''(e^{-\lambda}\phi) \in L^2$. Since this is a form of compact support, it also belongs to $L^2_{(p,q-1)}(\lambda)$. Thus for $\delta''_\lambda: L^2_{(p,q)}(\lambda) \to L^2_{(p,q-1)}(\lambda)$, $\tilde{\Lambda}_o^{p,q} \subseteq \mathcal{D}_{\delta''_\lambda}$

(= domain of definition of δ''_λ) and if $\phi \in \tilde{\Lambda}_o^{p,q}$, $\delta''_\lambda \phi = e^\lambda \delta''(e^{-\lambda}\phi)$. A simple computation using the formulas of §1 shows:

$$\delta''_\lambda \phi = -\sum_i i(\overline{X}_i) D_{X_i} \phi + \sum_i (X_i\lambda)\, i(\overline{X}_i)\phi$$

for every $\phi \in \tilde{\Lambda}_o^{p,q}$, where $\{X_i\}$ is an orthonormal basis of (1,0) vector fields as usual. Thus δ''_λ is a first order differential operator whose first order term has C^∞ coefficients and whose zero-th order term has locally bounded measurable coefficients. Hence if ψ is a form locally in L^2, then $\delta''_\lambda \psi$ makes sense as a distribution and

$$\delta''_\lambda \psi = e^\lambda \delta''(e^{-\lambda}\psi).$$

Now the argument in Hörmander's book [8], p. 109 (see also Vesentini [11]) shows in effect: If λ is continuous and M is complete, then the elements with compact support in $L^2_{(p,q)}(\lambda)$ are dense in $\mathcal{D}_{\delta''_\lambda}$ and $\mathcal{D}_{d''}$ in the graph norm. Furthermore, Friedrich's Lemma as formulated by Hörmander in [9], Proposition 1.2.3, says: If M is a Riemannian manifold and $P(D)$ is a first order differential operator on forms such that the first order coefficients of $P(D)$ are locally Lipschitzian and the zero-th order coefficients are locally bounded and measurable, then every locally L^2 form ψ with compact support such that $P(D)\psi$ is also locally L^2 can be approached by C_o^∞ forms in the graph norm. Combining these two facts, we see that on a complete Kähler manifold, if λ is locally Lipschitzian, then C_o^∞ is dense in $\mathcal{D}_{\delta''_\lambda}$ and $\mathcal{D}_{d''}$ in the graph norm. This justifies Lemma 3 of §2.

REFERENCES

1. Cheeger, J. and D. Gromoll, *On the structure of complete manifolds of nonnegative curvature*, Ann. of Math. 96 (1972), 413-443.

2. Greene, R. E. and H. Wu, *Curvature and complex analysis*, Bulletin Amer. Math. Soc., 77 (1971), 1045-1049.

3. __________, *Curvature and complex analysis* II, Bulletin Amer. Math. Soc. 78 (1972), 866-870.

4. __________, *Curvature and complex analysis* III, Bulletin Amer. Math. Soc. 79 (1973), 606-608.

5. _________, *On the subharmonicity and plurisubharmonicity of geodesically convex functions*, Indiana Univ. Math. J. 22 (1973), 641-653.

6. _________, *Integrals of subharmonic functions on manifolds of nonnegative curvature*, (to appear).

7. Gromoll, D. and W. Meyer, *On complete open manifolds of positive curvature*, Ann. of Math., 90 (1969), 75-90.

8. Hörmander, L., *An Introduction to Complex Analysis in Several Variables*, D. van Nostrand Co., Princeton, New Jersey, 1966.

9. _________, L^2 *estimates and existence theorems for the* $\bar{\partial}$ *operator*, Acta Math. 113 (1965), 89-152.

10. Narasimhan, R., *The Levi problem for complex spaces* II, Math. Annalen 146 (1962), 195-216.

11. Vesentini, E., *Lectures on Levi Convexity of Complex Manifolds and Cohomology Vanishing Theorems*, Tata Institute of Fundamental Research, Bombay, India, 1967.

REFINED RESIDUES, CHERN FORMS, AND INTERSECTIONS

James R. King
Massachusetts Institute of Technology

INTRODUCTION

Let $W \subset V$ be a complex submanifold of a complex manifold, and let η be a residue form for W. That is, η is a C^∞ form on $V - W$ with a singularity of residue type along W, and η satisfies the equation of currents: $W = d[\eta] - [d\eta]$. In other words the residue of η along W is 1. Then for any complex subvariety $X \subset V$ such that $\dim W + \dim X = \dim V + \dim W \cap X$ the intersection current (counted with multiplicities) $W \cdot X = d[X \wedge \eta] - [X \wedge d\eta]$. In more usual terminology, for any C^∞ form ψ with compact support on V,

$$\int_{W\cap X} m(W,X)\psi = \int_X \eta \wedge d\psi - d\eta \wedge \psi = \lim_{\varepsilon\to 0} \int_{X\cap S_\varepsilon} \eta \wedge \psi.$$

Here S_ε is the boundary of the ε-neighborhood of W in V and $m(W,X)$ is the intersection multiplicity at each point of $W \cap X$. This is discussed in more detail in this paper and was proved previously in [9], which is chiefly devoted to showing the existence of such η.

The purpose of this paper is to take the same approach to intersection theory when $\dim W \cap X > \dim W + \dim X - \dim V$. In [9], Proposition 5.3, it is observed that the form η is locally L^1 on X without the previous assumption on dimension. Therefore, the expression $d[X \wedge \eta] - [X \wedge d\eta]$ still defines a current supported on $W \cap X$. This current is no longer a locally constant function on $W \cap X$ but a "form" on $W \cap X$ whose degree is the "error" in the dimension of $W \cap X$. Then we have often $d[X \wedge \eta] - [X \wedge d\eta] = W \cap X \wedge \phi$ for some locally L^1 form ϕ, cf. 5.2.

It follows from the definition that if $d\eta$ is C^∞, the current $W \cap X \wedge \psi$ represents the cup product of the cohomology classes of W and of X. According to Theorem 2.13 of [9], this can be done with all forms preserving Hodge filtration and can also be done in Dolbeault cohomology.

In this paper we study the singularities of the residue forms more closely and show that ψ can be written as an expression in the Chern forms of the normal bundles of W and X. We can actually compute ψ when the intersection is nondegenerate.

The presence of the Chern forms is explained by studying the residue of η along the monoidal transform of W, which we call the refined residue. It is shown that given a metric in N_W, the normal bundle of W, that for any closed form ϕ on W, there is a residue form η on V with $d\eta$ a C^∞ form on all of V such that the refined residue of η is $p^*\phi \wedge \sum_{k=0}^{d-1} \omega^{d-k-1} \wedge p^*c_k$. Here the c_i are the Chern forms of N_W and ω is the Kähler metric in the fibers of $\mathbb{P}(N_W)$.

This theorem means that even though N_W may not be holomorphically imbeddable as a neighborhood of W, that by using residue calculus one may establish theorems in complex geometry that are proved in the C^∞ case by the use of tubular neighborhoods.

1. RESIDUES AND CURRENTS

If η is an r-form on a complex manifold V which is C^∞ on the complement V - W of a complex submanifold W of codimension $d = \dim_{\mathbb{C}} V - \dim_{\mathbb{C}} W$, then the operation of taking the *residue* of η associates to η a C^∞ r-2d form $\operatorname{res}\eta$ on W which satisfies the following residue formula:

$$\int_\gamma \operatorname{res}\eta = \lim_{\varepsilon\to 0} \int_{\tau_\varepsilon(\gamma)} \eta, \tag{1.1}$$

where γ is any differentiable r-2d chain on W and $\tau_\varepsilon(\gamma)$ is the inverse image of γ in the normal ε-sphere bundle bounding the ε-neighborhood of W in V. (We assume always that the normal bundle of W is imbedded as a neighborhood of W.) For this limit to exist, we must assume certain restrictions on the singularity of η. In this paper we will make no attempt to treat the most general singularities possible but will study a type rich enough in examples for our purposes.

Example 1.2: If f is meromorphic in $\mathbb{C}$, $\int_{|z|=\varepsilon} \frac{df}{f} = 2\pi i n$ for small ε, where n is the order of f at zero. Therefore, $\operatorname{res}(\frac{df}{f})$ is the zero form $2\pi i n$ on $\{0\}$.

Example 1.3: In $\mathbb{C}^n$, if $\mu = \frac{1}{2\pi i}\,\partial \log(|z_1|^2 + \ldots |z_n|^2)$ and $\omega = \bar{\partial}\mu$, the form $\eta = \mu \wedge \omega^{n-1}$ is the Bochner-Martinelli form. This form is invariant under the unitary group and $\int_{|z|=\varepsilon} \eta = 1$. Therefore, $\text{res}\,\eta = 1$. (Cf. [5], Prop. 1.10)

Example 1.4: If f is meromorphic in C^n and $W = \{f = 0\} \cup \{f = \infty\}$ is a complex manifold, then $\int_{\tau_\varepsilon(z_0)} \frac{df}{f} = 2\pi i n_{z_0}$, where $\tau_\varepsilon(z_0)$ is the normal ε-circle about z_0 and n_{z_0} is the order of f at z_0. Thus $\text{res}(\frac{df}{f}) = 2\pi i n_{z_0}$, which is a locally constant function on W. This Poincaré equation can be generalized to a singular divisor W. (Cf. [5], Prop. 1.1)

The residue formula (1.1) uniquely determines $\text{res}\,\eta$ because a C^∞ form is determined by its integrals over differentiable chains. A form $\text{res}\,\eta$ on W is also determined by all the integrals $\int_W (\text{res}\,\eta) \wedge \phi$, where ϕ is an arbitrary C^∞ form with compact support on W. We can interpret the residue formula in this context as well:

$$\int_W (\text{res}\,\eta) \wedge \phi = \lim_{\varepsilon \to 0} \int_{S_\varepsilon} \eta \wedge \pi^*\phi \tag{1.5}$$

for all C^∞ forms ϕ with compact support on W. S_ε is the imbedded normal ε-sphere bundle about W and $\pi\colon S_\varepsilon \to W$ is the bundle projection.

It is not difficult to see that 1.1 and 1.5 are equivalent, since γ may be approximated by C^∞ forms as a current and conversely (i.e., as a form with distribution coefficients, cf. [10] and [4]).

By Stokes Theorem we may rewrite the right hand side of 1.5. First, we may assume that the singularity of η is such that $\lim_{\varepsilon\to 0} \int_{S_\varepsilon} \eta \wedge \psi = 0$ if ψ is a form on V with $i^*\psi = 0$, where $i\colon W \to V$ is the inclusion. Roughly, if the singularity of η grows no faster than ε^{2d-1} in the normal direction this is true.

Under this assumption, if ψ is a compactly supported C^∞ extension of ϕ from W to V, i.e., $i^*\psi = \phi$, then $\lim_{\varepsilon\to 0} \int_{S_\varepsilon} \eta \wedge \pi^*\phi = \lim_{\varepsilon\to 0} \int_{S_\varepsilon} \eta \wedge \psi$. But $-S_\varepsilon$ is the oriented boundary of $V - B_\varepsilon$, where B_ε is the ε-normal neighborhood of W. By Stokes Theorem,

$$\int_{S_\varepsilon} \eta \wedge \psi - \int_{V-B_\varepsilon} d(\eta \wedge \Psi) =$$

$$- \int_{V-B_\varepsilon} d\eta \wedge \psi + (-1)^{r+1} \int_{V-B_\varepsilon} \eta \wedge d\psi.$$

If η and $d\eta$ are both locally L^1 on V, then $\lim_{\varepsilon\to 0} \int_{S_\varepsilon} \eta \wedge \psi =$

$$- \int_V d\eta \wedge \psi - (-1)^r \int_V \eta \wedge d\psi.$$

Thus we finally have for any compactly supported C^∞ ψ on V:

$$\int_W \mathrm{res}\,\eta \wedge \psi = \int_V (-1)^{\deg \eta+1} \eta \wedge d\psi - \int_V d\eta \wedge \psi. \tag{1.6}$$

Now equation 1.6 may be viewed as an equation of distributions, or more precisely, currents, which are dual to forms.

Definition 1.7: If η is a locally L^1 r-form on V, $\dim_C V = n$, the linear functional $[\eta]: A_C^{2n-r}(V) \to C$ on the space of compactly supported C^∞ forms on V is defined by $[\eta](\psi) = \int_V \eta \wedge \psi$. The exterior differential $d[\eta]$ is defined by $d[\eta](\psi) = (-1)^{r+1} \int_V \eta \wedge d\psi$.

Other linear functionals used in this paper will be of the form $[X \wedge \eta]$, where X is a complex analytic subvariety of V and η is a locally L^1 s-form on X (in the induced volume measure). Then $[X \wedge \eta](\psi) = \int_X \eta \wedge \psi$ for any C^∞ form ψ on V with compact support. Again, $d[X \wedge \eta](x) = (-1)^{s+1}[X \wedge \eta](d\psi)$.

We need to know nothing of the theory of distributions to use these definitions as a shorthand notation. Thus 1.6 takes the following form as an equation of currents:

$$W \wedge \mathrm{res}\,\eta = d[\eta] - [d\eta]. \tag{1.8}$$

If λ is any k-form on the m-dimensional submanifold W, we define the current $W \wedge \lambda: A_C^{2m-k}(V) \to C$ by $W \wedge \lambda(\psi) = \int_W \lambda \wedge \psi$.

Notice that 1.8 shows that $d[\eta]$ and $[d\eta]$ need not be the same current. In fact the difference is precisely the residue. The residue is a measure of the failure of Stokes Theorem to hold with respect to a form η which is not C^∞.

Definition 1.9: If η is a locally L^1 form on V, C^∞ on $V - W$, and $d\eta$ is also locally L^1, then we define the *residue*

$$\mathrm{Res}\,\eta = d[\eta] - [d\eta].$$

This is a current supported on W. If $\mathrm{Res}\,\eta = W \wedge \phi$ for a form on W, we set $\mathrm{res}\,\eta = \phi$. Thus $\mathrm{Res}\,\eta = W \wedge \mathrm{res}\,\eta$.

Example 1.2 (bis): If $\eta = \frac{df}{f}$, $[d\eta] = 0$ by $d[\eta] = \{0\} \wedge 2\pi i n$.

Example 1.10: $d[\log |z|^2] = [d \log |z|^2]$ even though there is a singularity at 0.

It is in the form 1.8 (or Definition 1.9) that we will discuss residue formulas. The characterization in terms of chains can be recovered from this. In this form the properties of residues with respect to cohomology are easily expressed and it is easier to consider the case of a W with singularities.

2. CHERN CLASSES AND THE FUNDAMENTAL RESIDUE FORMULA

Let E be a holomorphic k-dimensional vector bundle over the complex manifold W. Let h be a C^∞ hermitian metric on E. The function $|s|^2 = h(s,s)$ is a non-negative C^∞ function on E which defines a Euclidean length in each fiber E_x, $x \in W$.

We will make great use of the forms

$$\mu = \frac{1}{2\pi i} \partial \log |s|^2 \tag{2.1}$$

$$\omega = \bar{\partial}\mu . \tag{2.2}$$

If we choose an orthonormal C-basis $e_1,\ldots,e_k$ in E_x and set $s = \sum_{i=1}^{k} s_i e_i$ for $s \in E_x$, then $|s|^2 = \sum_{i=1}^{k} |s_i|^2$ and, restricted to E_x, $\omega = \frac{i}{2\pi} \partial\bar{\partial} \log (|s_1|^2 + \ldots + |s_k|^2)$, which is the pullback to E_x of the Kähler form $\widetilde{\omega}$ of the Fubini-Study metric on $\mathbb{P}(E_x)$. Furthermore, the form $\mu \wedge \omega^{k-1}$ restricted to E_x is just the Bochner-Martinelli form defined in Example 1.3. This form satisfies the following residue formula:

$$Z = d[\mu \wedge \omega^{k-1}] - [\omega^k] = \bar{\partial}[\mu \wedge \omega^{k-1}] - [\omega^k], \tag{2.3}$$

where Z is the zero section of E, a complex manifold biholomorphic to W. Proofs of this may be found in [7] and [9]; we will also give an indication of the proof in the next section.

The only trouble with this residue formula is that the form $[\omega^k]$ is not in general a C^∞ form. In the case of the product bundle $W \times \mathbb{C}^k$ with the flat metric, $\omega^k \equiv 0$ and this problem does not arise. For an arbitrary holomorphic bundle, however, if we seek an equation $Z = d[\eta] - [d\eta]$, where $d\eta$ is C^∞ we must look further. This is where the Chern forms come in.

Given a holomorphic k-vector bundle E over W and a hermitian metric h on E, the Chern forms of this metric are usually defined by taking invariant polynomials in the curvature forms to yield 2r-forms $c_r(E)$, $r = 1, \ldots, k$, on W (cf. [1] or [2] Chapter 5). We will not indicate the dependence on h since we assume the metric is chosen once and for all. When it is unambiguous we write c_r instead of $c_r(E)$.

If $\pi: E \to W$ is the projection, we have the following polynomial relation of forms on $E - Z$:

$$\omega^k + \omega^{k-1} \wedge \pi^*c_1 + \ldots + \omega \wedge \pi^*c_{k-1} + \pi^*c_k \equiv 0 \tag{2.4}$$

This may be taken as a definition of the c_i, since this relation defines them uniquely, as we will see below. For a proof of 2.4, see [1].

Equation 2.4 yields a residue formula of the desired type. Let $\eta = \mu \wedge (\omega^{k-1} + \omega^{k-2} \wedge \pi^*c_1 + \ldots + \pi^*c_{k-1})$. Then $d\eta = \omega \wedge \left(\sum_{r=0}^{k-1} \omega^{k-1-r} \wedge \pi^*c_r\right) = -\pi^*c_k$. We have the residue formula

$$Z = d[\eta] - [d\eta] = d[\eta] + \pi^*c_k , \tag{2.5}$$

which we will refer to as the *Fundamental Residue Formula*.

To prove this formula, we first observe that

$$d[\mu \wedge \omega^{k-1-r} \wedge \pi^*c_r] - [\omega^k \wedge \pi^*c_r] = (d[\mu \wedge \omega^{k-1-r}] - [\omega^k]) \wedge \pi^*c_r .$$

This follows immediately by applying each side to a form ϕ with compact support on E. $(d[\mu \wedge \omega^{k-1-r} \wedge \pi^*c_r](\phi) = \int_E \mu \wedge \omega^{k-1-r} \wedge \pi^*c_r \wedge d\phi =$ $= \int_E \mu \wedge \omega^{k-1-r} \wedge d(\pi^*c_r \wedge \phi) - \mu \wedge \omega^{k-1-r} \wedge (d\pi^*c_r) \wedge d\phi$, etc. We will see below that $dc_r = 0$ and simplify the computation.).

Using this fact it suffices to show that $d[\mu \wedge \omega^s] = [\omega^{s+1}]$ for $s < k - 1$ and that $d[\mu \wedge \omega^{k-1}] - [\omega^k] = Z$. This is proved in [7] and [9]; we shall also indicate the proof in the next section.

3. INTEGRATION ALONG THE FIBER AND THE MONOIDAL TRANSFORM

For proving the residue formulas in this paper, a simple method is to analyze the singularity of the residue form by means of a monoidal transform and then to show the residue formula by integration along the fiber.

In the language of currents integration along the fiber is the same as the push-forward map of chains. If $f\colon M_1 \to M_2$ is a proper (f^{-1} of a compact is compact) smooth map between oriented manifolds and u is a current on M_1, then we define the current f_*u on M_2 by setting $f_*u(\phi) = u(f^*\phi)$ for any compactly supported form ϕ on M_2. Suppose $u = [\psi]$, the current defined by a form ψ. It will happen in certain cases that $f_*[\psi]$ will also be defined by a form, which we will denote by $f_*\psi$. By definition $[f_*\psi] = f_*[\psi]$ if such an $f_*\psi$ exists. Here are two important examples:

Example 3.1: f is a smooth fibration. In this case if ψ is C^∞, then $f_*\psi$ is also a C^∞ form obtained by integrating out the differentials of the fiber coordinates, cf. [10].

Example 3.2: f is a diffeomorphism off of a set of measure zero. This means that there is a closed set $A \subset M_1$ of measure zero such that $B = f(A)$ has measure zero in M_2 and $f\colon M_1 - A \to M_2 - B$ is a diffeomorphism. In this case, if ψ is a locally L^1 form on M_1, then $f_*[\psi] = [f^{-1*}\psi]$, where $f^{-1*}\psi$ is well-defined on $M_2 - B$, and $f_*\psi = f^{-1*}\psi$. To see this it suffices to show that the two measures $f_*[\psi]$ is $[f^{-1*}\psi]$ on M_2. It is enough to check this on $M_2 - B$, since $f_*[\psi]$ is absolutely continuous with respect to Lebesgue measure (locally) and so is defined by a locally L^1 function. The equality on $M_2 - B$ is just the usual invariance under coordinate change.

Both of these cases come into play when we use the *monoidal transform.* If $W \subset V$ is a complex m-submanifold of the complex n-manifold V, then we may construct a second complex n-manifold $\hat{V}$ called the monoidal transform of V along W. There is a proper holomorphic map $\sigma\colon \hat{V} \to V$ with the following properties:

(a) $\sigma\colon \hat{V} - \sigma^{-1}(W) \to V - W$ is biholomorphic.

(b) $\sigma\colon \sigma^{-1}(W) \to W$ is a holomorphic fiber bundle with fiber $\mathbb{P}^{n-m-1}$. In fact this bundle is $\mathbb{P}(N_W)$, where N_W is the normal bundle of W in V.

The submanifold $\hat{W} = \sigma^{-1}(W) \subset \hat{V}$ has dimension $n - 1$; it is a divisor. We will write $p\colon \hat{W} \to W$ for the restriction of σ to $\hat{W}$. These relationships are summarized by this commutative diagram:

$$\begin{array}{ccccccc} \mathbb{P}(N_W) \simeq \hat{W} & \subset & \hat{V} & \supset & \hat{V} - \hat{W} \\ \downarrow p & & \downarrow \sigma & & \wr\! \wr \\ W & \subset & V & \supset & V - W \end{array} \tag{3.3}$$

Having chosen local coordinates $(w_1,\ldots,w_m, z_1,\ldots,z_{n-m})$ at $y \in W$

such that $W = \{z_1 = \ldots = z_{n-m} = 0\}$ locally, we may choose coordinates near any $x \in \hat{W}$ with $\sigma(x) = y$ of a special form. In these coordinates $(u_1,\ldots,u_m,v_1,\ldots,v_{k-1},t,v_{k+1}),\ldots,v_{n-m})$ $\hat{W} = \{t = 0\}$ and the mapping σ is given by $w_i = u_i$, $z_j = tv_j$ if $j \neq k$ and $z_k = t$. The choice of k depends on x. For the precise definition of $\hat{V}$, see [9].

Now following [9] we say that a form η on V is of *residue type along W* if η is C^∞ on $V - W$ and $\sigma^*\eta = \frac{dt}{t} \wedge \psi_1 + \psi_2$ near $\hat{W}$ (in the local coordinates above), where the ψ_i are C^∞. In this case $[\eta] = \sigma_*[\sigma^*\eta]$ and $[d\eta] = \sigma_*[d\sigma^*\eta]$ by 3.2.

Thus by definition 1.9

$$\begin{aligned} W \wedge \operatorname{res}\eta = \operatorname{Res}\eta &= d[\eta] - [d\eta] = d\sigma_*[\sigma^*\eta] - \sigma_*[d\sigma^*\eta] \\ &= \sigma_*(d[\sigma^*\eta] - [d\sigma^*\eta]) = \sigma_*(\hat{W} \wedge \operatorname{res}\sigma^*\eta). \end{aligned}$$

But for any C^∞ form λ on $\hat{W}$, the push-forward $\sigma_*(\hat{W} \wedge \lambda) = W \wedge p_*\lambda$. By case 3.1 the fiber integral $p_*\lambda$ is C^∞. We have shown

$$\operatorname{res}\eta = p_*\operatorname{res}\sigma^*\eta.$$

In this case the form $\operatorname{res}\sigma^*\eta$ is given locally on $\hat{W}$ by the restriction to $\hat{W}$ of $2\pi i\psi_1$.

To prove the residue formulas of Section 2, we simply observe that in the case of the zero section $Z \subset E$, $\sigma^*\eta = \frac{1}{2\pi i}\partial \log|s\circ\sigma|^2 = \frac{1}{2\pi i}\frac{dt}{t} + \rho$, where ρ is C^∞. Then $\sigma^*\omega = \bar{\partial}\rho$, a C^∞ form whose restriction to $\hat{Z} = \mathbb{P}(E)$ is the Fubini-Study Kähler form on the fibers.

Consequently, by 3.4, $\operatorname{res}\mu \wedge \omega^r = p_* \operatorname{res}\sigma^*(\mu \wedge \omega^r) = p_*\omega^r = 0$ if $r < k - 1$ and $= 1$ if $r = k - 1$. This proves the residue formulas of Section 2; a more detailed account is found in [7] or [9].

We may also use this procedure with 2.4 to compute $\operatorname{res}\mu \wedge \omega^r$ for $r \geq k$. For example, $\operatorname{res}\mu \wedge \omega^k = p_*\omega^k = p_*(-\omega^{k-1} \wedge p^*c_1 - \ldots - p^*c_k) = -c_1$.

Proposition 3.5 (after Bott [1]): *If* E *is a holomorphic k-vector bundle over* W, *the formal power series of forms*

$$\frac{\mu}{1-t\omega} = \mu \wedge (1 + t\omega + t^2\omega^2 + \ldots) \text{ on } E$$

and

$$\frac{1}{1-t\omega} = 1 + t\omega + t^2\omega^2 + \ldots \text{ on } \mathbb{P}(E)$$

satisfy $\operatorname{res}\left(\frac{\mu}{1-t\omega}\right) = p_*\left(\frac{1}{1-t\omega}\right) = \frac{t^{k-1}}{c_t(E)}$. *Here,* $c_t(E) = 1 + tc_1(E) + \ldots + t^k c_k(E)$, *the Chern polynomial, which has a formal power series inverse.*

This equation is an equation of formal power series in t.

Proof: Just iterate the results above using 2.4 to reduce powers of ω. For example $\omega^{k+1} = \omega(-\omega^{k-1} \wedge p^*c_1 - \omega^{k-2} \wedge p^*c_2 - \ldots) = -\omega^k \wedge p^*c_1 - \omega^{k-1} \wedge p^*c_2 - \ldots = (\omega^{k-1} \wedge p^*c_1 + \omega^{k-2} \wedge p^*c_2 + \ldots) \wedge p^*c_1 - \omega^{k-1} \wedge p^*c_2 = \omega^{k-1} \wedge p^*(c_1^2 - c_2)$ + lower order terms in ω. Thus $p_*\omega^{k+1} = c_1^2 - c_2$.

Q.E.D.

4. REFINED RESIDUES ALONG A COMPLEX SUBMANIFOLD

In [9], Theorem 2.13, it was shown that for any complex submanifold W of codimension $k > 0$ of a complex manifold V and for any closed smooth form ϕ on W there is a residue form η such that $W \wedge \phi = W \wedge \text{res}\,\eta = d[\eta] - [d\eta]$ and that $d\eta$ extends to a C^∞ form on all of V. Given the (p,q) type of ϕ one can say something about the (p,q) type of η as well.

Definition 4.1: A form η, C^∞ on $V - W$, with a singularity of residue type along W is said to be *smoothing* if $d\eta$ extends to a C^∞ form on V. If $\bar{\partial}\eta$ is C^∞ on V, we say that η is *$\bar{\partial}$-smoothing*.

Smoothing residue forms may be used to relate the intersection of two varieties to the wedge product of cohomologous smooth forms. For example, suppose W_1 and W_2 are complex submanifolds with $\dim(W_1 \cap W_2) = \dim W_1 + \dim W_2 - \dim V$. If $W_i = d[\eta_i] - [d\eta_i]$, $i = 1,2$, then the intersection current $W_1 \cdot W_2 = [d\eta_1 \wedge d\eta_2] + d([W_1 \wedge \eta_2] - [\eta_2 \wedge d\eta_2])$. If the η_i are smoothing, this says the current $W_1 \cdot W_2$ is cohomologous to $d\eta_1 \wedge d\eta_2$. The truth of this equation depends on the fact that the current $W_1 \wedge \eta_2$ is well defined, but it is proved in Proposition 5.3 of [9] that η_2 is locally L^1 on W_1. If $\dim(W_1 \wedge W_2) = 0$ and V is compact, we can apply both sides of this equation of currents to the constant function 1; this implies that the intersection number of W_1 and W_2 is $\int_V d\eta_1 \wedge d\eta_2$.

If $\dim(W_1 \cap W_2) > \dim W_1 + \dim W_2 - \dim V$ but no component of W_1 is contained entirely within W_2, then η_2 is still locally L^1 on W_1. Therefore, the current $[d\eta_1 \wedge d\eta_2] + d([W_1 \wedge \eta_2] - [\eta_1 \wedge d\eta_2])$ is still defined. The support of this current is still contained in $W_1 \wedge W_2$, but the (real) dimension of the current itself is still

$2(\dim W_1 + \dim W_2 - \dim V)$. We will see that this is a current of the type $W_1 \cap W_2 \wedge \psi$, where ψ is a locally L^1 form on $W_1 \cap W_2$. To determine ψ more precisely, we need to study the singularity of ψ more carefully; that is the subject of this section.

As in the previous section we will use the monoidal transform $\sigma: \hat{V} \to V$ to study these singularities. Recall $\hat{W} = \sigma^{-1}(W) \simeq \mathbb{P}(N_W)$, where N_W is the normal bundle to W. If the divisor $\hat{W}$ is defined locally by $\{t = 0\}$, a form μ on $\hat{V}$ has residue type on $\hat{V}$ along $\hat{W}$ if locally $\mu = \frac{dt}{t} \wedge \psi_1 + \psi_2$ where the ψ_i are C^∞. A form η on V is of residue type with singularity along W if $\sigma^*\eta$ is of residue type with singularity on $\hat{W}$.

By the discussion in Section 3 we know that $\operatorname{res} \eta = p_* \operatorname{res} \sigma^*\eta$ as forms on W, where $p: \mathbb{P}(N_W) \to W$ is defined by σ. Thus if η_1, η_2 are forms of residue type and $\operatorname{res} \sigma^*\eta_1 = \operatorname{res} \sigma^*\eta_2$, then $\operatorname{res} \eta_1 = \operatorname{res} \eta_2$ but not conversely as we see in Example 4.3.

Definition 4.2: If η is a form on V of residue type along W, the *refined residue* of η is $\operatorname{res} \sigma^*\eta$, a C^∞ form on $\hat{W}$.

Example 4.3: Let $Z \subset E$ be the zero section of the holomorphic k-vector bundle E and $\mu = \frac{1}{2\pi i} \partial \log |s|^2$ and $\omega = d\mu$. It was proved in Section 3 that if $\eta_1 = \mu \wedge \omega^{k-1}$ and $\eta_2 = \mu \wedge \sum_{j=0}^{k-1} \omega^{k-1-j} \wedge \pi^* c_j$,

$\operatorname{res} \eta_1 = \operatorname{res} \eta_2 = 1$. But $\operatorname{res} \sigma^*\eta_1 = \omega^{k-1}$, $\operatorname{res} \sigma^*\eta_2 = \sum \omega^{k-1-j} \wedge \pi^* c_j$.

Example 4.3 is the archetype for the discussion that follows. Observe that η_2 is smoothing and $\bar{\partial}$-smoothing but η_1 is not. We will see in Corollary 6.5 that in general ω^{k-1} is not the refined residue of a smoothing form. Since this form will recur, we will establish the notation:

$$q(E) = \sum_{j=0}^{k-1} \omega^{k-1-j} \wedge p^* c_j(E). \tag{4.4}$$

In this case $p: E \to W$ is a holomorphic k-vector bundle and the forms c_j and ω are defined in 2.4 and 2.2. The form $q(E)$ depends on the choice of hermitian metric in E, but the cohomology class of $q(E)$ does not.

The following theorem is the central result of this section; its proof will occupy us for the remainder of this section.

Theorem 4.5 (Existence of Residue Form): *Suppose* W *is a (closed) complex submanifold of* V *of complex codimension* $k > 0$. *Having chosen a metric on the normal bundle* N_W, *then there exists a smoothing residue form* η *and* $\bar\partial$- *smoothing residue form* ζ *such that the refined residues of* η *and* ζ *both equal* $q(N_W)$. *More generally, such forms* η (ζ) *may be found with refined residues* $q(N_W) \wedge p^*\phi$, *where* ϕ *is any smooth closed* ($\bar\partial$-*closed) form on* W. *Furthermore, if* ϕ *has Hodge filtration* p, η *may be chosen with Hodge filtration* $p + k$ *and if* ϕ *is of type* (p,q), ζ *may be chosen of type* $(p + k, q + k - 1)$.

Remark: This asserts that $d\eta$ and $\bar\partial\zeta$ are C^∞ and $\hat{W} \wedge (q(N_W) \wedge p^*\phi) = d[\eta] - [d\eta]$ if $d\phi = 0$ and $= -\bar\partial[\zeta] - [\bar\partial\zeta]$ if $\bar\partial\phi = 0$. Hodge filtration p means that a form has zero (r,s) components for $r < p$.

In the case that V is a compact Kähler manifold the form η can also be chosen to be of type $(p + k, q + k - 1)$ if ϕ is a closed form of type (p,q). This follows because $-d\eta = d\lambda + \zeta$, where λ is C^∞ and ζ is the harmonic $(p + k, q + k - 1)$ form cohomologous to $W \wedge \phi$ and hence $-d\eta$.

In order to prove Theorem 4.5 we procede as in [9] to compute DeRham and Dolbeault cohomology with complexes of residue forms.

Definition 4.6: Let $\mathcal{B}^{p,q}_{\hat{V},\hat{W}}$ be the sheaf of germs of C^∞ forms on $\hat{V} - \hat{W}$ of residue type along W *which have zero residue*. We will write $\mathcal{B}^{p,q}$ when this is not ambiguous; we write $B^{p,q}_{\hat{V},\hat{W}}$ for $\Gamma(\hat{V}, \mathcal{B}^{p,q}_{\hat{V},\hat{W}})$.

The $\bar\partial$ operator maps $\mathcal{B}^{p,q}$ into $\mathcal{B}^{p,q+1}$. If $\eta \in \mathcal{B}^{p,q}_x$, $x \in \hat{V}$, $q > 0$ and $\bar\partial\eta = 0$ then since $\operatorname{res}\eta = 0$, $\bar\partial[\eta] = 0$. Thus by 3.1 of [9] there is a (p,q) form μ of residue type with $\bar\partial\mu = \eta$ near x. If $\operatorname{res}\mu \neq 0$, at least $\bar\partial \operatorname{res}\mu = 0$ since $\hat{W} \wedge \operatorname{res}\eta = \bar\partial[\mu] - [\eta]$. Writing $\hat{W}$ locally as $\{t = 0\}$ and extending $\operatorname{res}\mu$ from $\hat{W}$ to a $\bar\partial$ closed form ϕ in a neighborhood of x in $\hat{V}$, we define $\zeta = \mu - \frac{dt}{t} \wedge \phi$. Then $\bar\partial\zeta = \eta$ and $\operatorname{res}\zeta = \operatorname{res}\mu - \phi = 0$ on $\hat{W}$.

Lemma 4.7: *If* $\eta \in \mathcal{B}^{p,q}_{\hat{V},\hat{W},x}$, $q > 0$, $\bar\partial\eta = 0$, *there is a* $\zeta \in \mathcal{B}^{p,q-1}_{\hat{V},\hat{W},x}$ *such that* $\bar\partial\zeta = \eta$. *Consequently*

$$H^q(\hat{V},\Omega^p_{\hat{V}}) \simeq \frac{\ker(\bar\partial : B^{p,q} \to B^{p,q+1})}{\operatorname{im}(\bar\partial : B^{p,q-1} \to B^{p,q})} .$$

Proof: The first part, a Dolbeault lemma for $\mathcal{B}^{p,q}$ is proved in the preceding paragraph. Furthermore, if $\eta \in B^{p,0}$ and $\bar\partial\eta = 0$, η is holomorphic on $\hat{V} - \hat{W}$. Since $\text{res}\,\eta = 0$, η is actually holomorphic on all of $\hat{V}$. Thus the $\mathcal{B}^{p,q}$ form a fine resolution of $\Omega^p_{\hat{V}}$, the sheaf of holomorphic p-forms. Q.E.D.

Let $A^{p,q}_{\hat{W}}$ be the C^∞ (p,q) forms on $\hat{W}$. Then we may define a *restriction map* $r: \mathcal{B}^{p,q}_{\hat{V},\hat{W}} \to A^{p,q}_{\hat{W}}$. If $y \in \hat{W}$, near y the divisor $\hat{W} = \{t = 0\}$ for a holomorphic function t. By definition $\alpha = \frac{dt}{t} \wedge \psi_1 + \psi_2$ near y for any $\alpha \in \mathcal{B}^{p,q}$. Set $r(\alpha) = i^*\psi_2$, where $i: \hat{W} \to \hat{V}$ is the inclusion. The map r is well-defined since ψ_2 is uniquely determined modulo dt. For a different local defining equation, $\hat{W} = \{s = 0\}$, $s = at$, $a \neq 0$, we have $\alpha = \frac{ds}{s} \wedge \psi_1 - \frac{da}{a} \wedge \psi_1 + \psi_2$. But $i^*(\frac{da}{a} \wedge \psi_1) = 0$ since $\text{res}\,\alpha = i^*\psi_1 = 0$.

Definition 4.8: Let the sheaf $C^{p,q}_{V,W}$ be the sheaf of germs of residue forms η on V along W such that $\sigma^*\eta \in \mathcal{B}^{p,q}_{\hat{V},\hat{W}}$ and $r(\sigma^*\eta) \in p^*A^{p,q}_{W}$. That is, the restriction of $\sigma^*\eta$ to $\hat{W}$ is the pull-back by $p: \hat{W} \to W$ of a C^∞ form on W.

Lemma 4.9 (Dolbeault Lemma): *If* $\eta \in C^{p,q}_{V,W,x}$, $q > 0$ *and* $\bar{\partial}\eta = 0$, *there exists a* $\mu \in C^{p,q-1}_{V,W,x}$ *such that* $\bar{\partial}\mu = \eta$.

Proof: We may assume that η is defined in a neighborhood U of x which is biholomorphic to $\Delta^{n-k} \times \Delta^k$, a product of polydiscs. Under this biholomorphism, $W \cap U$ is identified with $\Delta^{n-k} \times \{0\}$; let $(z_1,\ldots,z_{n-k}, w_1,\ldots,w_k)$ be the standard coordinates.

Under this identification $\sigma^{-1}(U) \simeq \Delta^{n-k} \times \hat{\Delta}^k$, where $\hat{\Delta}^k$ denotes the k-polydisc with the origin blown up to a $\mathbb{P}^{k-1}$, and $W \cap \sigma^{-1}(U) \sim \Delta^{n-k} \times \mathbb{P}^{k-1}$. According to 3.5 of [9], the Dolbeault cohomology group $H^q(\Delta^{n-k} \times \hat{\Delta}^k, \Omega^p) \simeq H^0(\Delta^{n-k}, \Omega^{p-q}_{\Delta^{n-k}})$ if $q > 0$. (We agree that this = 0 if $p < q$). If we write this cohomology in terms of C^∞ forms, every class is represented by a unique form $(\sigma^*\zeta) \wedge \omega^q$, where ζ is a holomorphic $(p - q)$ form on Δ^{n-k} and $\omega = \frac{1}{2\pi i}\,\bar{\partial}\partial \log(|w_1|^2+\ldots+|w_k|^2)$.

Lemma 4.7 tells us that the Dolbeault cohomology can also be computed by the forms in $B^{p,q}$. Since $\sigma^*\eta \in B^{p,q}$ and $\bar{\partial}\sigma^*\eta = 0$, $\sigma^*\eta = \bar{\partial}\alpha + (\sigma^*\zeta) \wedge \omega^q$, where $\alpha \in B^{p,q-1}$ and ζ and ω are as described above. If $q \geq k$, $\omega^q = 0$ and we are done, so we assume $q < k$. Apply the restriction map r to get $0 = r(\sigma^*\eta) = \bar{\partial}r(\alpha) + p^*\zeta \wedge r(\omega^q)$. We wedge with ω^{k-1-q} and integrate along the fiber of $p: \Delta^{n-k} \times \mathbb{P}^{k-1} \to \Delta^{n-k}$; $0 = \bar{\partial}p_*(r(\alpha) \wedge \omega^{k-1-q}) + \zeta$. This is true since $p_*r(\omega^{k-1}) = 1$. This says that the holomorphic (p-q) form ζ equals $\bar{\partial}$ of some form.

Therefore, $\zeta \equiv 0$.

As a result of this, we now have $\sigma^*\eta = \bar{\partial}\alpha$. It only remains to show that α can be chosen with the restriction $r(\alpha) = p^*\lambda$. It will be useful to write $\pi\colon \Delta^{n-k} \times \hat{\Delta}^k \to \Delta^{n-k} \times \mathbb{P}^{k-1}$ for the natural projection. By hypothesis, $r(\sigma^*\eta) = p^*\phi$ for a form ϕ on Δ^{n-k}. Since $\bar{\partial}\sigma^*\eta = 0$, $\bar{\partial}\phi = 0$ and $\phi = \bar{\partial}\psi$ for a $(p,q-1)$ form ψ on Δ^{n-k}.

Therefore, $\sigma^*\eta = \bar{\partial}\pi^*p^*\psi + \bar{\partial}(\alpha - \pi^*p^*\psi)$. But $r(\sigma^*\eta) = p^*\phi = \bar{\partial}p^*\psi = \bar{\partial}p^*\psi + \bar{\partial}r(\alpha - \pi^*p^*\psi)$. If we set $\lambda = \pi^*r(\alpha - \pi^*p^*\psi)$, then $\bar{\partial}\lambda = 0$. Thus $\sigma^*\eta = \bar{\partial}(\alpha - \lambda)$ and $r(\alpha - \lambda) = r(\pi^*p^*\psi + (\alpha - \pi^*p^*\psi) - \lambda) = p^*\psi + \lambda - \lambda = p^*\psi$.

Q.E.D.

Lemma 4.10 (Poincaré Lemma): *If* $\eta \in F^p C^k_{V,W,x} = \sum_{\substack{r+s=k \\ r\geq p}} C^{r,s}_{V,W,x}$, *the forms in* C^k *of Hodge filtration* $p < k$, *then* $d\eta = 0$ *implies that there exists a form* $\mu \in F^p C^{k-1}_{V,W,x}$ *such that* $d\mu = \eta$.

Proof: This is a formal consequence of Lemma 4.9. The form $\eta = \eta_1 + \eta_2$, where $\eta_1 \in C^{k,o}$ and $\eta_2 \in \sum_{\substack{r+s=k \\ k>r\geq p}} C^{r,s}$. Then $d\eta = (d\eta_1 + \partial\eta_2) + \bar{\partial}\eta_2 = 0$; this implies $\bar{\partial}\eta_2 = 0$ since $d\eta_1 + \partial\eta_2 \in F^k C^{k+1}$. By 4.9 $\eta_2 = \bar{\partial}\mu_2$ for some $\mu_2 \in F^p C^{k-1}$. Then $\eta - d\mu_2 \in C^{k,o}$ and $d(\eta - d\mu_2) = 0$. This form is a closed holomorphic k-form; by the Poincaré lemma for holomorphic forms there is a holomorphic (k-1)-form μ_1 such that $\eta - d\mu_2 = d\mu_1$.

Q.E.D.

We will denote by $C^{p,q}$, $F^p C^k$, $A^{p,q}$, $F^p A^k$, sections in $\Gamma(V, C^{p,q}_{V,W})$, etc.

Corollary 4.11: *If* $\eta \in F^p C^k$ *and* $d\eta = 0$, *there is a form* $\mu \in F^p C^k$ *such that* $\eta = d\mu + \alpha$, *where* $\alpha \in F^p A^k$. *Similarly, if* $\eta \in C^{p,q}$ *and* $\bar{\partial}\eta = 0$, *there is a* $\nu \in C^{p,q-1}$ *such that* $\eta = \bar{\partial}\nu + \beta$, $\beta \in A^{p,q}$. *(We set* $\mu = \nu = 0$ *if* $p = k$ *or* $q = 0$.*)*

Proof: Let Ω^p and $\hat{\Omega}^p$ be the sheaves of germs of holomorphic and closed holomorphic p-forms on V. Then by Lemmas 4.10 and 4.9 the sequences

$$0 \to \hat{\Omega} \to F^p C^p \xrightarrow{d} F^p C^{p+1} \xrightarrow{d} \cdots \quad \text{and}$$
$$0 \to \Omega^p \to C^{p,o} \xrightarrow{\bar{\partial}} C^{p,1} \xrightarrow{\bar{\partial}} \cdots$$

are fine resolutions of $\hat{\Omega}^p$ and Ω^p. (We observe that any form $\gamma \in C^{p,o}$ which has $\bar{\partial}\gamma = 0$ is actually holomorphic.) Consequently, the corresponding DeRham and Dolbeault cohomology groups are isomorphic to the usual ones defined by C^∞ forms (cf. [9], sections 2 and 3).

Therefore, the inclusion $A^{p,q} \subset C^{p,q}$ defines isomorphisms:

$$\frac{\ker d\colon F^p A^k \to F^p A^{k+1}}{d(F^p A^{k-1})} \simeq \frac{\ker F^p C^k \to F^p C^{k+1}}{d(F^p C^{k-1})}$$

and

$$\frac{\ker \bar\partial\colon A^{p,q} \to A^{p,q+1}}{\bar\partial(A^{p,q-1})} \simeq \frac{\ker \bar\partial\colon C^{p,q} \to C^{p,q+1}}{\bar\partial(C^{p,q-1})} .$$

This is precisely what is stated in the Corollary. Q.E.D.

Since $\hat{W} \subset \hat{V}$ is a submanifold of codimension one, there is a locally finite open covering $\{U_i\}$ of V such that for each i, $\hat{W} \cap U_i = \{t_i = 0\}$, where t_i is a holomorphic function in U_i vanishing to first order on $\hat{W} \cap U_i$. The non-vanishing functions in $U_i \cap U_j$, $f_{ij} = t_i t_j^{-1}$ define a holomorphic line bundle $L_{\hat{W}}$. There is a section τ of $L_{\hat{W}}$ defined in U_i by t_i, for $t_i = f_{ij} t_j$. Of course the set $\{\tau = 0\} = \hat{W}$.

The restriction of the bundle $L_{\hat{W}}$ to $\hat{W}$ is just the normal bundle $N_{\hat{W}}$ of $\hat{W}$ in $\hat{V}$. Any Hermitian metric on $N_{\hat{W}}$ can be extended to a Hermitian metric on $L_{\hat{W}}$. If $|\tau|^2$ is the square of the length of τ with respect to a Hermitian metric on $L_{\hat{W}}$, the 1-form $\frac{1}{2\pi i}\partial \log |\tau|^2$ is a form of residue type on $\hat{V}$ along $\hat{W}$ whose residue is the constant function 1. This is easy to see, for locally $|\tau|^2 = g_i|t_i|^2$ and $\frac{1}{2\pi i}\partial \log |\tau|^2 = \frac{1}{2\pi i}\frac{dt_i}{t_i} + \frac{1}{2\pi i}\frac{dg_i}{g_i}$ ($g_i \neq 0$ so dg_i/g_i is C^∞.)

For any Hermitian metric on N_W, the normal bundle of $W \subset V$, there is a corresponding metric on $N_{\hat{W}}$ induced by the natural map $N_{\hat{W}} \to N_W$, which is linear on the fibers (recall that $N_{\hat{W}}$ along the zero section). If we extend this metric to $L_{\hat{W}}$, the form $\tilde{\omega} = -c_1(L_{\hat{W}}) = \frac{1}{2\pi i}\bar\partial\partial \log |\tau|^2$ is a closed (1,1) extension of the ω from $\hat{W}$ to all of $\hat{V}$.

Proof of Theorem 4.5: Let us begin with the case of a C^∞ r-form ϕ on W of Hodge filtration p, with $d\phi = 0$. Let α be a C^∞ form of Hodge filtration $p + k - 1$ on $\hat{V}$ which extends the form $q(N_W) \wedge p^*\phi$ on $\hat{W}$. This may be done by extending locally and patching by a partition of unity, since α need not be closed.

The form $\mu = \frac{1}{2\pi i}\partial \log |\tau|^2 \wedge \alpha$ is a residue form of Hodge filtration $p + k$, and $\operatorname{res} \mu = q(N_W) \wedge p^*\phi = i^*\alpha$, for $i\colon \hat{W} \subset \hat{V}$ since $\operatorname{res} \frac{1}{2\pi i}\partial \log |\tau|^2 = 1$ and α is C^∞. Thus, $\hat{W} \wedge q(N_W) \wedge p^*\phi = d[\mu] - [d\mu]$.

But $d\mu = \tilde{\omega} \wedge \alpha - \frac{1}{2\pi i} \partial \log |\tau|^2 \wedge d\alpha$. Thus $\operatorname{res} d\mu = -i^* d\alpha = d(q(N_W) \wedge p^*\phi) = 0$, and the restriction map r is defined. First, $r(\tilde{\omega} \wedge \alpha) = \left(\sum_{j=0}^{k-1} \omega^{k-j} \wedge p^* c_j \right) \wedge p^*\phi = -p^*(c_k \wedge \phi)$ by 2.4. Second, if locally $\hat{W} = \{t_i = 0\}$, $r(\partial \log |\tau|^2 \wedge d\alpha) = i^*(\frac{dg_i}{g_i} \wedge d\alpha) = 0$. Consequently, the form $\sigma_* d\mu$ on V is a closed section of $F^p C^{2k+r}$.

As a consequence of Corollary 4.11; there is a form $\nu \in F^p C^{2k+r}$ such that $\sigma_* d\mu = d\nu + \alpha$, $\alpha \in F^p A^{2k+r}$. Therefore, $\sigma_*\mu - \nu$ is a residue form on V with refined residue $q(N_W) \wedge p^*\phi$ and $d(\sigma_*\mu - \nu)$ is C^∞:

$$\hat{W} \wedge q(N_W) \wedge p^*\phi = d[\sigma_*\mu - \nu] - \alpha.$$

The same argument applied to $\bar{\partial}\mu$ instead of $d\mu$, using the other part of Corollary 4.11, completes the proof of the theorem.

5. REFINED RESIDUES AND INTERSECTIONS

If V is a complex n-manifold and $Y \subset V$ is a complex subvariety of pure dimension r, then $Y = \bigcup Y_i$, a union of irreducible subvarieties of dimension r. Each subvariety defines a 2r-dimensional current of integration Y_i. Then we will define a *variety with multiplicities* with support Y to be a current of the form $\sum n_i[Y_i]$, where the n_i are nonnegative integers. In other words multiplicity is defined for each point of the regular locus of Y and this multiplicity is locally constant on the regular locus.

For example, if $W \subset V$ is a complex submanifold of dimension m and $X \subset V$ is a subvariety of pure dimension k, then if $W \cap X$ has dimension $m + k - n$, the intersection $W \cdot X$ is defined. $W \cdot X$ is a variety with multiplicities with support $W \cap X$. The multiplicities of $W \cdot X$ may be defined in an algebraic or an analytic manner (see [3] and [8], section 4).

The integration currents $[W]$ and $[X]$ define classes in $H^{2n-2m}(V,\mathbb{C})$ and $H^{2n-2k}(V,\mathbb{C})$ by means of the DeRham cohomology with distribution coefficients. The current $W \cdot X$ represents the cup product of the classes in $H^{4n-2m-2k}(V,\mathbb{C})$. This is proved using residue forms in [9]. In this section we wish to define a current supported on $W \wedge X$ which represents this cup product even when $\dim(W \wedge X) > m + k - n$. This method, also using residue forms, preserves the Hodge filtration as well.

First, we make some remarks about intersections and blowing-up. If

$\sigma: \hat{V} \to V$ is the monoidal transform along W, then we define the *transform* $\hat{X}$ of X as the closure $\overline{\sigma^{-1}(X - W)}$ in $\hat{V}$. This is a subvariety of $\hat{V}$ of dimension k because $\sigma^{-1}(X \cup W)$ is a subvariety and $\hat{W} = \sigma^{-1}(W)$ is an irreducible subvariety. Thus $\overline{\sigma^{-1}(X \cup W) - \hat{W}}$ is also a subvariety [6].

Definition 5.1: The *refined intersection of* X *along* W is the subvariety $\hat{W} \cdot \hat{X}$ of $\hat{V}$.

This variety has dimension $k - 1$ since $\hat{W}$ has codimension one in $\hat{V}$ and contains no component of $\hat{X}$. We remark that this definition is not symmetric with respect to W and X.

Since $\hat{W} \simeq \mathbb{P}(N_W)$, the projective normal bundle of W in V, the variety $\hat{W} \cdot \hat{X}$ is a subvariety of the restriction of $\mathbb{P}(N_W)$ to $W \cap X$. Let dimension $W \cap X = m + k - n + e$. Since the projection $p: \hat{W} \cap \hat{X} \to W \cap X$ is onto, the dimension of the generic fiber is $k - 1 - \dim(W \wedge X) = d - 1 + e = n - m - 1 + e$. For an intersection of the "right" dimension, $e = 0$ and the fiber at x is all of $\mathbb{P}(N_W)_x \simeq \mathbb{P}^{d-1}$.

Theorem 5.2: *Suppose* η *is a residue form along* W *with refined residue* ψ. *Then for any subvariety* $X \subset V$, *none of whose components are contained in* W,

$$d[X \wedge \eta] - [X \wedge d\eta] = \sigma_*(\hat{W} \cdot \hat{X} \wedge \psi).$$

Proof: We know that $\sigma_* d[\hat{X} \wedge \sigma^*\eta] = d[X \wedge \eta]$ and $\sigma_*[\hat{X} \wedge d\sigma^*\eta] = [X \wedge d\eta]$ since σ is biholomorphic on the complement of a set of measure zero on $\hat{X}$ by 3.2. Consequently, $d[X \wedge \eta] - [X \wedge d\eta] = \sigma_*(d[\hat{X} \wedge \sigma^*\eta] - [\hat{X} \wedge d\sigma^*\eta])$. Locally, if $\hat{W} = \{t = 0\}$, $\sigma^*\eta = \frac{dt}{t} \wedge \psi + \rho$. Since $d[\hat{X} \wedge \frac{dt}{t}] = 2\pi i\ \hat{W} \cdot \hat{X}$ by 6.3 of [9], $d[\hat{X} \wedge \sigma^*\eta] = 2\pi i\ \hat{W} \cdot \hat{X} \wedge \psi + [\hat{X} \wedge d\sigma^*\eta]$ and the theorem is proved.

By restriction the map σ defines the map $\pi: \hat{W} \wedge \hat{X} \to W \wedge X$. If we set the form $\phi = p_*(\hat{W} \cdot \hat{X} \wedge \psi)$, ϕ is a locally L^1 form on $W \cap X$ by 5.1 of [9] and $W \cap X \wedge \phi = d[X \wedge \eta] - [X \wedge d\eta]$ by definition.

If we restrict our attention to the case where $\operatorname{res} \eta = 1$, we know by the results of Section 4 that we can choose the refined residue $\psi = \sum_{r=0}^{d-1} \omega^{d-r-1} \wedge p^*c_r$. The forms ω and c_r depend on some Hermitian metric chosen in N_W. Thus given the submanifold W, there is an η with $\operatorname{Res} \eta = W$, $d\eta$ is C^∞ on V and for any X as in 5.2,

$$d[X \wedge \eta] - [X \wedge d\eta] = \sum_{r=0}^{d-1} \sigma_*(\hat{W} \cdot \hat{X} \wedge \omega^{d-r-1}) \wedge c_r. \tag{5.3}$$

In the case of the "right" dimension, $e = 0$, the fiber of $\hat{W} \cap \hat{X} \to W \wedge X$ is $\mathbb{P}^{d-1}$ and the integration along the fiber of ω^r is zero for $r < d - 1$ and is one for $r = d - 1$. Thus for $e = 0$, $W \cdot X = d[X \wedge \eta] - [X \wedge d\eta]$. This is the proof of Theorem 6.3 of [9].

In this case 5.3 gives a formula for $W \cdot X$.

Since 5.3 is true for any X with no components contained in W we may define a generalized intersection by that equation.

Definition 5.4: If W and X are as in 5.2 and h is a Hermitian metric in N_W, the *generalized intersection current* $W *_h X$ is defined to be $\sigma_*(\hat{W} \cdot \hat{X} \wedge \sum_{r=0}^{d-1} \omega^{d-1-r} \wedge p^* c_r)$, $d = \operatorname{codim} W$.

The forms ω and c_r are defined in 2.4 and 2.2; we write $W * X$ without the h when this is clear. If X is also a manifold $W * X \neq X * W$ necessarily.

Proposition 5.5: *The current* $W * X$ *represents the cup product of* W *and* X *in cohomology, with Hodge filtration preserved. More precisely, if* $W = \phi_W + dT_W$, $X = \phi_X + dT_X$, ϕ_W, T_W *(resp.* ϕ_X, T_X*) being a form and current of Hodge filtration* d *(resp.* $n - k$*), then* $W * X = \phi_W \wedge \phi_X + ds$, s *being a current of filtration* $d + n - k$.

Proof: Choose a residue form η of Hodge filtration d such that $W = d[\eta] - [d\eta]$ and $d\eta$ is C^∞. Also choose η with refined residue $\Sigma\omega^{d-r-1} \wedge p^*c_r$. This is possible by 4.5. Then since they are cohomologous, $\phi_W = d\eta + d\lambda$, where λ is C^∞ on all of V and of Hodge filtration d. Thus $W = \phi_W + d[\eta - \lambda]$ and consequently $\phi_W = -d(\eta - \lambda)$. Now the refined residue of $\eta - \lambda$ is the same as that of η so by 5.3 $W * X = d[X \wedge (\eta - \lambda)] - [X \wedge d(\eta - \lambda)] = d[X \wedge (\eta - \lambda)] + X \wedge \phi_W$.

Now if $X = \phi_X + dT_X$, $X \wedge \phi_W = \phi_W \wedge \phi_X + d(T_X \wedge \phi_W)$ and thus $W * X = \phi_W \wedge \phi_X + d(X \wedge (\eta - \lambda) + T_X \wedge \phi_W)$. Q.E.D.

Remark: If $W = \Psi_W + \bar{\partial} S_W$ and $X = \psi_X + \bar{\partial} S_X$, it is also true that $W * X = \psi_W \wedge \psi_X + \bar{\partial} T$ with ψ_W, S_W, ψ_X, S_X, T of (p,q) types (d,d), $(d, d-1)$, $(n-m, n-m)$, $(n-m, n-m-1)$, $(d+n-m, d+n-m-1)$.

6. COMPUTATION OF THE GENERALIZED INTERSECTION

In the last section we have defined generalized intersection currents for an m-dimensional submanifold $W \subset V$ and an r-dimensional subvariety $X \subset V$, where V is an n-dimensional complex manifold. In this section we will observe that $W * X$ can actually be computed, at least in special cases.

We will denote the tangent bundle of a complex manifold V by T_V and the normal bundle of a submanifold $W \subset V$ by $N_{W;V}$.

Definition 6.1: If W, X above are *both* complex submanifolds of V and so is $W \cap X$, we say that this intersection is *nondegenerate* if for each $x \in W \cap X$, the intersection of the tangent spaces $T_{W,x} \cap T_{X,x}$ equals $T_{W\cap X,x} \subset T_{V,x}$. This is equivalent to saying that if $W = \{F = (f_1,\ldots,f_{n-m}) = 0\}$ and $X = \{G = (g_1,\ldots,g_{n-r}) = 0\}$, where rank $F = n - m$ and rank $G = n - r$, then rank $(F,G) = n - \dim(W \cap X)$. We also require that no component of X is contained in W.

Proposition 6.2: *Suppose* W *and* X *intersect nondegenerately. If* $\sigma\colon \hat{V} \to V$ *is the monoidal transform of* V *along* W, *then* $\hat{W} \cap \hat{X}$ *can be identified biholomorphically with* $\mathbb{P}(N_{W\cap X;X})$.

Proof: Obvious when written in local coordinates.

For any formal power series $A(t) = \sum_{\alpha=0}^{\infty} a_\alpha t^\alpha$ we denote by $A(t)\{k\}$ the kth coefficient a_k. If $a_0 \neq 0$, then $1/A(t)$ is also a formal power series. In particular, $1/1-t\omega = 1 + t\omega + t^2\omega^2 + \ldots$. For a holomorphic vector bundle E with a Hermitian metric, we let $C(E)(t) = 1 + tc_1(E) + t^2c_2(E)+\ldots$ be the (finite) series of Chern forms. Then the form $q(E)$ of 4.4 equals $\dfrac{p^*C(E)(t)}{1 - t\omega}\{k - 1\}$.

Theorem 6.3: *If* W *and* X *intersect nondegenerately and* $\dim(W \cap X) = \dim W + \dim X - \dim V + e$, *then the generalized intersection current* $W * X = [W \cap X] \wedge \psi$, *with*

$$\psi = \frac{C(N_{W;V})(t)}{C(N_{W\cap X;X})(t)}\{e\} .$$

The form ψ *depends on the Hermitian metric in* $N_{W;V}$ *but its cohomology class does not; the metric in* $N_{W\cap X;X}$ *is induced from* $N_{W;V}$. *(If* $W \cap X$ *is not pure dimensional,* e *varies for every component of* $W \cap X$.)

Proof: From 6.2 we see that $\sigma\colon \hat{W} \cap \hat{X} \to W \cap X$ may be identified with $\pi\colon \mathbb{P}(N_{W\cap X;X}) \to W \cap X$. By Definition 5.4 we have that $W * X = \sigma_*(\hat{W} \cdot \hat{X} \wedge q(N_{W;V}))$. Therefore, $W * X = [W \cap X] \wedge \pi_* q(N_{W;V})$ in the non-degenerate case. But $\pi_* q(N_{W;V}) = \pi_* \dfrac{\pi^* C(N_{W;V})(t)}{1 - t\omega}\{k - 1\}$

$$= C(N_{W;V})(t)\ \pi_* \frac{1}{1-t\omega}\{k - 1\} = \frac{t^{d-1}\ C(N_{W;V})(t)}{C(N_{W\cap X;X})(t)}\{k - 1\},$$

where $k = \dim V - \dim W$ and $d = \dim X - \dim W \cap X$. The last equality is true because the restriction of the form ω of 2.2 on $\mathbb{P}(N_{W;V})$ restricts to the corresponding form ω on $\mathbb{P}(N_{W\cap X;X}) \subset \mathbb{P}(N_{W;W})|_{W\cap X}$. Finally, since $k - d = e$, we obtain $\pi_* q(N_{W;V}) = \dfrac{C(N_{W;V})(t)}{C(N_{W\cap X;X})(t)}\{e\}$. Q.E.D.

If τ is a holomorphic section of the k-vector bundle $\pi\colon E \to V$, then we say τ is nondegenerate if the intersection of the graph $\tau(V) \subset E$ with the zero section is nondegenerate. If $\tau = \sum_{i=1}^{k} t_i e_i$ for holomorphic functions t_i and a local holomorphic frame $\{e_i\}$, this is equivalent to saying $\tau \not\equiv 0$ and on $T = \{\tau = 0\} \subset V$, $\operatorname{rank}(t_1,\ldots,t_k) = \dim V - \dim T = k - e$.

Corollary 6.4: *If* τ *is a nondegenerate section of* E,

$$Z * \Gamma_\tau = T \wedge \left(\frac{C(E)(t)}{C(N_{T;V})(t)}\{e\}\right),$$

where $\Gamma_\tau = \tau(V)$, Z = zero section, *and* $T = \{\tau = 0\}$. *Considered as a current in* V, $T \wedge \left(\dfrac{C(E)(t)}{C(N_{T;V})(t)}\{e\}\right)$ *is cohomolgous to the form* $c_k(E)$.

Proof: The first part is immediate from 6.3. The last part follows from 2.5

$$\begin{aligned} Z * \Gamma_\tau &= d[\Gamma_\tau \wedge \mu \wedge q(E)] - [\Gamma_\tau \wedge \omega \wedge q(E)] \\ &= d[\Gamma_\tau \wedge \mu \wedge q(E)] + \Gamma_\tau \wedge c_k(E). \end{aligned}$$

If we apply π_* to both sides we obtain the result. Q.E.D.

Corollary 6.5: *If* $W \subset V$ *is a submanifold, in general there is no residue form* $\tilde{\eta}$ *with singularity along* W *such that* $d\eta$ *is* C^∞ *and the refined residue is* ω^{d-1}.

Proof: If one substitutes ω^{d-1} for $q(N_W)$ in 6.3, one obtains different

Chern forms. Therefore, since $-X \wedge d\tilde{\eta}$ is cohomologous to $W * X$, if $\tilde{\eta}$ exists, this leads to a contradiction for a general N_W, if there exist varieties X in sufficient number.

7. THE DIAGONAL AND SELF INTERSECTION

A generalized intersection can be defined in the case that $X \subset W$ and in the case that neither X nor W is a manifold by using the trick of intersection with the diagonal. Let us assume in this section that V is a compact complex n-manifold.

Let the diagonal $\Delta = \{(x,y): x = y\} \subset V \times V$. The two projections $p_i: V \times V \to V$, $i = 1,2$, are biholomorphic restricted to Δ.

If ϕ_i, $i = 1,\ldots,r$, are closed forms on V of degree d_i which furnish a basis for the DeRham cohomology $H^*(V,C) = \sum_{j=0}^{2n} H^j(V,\mathbb{C})$, then let $\psi_1,\ldots,\psi_r$ represent the dual basis. This means that ψ_j is a closed form of degree $2n - d_j$ such that $\int_V \phi_i \wedge \psi_j = \delta^i_j$. By the Künneth formula, every element of $H^*(V \times V, \mathbb{C})$ may be represented by a form of the type

$$\sum_{i,j=1}^{r} c_{ij} p_1^* \phi_i \wedge p_2^* \psi_j .$$

The current Δ is cohomologous to a closed C^∞ form Φ on $V \times V$ if for any closed form α, $\int_\Delta \alpha = \Delta(\alpha) = [\Phi](\boldsymbol{\alpha}) = \int_{V\times V} \Phi \wedge \alpha$. This is a consequence of Poincaré duality. By the remark above it suffices to check for $\alpha = p_1^* \phi_i \wedge p_2^* \psi_j$, $i,j = 1,\ldots,r$. We observe that for the form $\Phi = \sum_{k=1}^{r} (-1)^{\deg \phi_k} p_1^* \phi_k \wedge p_2^* \psi_k$, $\int_{V\times V} \Phi \wedge p_1^* \phi_i \wedge p_2^* \psi_j = \delta^i_j$. But

$$\Delta(p_1^* \phi_i \wedge p_2 \psi_j) + \int_V \phi_i \wedge \psi_j = \delta^i_j$$

as well, so Δ is cohomologous to this Φ.

Now take any complex subvarieties $X, Y \subset V$. If the currents defined by X and Y are cohomologous to the forms ϕ_X and ϕ_Y, respectively, the current defined by $X \times Y$ is cohomologous in $V \times V$ to $p_i^* \phi_X \wedge p_2^* \phi_Y$. This follows from Poincaré duality as above: apply both currents to $\alpha = p_1^* \phi_i \wedge p_2^* \psi_j$.

Therefore, by 5.5, the generalized intersection $\Delta * (X \times Y)$ is cohomologous to $\Delta \wedge p_1^* \phi_X \wedge p_2^* \phi_Y$. Then the currents $p_{1*}(\Delta * (X \times Y))$ and

$p_{2*}(\Delta * (X \times Y)$ are both cohomologous to $p_{1*}(\Delta \wedge p_1^* \phi_X \wedge p_2^* \phi_Y)$ $= p_{2*}(\Delta \wedge p_1^* \phi_X \wedge p_2^* \phi_Y) = [\phi_X \wedge \phi_Y]$. This is a C^∞ form representing the cup product of the two classes. Thus, either of these currents gives a generalized intersection current supported in $X \cap Y$ and representing the cup product.

Example 7.1: If $X = Y$ this defines a self-intersection current. In particular, if X is a submanifold of V we compute:

$$p_{1*}(\Delta * X \times X) = X \wedge \frac{C(N_{\Delta;V\times V})(t)}{C(N_{\Delta\cap X\times X;X\times X})(t)}\{e\}$$

$$= X \wedge \frac{C(T_V)(t)}{C(T_X)(t)}\{e\}$$

which is cohomologous to $X \wedge C_e(N_{X;V})$, where $e = \operatorname{codim} X$. (We use the fact that $N_{\Delta;V\times V} = T_V$ and $N_{X;V} = TV/TX$.)

Another place where the diagonal appears is in the Lefschetz fixed point theorem. If $f: V \to V$ is a map, say a holomorphic map in this context, then the *Lefschetz number* $L(f) = \sum (-1)^q \operatorname{trace} f_q^*$, where $f_q^*: H^q(V,\mathbb{C})$ $H^q(V,\mathbb{C})$ is the map induced by f^* on the qth DeRham cohomology. In terms of the basis chosen earlier,

$$L(f) = \int_V \sum_{k=1}^{r} (-1)^{\deg \phi_k} \phi_k \wedge f^*\psi_k .$$

If $\Gamma_f = \{(x,f(x)\} \subset V \times V$ is the *graph* of f, Γ_f is a complex n-submanifold of $V \times V$. Then $L(f) = \int_{\Gamma_f} \Phi$, where Φ is the form cohomologous to Δ defined above. Then by 5.5 $\Gamma_f \wedge \Phi$ is cohomologous to $\Delta * \Gamma_f$. Thus $\Delta * \Gamma_f(1) = [\Gamma_f \wedge \Phi](1) = \int_{\Gamma_f} 1\Phi = L(f)$. Since the current $\Delta * \Gamma_f =$ $\Delta \cap \Gamma_f \wedge \psi$ for some L^1 form ψ by the remark after 5.2, $L(f) =$ $\Delta * \Gamma_f(1) = \int_{\Delta\cap\Gamma_f} \psi$.

If we identify Δ with V by means of p_1, the set $\Delta \cap \Gamma_f$ is identified with the fixed-point set S of f. If $\dim(\Delta \cap \Gamma_f) = 0$, $\Delta * \Gamma_f =$ $\Delta \cdot \Gamma_f$; and $L(f) = \Delta \cdot \Gamma_f(1) =$ the number of fixed points of f, counted with multiplicity. If Δ intersects Γ_f nondegenerately, this implies that $L(f)$ equals $\int_S \frac{C(T_V)(t)}{C(N_{S;V})(t)}\{\dim S\}$.

Finally, we observe that it would be more interesting to study the Atiyah-Bott Lefschetz fixed point theorem for the $\bar{\partial}$ complex, since we are in the category of complex manifolds here. To do this it will be necessary to split the residue form η for Δ into parts η_p, $p = 0,\ldots,n$ corresponding to the portion of η with exactly p dz's from the first factor of $V \times V$. It is not difficult to compute the residue of this η_p, and this yields the Lefschetz theorem for isolated zeroes, but in order to study higher dimensional fixed point sets, it would be necessary to compute the generalized residue; and this seems to be more difficult.

REFERENCES

1. Bott, R., *A residue formula for holomorphic vector fields,* J. Diff. Geom. 1(1967), 311-330.

2. Chern, S. S., *Complex Manifolds Without Potential Theory,* Van Nostrand, Princeton, N.J., 1967.

3. Draper, R. N., *Intersection theory in analytic geometry,* Math. Ann. 180(1969), 175-204.

4. Federer, H., *Geometric Measure Theory,* Springer-Verlag, New York, 1969.

5. Griffiths, P. and King, J., *Nevanlinna theory and holomorphic mappings between algebraic varieties,* Acta Math. 130:3-4(1973), 145-220.

6. Gunning, R. C. and Rossi, H., *Analytic Functions of Several Complex Variables,* Prentice-Hall, Englewood Cliffs, N.J., 1965.

7. King, J., *A residue formula for complex subvarieties,* Carolina Conference Proc.: Holomorphic Mappings and Minimal Surfaces, Chapel Hill, N.C., 1970.

8. __________, *The currents defined by analytic varieties,* Acta Math. 127(1971), 185-220.

9. __________, *Global residues and intersections on a complex manifold,* To appear, Trans. AMS.

10. Schwartz, L., *Théorie des distributions,* Hermann, Paris, (New edition), 1966.

FRENET FRAMES ALONG HOLOMORPHIC CURVES

Shiing-Shen Chern
Michael J. Cowen
Albert L. Vitter III

University of California at Berkeley
Princeton University
Tulane University

1. INTRODUCTION

The study of holomorphic curves has received in recent years a revival of interest and new impetus, mainly as a special case in the general framework of holomorphic mappings. The most refined results are for holomorphic curves in complex projective space. This theory was initiated by H. and J. Weyl in 1938 [7] and its most important results, the defect relations, are due to Ahlfors [1]. A principal tool in both the original and modern (see [5] and [8]) contributions to this subject has been the systematic use of Frenet frames along holomorphic curves in projective space. In this paper we give the concept of a Frenet frame a differential geometric definition in the context of holomorphic curves on general hermitian manifolds. We characterize those hermitian manifolds possessing a Frenet frame for every holomorphic curve by curvature conditions. In particular, we show that if such a manifold is Kähler then it has constant holomorphic sectional curvature. The classical definition of Frenet frames on projective spaces is then examined. Finally, we consider the application of Frenet frames to the study of holomorphic sections of line bundles. Here too, the classical case (hyperplane bundle) is examined in terms of the general theory.

2. REVIEW OF HERMITIAN GEOMETRY (cf. [2])

Let M be a hermitian manifold of dimension n. Locally its metric can be written

$$<,> = \sum_k \omega_k \wedge \overline{\omega}_k$$

where $\omega_1, \ldots \omega_n$, called the coframe, are forms of bidegree (1,0) and are determined up to unitary transformation (we will agree that all small Latin letters in this section have range $1,\ldots n.$). Let $e_1, \ldots e_n$ be the unitary frame dual to $\omega_1, \ldots \omega_n$. It is well-known that there is a unique connection

$$De_i = \sum_k \omega_{ik} e_k$$

on M characterized by the conditions

$$\omega_{ik} + \overline{\omega}_{ki} = 0 \tag{1}$$

$$T_i \equiv d\omega_i - \sum_k \omega_k \wedge \omega_{ki} \text{ of bidegree } (2,0). \tag{2}$$

The first condition means that the connection preserves the hermitian structure. $T_1, \ldots T_n$ are called the torsion forms. From (1) and (2) it follows that

$$d\omega_{ik} = \sum_j \omega_{ij} \wedge \omega_{jk} + \Omega_{ik} \tag{3}$$

where the curvature forms Ω_{ik} are of bidegree (1,1) and satisfy

$$\Omega_{ik} + \overline{\Omega}_{ki} = 0\,. \tag{4}$$

More explicitly we have

$$\Omega_{ij} = \sum_{k,\ell} R_{ijk\ell}\, \omega_k \wedge \overline{\omega}_\ell$$

where

$$R_{ijk\ell} = \overline{R}_{ji\ell k}\,. \tag{5}$$

For each $p \in M$, $X,Y \in TM_p$ the curvature forms define a linear transformation $\Omega_p(X,\overline{Y}) : TM_p \to TM_p$ determined by

$$<\Omega_p(X,\overline{Y})e_i, e_j> = \Omega_{ij}(X,\overline{Y})$$

Let $L \xrightarrow{\pi} M$ be a holomorphic line bundle with hermitian norm. Relative to an open covering $\{U,V\ldots\}$ of M, the points of $\pi^{-1}(U)$ will have local coordinates (x,y_U), $x \in U$, $y_U \in \mathbb{C}$, which are related by the equations of transition

$$y_U g_{UV} = y_V \qquad \text{in } \pi^{-1}(U \cap V)$$

where g_{UV} is a holomorphic function without zero in $U \cap V$. The hermitian norm is given in terms of coordinates by

$$||y||^2 = h_U|y_U|^2$$

where $h_U > 0$ is a C^∞ function in U. In $U \cap V$ we have therefore

$$h_U|y_U|^2 = h_V|y_V|^2$$

or

$$h_U = h_V|g_{UV}|^2 .$$

A connection on L of bidegree (1,0), uniquely determined by the condition that it leaves the hermitian norm invariant is given by the form

$$\phi_U = \partial \log h_U .$$

Its curvature form is

$$\Phi = - \partial\bar{\partial} \log h_U.$$

For a section y_U its covariant differential is

$$Dy_U = dy_U + \phi_U y_U.$$

When y_U is holomorphic, this is a form of bidegree (1,0) with values in L.

3. FRENET FRAMES

A holomorphic curve in M is a holomorphic mapping $f:X \to M$, where X is a simply connected complex manifold of dimension one (the most important case is $X = \mathbb{C}$). We assume f is not the constant mapping. Let ζ and z_i, $i = 1$ to M, be the local coordinates of X and M respectively and consider a neighborhood of $\zeta = 0$. Then f is given by $z_i = z_i(\zeta)$ and its tangent vector $f'(0) \in TM_{z(0)}$ is given by

$$\sum_i \frac{\partial z_i}{\partial \zeta}(0)\frac{\partial}{\partial z_i} = \zeta^p V$$

where p is a non-negative integer and V a non-zero vector. The isolated points where $p > 0$ are called the *stationary points of* f *of order zero.*

By a *unitary frame for* M *along* f we mean C^∞ mappings $e_i: X \to TM$ such that $e_1(p), \ldots e_n(p)$ is a unitary basis for $TM_{f(p)}$ for each $p \in X$. The covariant differential of the field e_i is given by $De_i = \sum_j f^*(\omega_{ij})e_j$ which will be denoted simply by $De_i = \sum_j \omega_{ij} e_j$. A unitary frame for M along f is called a *Frenet frame* if it has the additional properties at each point of X:

(i) $e_1 = ||v||^{-1} v$

(ii) $De_i = \omega_{ii-1} e_{i-1} + \omega_{ii} e_i + \omega_{ii+1} e_{i+1}$
for $i = 1$ to $n - 1$ where ω_{ii+1} is a (1,0) form.

The points of X where ω_{ii+1} vanishes are called the *stationary points of* f *of order* i.

In attempting to construct a Frenet frame along f one begins by defining e_1 locally according to (i) and completing to a local unitary frame $e_1, \ldots e_n$. We have

$$\begin{aligned} \omega_r &= 0 \qquad (\text{i.e., } f^*(\omega_r) = 0)\, r = 2 \text{ to } n \\ \omega_1 &= h_1\, d\zeta . \end{aligned} \tag{6}$$

We then need the following result [3]: Let $w_\lambda(\zeta)$ be complex-valued functions which satisfy the differential system

$$\frac{\partial w_\lambda}{\partial \bar{\zeta}} = \sum_\mu a_{\lambda\mu}(\zeta) w_\mu \qquad 1 \le \lambda, \mu \le n \tag{7}$$

in a neighborhood of $\zeta = 0$, where $a_{\lambda\mu}$ are complex valued C^1-functions. Suppose the w_λ do not all vanish identically. Then w_λ are of the form

$$w = \zeta^p \tilde{w}_\lambda, \tag{8}$$

where p is an integer ≥ 0 and $\tilde{w}_\lambda(0)$ are not all zero.

Theorem 1: *Let* M *be a hermitian manifold. consider the following two curvature conditions:*

$$\Omega_p(X,\bar{X})X = a(X)X \quad \textit{where} \quad a(X) \in \mathbb{R}, \quad X \in TM_p \quad \forall p \in M \tag{9A}$$

$$\Omega_p(X,\bar{X})Y = b(X)Y \quad \textit{where} \quad b(X) \in \mathbb{R}, \quad \forall X,Y \in TM_p,\ X \perp Y,\ \forall p \in M \tag{9B}$$

M has a Frenet frame for every holomorphic curve iff (i) dimension $M = 2$ or (ii) dimension $M = 3$ and (9A) holds or (iii) dimension $M \geq 4$ and both (9A) and (9B) hold.

Proof: First, the existence of a Frenet frame along every holomorphic curve will be shown to follow from the curvature and dimension conditions. From (2) and (6) we have

$$\omega_1 \wedge \omega_{1r} = 0 \qquad r = 2 \text{ to } n. \tag{10}$$

This implies that at a non-stationary point of f, ω_{1r} is a multiple of ω_1, i.e.

$$\omega_{1r} = h_r d\zeta \qquad r = 2 \text{ to } n. \tag{11}$$

By continuity, this is also true at a stationary point and hence at all points. If $n = 2$, we are done. If $n \geq 3$, Ω_{1r} along f is

$$\Omega_{1r}(e_1, \bar{e}_1)\|f'\|^2 d\zeta \wedge d\bar{\zeta} = 0$$

by (9A) and so differentiation of (11) gives

$$(dh_r - h_r \omega_{11} + \sum_q h_q \omega_{qr}) \wedge d\zeta = 0 \qquad 2 \leq q, r \leq n.$$

The expression in parenthesis is therefore a multiple of $d\zeta$ which means that $h_r(\zeta)$, $r = 2$ to n, satisfy a differential system of type (7). Hence the conclusion (8) implies $De_1 = \omega_{11}e_1 + d\zeta \wedge \sum_r h_r e_r$ where either $h_r \equiv 0$ or $h_r(\zeta) = \zeta^p \tilde{h}(\zeta)$ such that the $\tilde{h}_r(0)$ are not all zero. If $h_r \equiv 0$ for all r, $e_1, \ldots e_n$ is trivially a Frenet frame; if not we can make a unitary change of $e_2, \ldots e_n$ so that $De_1 = \omega_{11}e_1 + \omega_{12}e_2$ where ω_{12} is a (1,0) form. Now since $\omega_{1r} = 0$ for $r \geq 3$ and $\Omega_{ij}(e_1, \bar{e}_1) = 0$ for $i \neq j$, (3) yields $0 = \omega_{12} \wedge \omega_{23}$ and so ω_{23} is a (1,0) form. If $n = 3$ we are done. If $n \geq 4$, continuing the above procedure produces a Frenet frame locally on X. If $e_1, \ldots e_n$ is a Frenet frame on an open set $U \subset X$ then the e_j are defined up to $e_j \to \exp(i\tau_j)e_j$ where τ_j is a real valued C^∞ function on U. A standard patching arguement, using the simple connectivity of X, now yields a Frenet frame globally on X.

Now assume M has a Frenet frame along every holomorphic curve and that the dimension of M is ≥ 3. Using local coordinates one sees

$\forall p \in M$ and X, Y unit vectors in TM_p such that $X \perp Y$ there is a holomorphic curve on M through p whose Frenet frame at p is of the form $e_1 = X, e_2 = Y, e_3, \ldots e_n$. Then

$$\Omega(X,\overline{X})X = \sum_j \Omega_{ij}(X,\overline{X})\ e_j$$
$$= \Omega_{11}(X,\overline{X})X + \Omega_{12}(X,\overline{X})Y$$

($\Omega_{1r}(X,\overline{X}) = 0$ for $r \geq 3$ because of the Frenet frame conditon and (3)). Since the left hand side is independent of Y and since $n \geq 3$, $\Omega_{12}(X,\overline{X}) = 0$ and so $\Omega(X,\overline{X})X = \Omega_{11}(X,\overline{X})X$.

$$\Omega(X,\overline{X})Y = \Omega_{21}(X,\overline{X})X + \Omega_{22}(X,\overline{X})Y + \Omega_{23}(X,\overline{X})e_3$$

$\Omega_{21}(X,\overline{X}) = -\overline{\Omega}_{12}(X,\overline{X}) = 0$ and e_3 can be chosen to be an arbitrary unit vector in TM_p orthogonal to X and Y. Since the left hand side is independent of e_3, $n \geq 4$ implies that $\Omega_{23}(X,\overline{X}) = 0$ and so $\Omega(X,\overline{X})Y = \Omega_{22}(X,\overline{X})Y$.

Q.E.D.

The next theorem exploits equations (9A) and (9B) to give a precise description of the curvature of M.

Theorem 2: *A hermitian manifold* M *of dimension* ≥ 4 *has a Frenet frame along every holomorphic curve iff the curvature of* M *at each point has the following form:*

$$\Omega_{ij} = \rho\, \omega_j \wedge \overline{\omega}_i \qquad \forall i \neq j$$
$$\Omega_{ii} = \rho\, \omega_i \wedge \overline{\omega}_i + \sum_j b_j \omega_j \wedge \overline{\omega}_j \qquad (12)$$
$$+ \sum_{j<k} [c_{jk}\omega_j \wedge \overline{\omega}_k + \overline{c}_{jk}\omega_k \wedge \overline{\omega}_j]$$

where $\rho \in \mathbb{R}$ *is independent of* i *and* j, $b_j \in \mathbb{R}$, *and* $c_{jk} \in \mathbb{C}$ *is independent of* i.

A hermitian manifold M *of dimension* $= 3$ *has a Frenet frame for every holomorphic curve iff the curvature of* M *at each point has the following form:*

$$\Omega_{ij} = \rho_{ij}\omega_j \wedge \overline{\omega}_i \qquad \forall i \neq j$$
$$\Omega_{ii} = (b_{ji} + \rho_{ij})\omega_i \wedge \overline{\omega}_i + \sum_{j|j\neq i} b_{ij}\omega_j \wedge \overline{\omega}_j \qquad (13)$$
$$+ \sum_{j<k} [c_{jk}\omega_j \wedge \overline{\omega}_k + \overline{c}_{jk}\omega_k \wedge \overline{\omega}_j]$$

where $\rho_{ij} = \rho_{ji} \in \mathbb{R}$, $b_{ij} \in \mathbb{R}$, *and* $c_{jk} \in \mathbb{C}$ *is independent of* i.

Proof: Equation (9A) for the ith and jth components of $\Omega(X,\overline{X})X$ $(i \neq j)$ yields

$$a(X)\,X_i = \sum_{\mu,k,\ell} R_{\mu ik\ell}\, X_\mu\, X_k\, \overline{X}_\ell$$

$$a(X)\,X_j = \sum_{\mu,k,\ell} R_{\mu jk\ell}\, X_\mu\, X_k\, \overline{X}_\ell\,.$$

Multiplying the first equation by X_j and the second by X_i we get

$$X_j \sum_{\mu,k,\ell} R_{\mu ik\ell}\, X_\mu X_k \overline{X}_\ell = X_i \sum_{\mu,k,\ell} R_{\mu jk\ell}\, X_\mu X_k \overline{X}_\ell\,. \tag{14}$$

Comparing coefficients one sees that

$$R_{\mu ik\ell} = 0 \quad \text{unless} \quad k \text{ or } \mu = i \tag{15}$$

and so by (5),

$$R_{i\mu\ell k} = 0 \quad \text{unless} \quad k \text{ or } \mu = i\,. \tag{16}$$

Considering the other terms in (14) with the aid of (15) and (16) yields

$$R_{iiii} = R_{jjii} + R_{ijji} \tag{17}$$

$$R_{iik\ell} \quad \text{is independent of} \quad i \quad \text{for} \quad k \neq \ell. \tag{18}$$

Applying (5) again we get

$$R_{ijji} = R_{jiij} \quad \in \mathbb{R}. \tag{19}$$

Furthermore it is clear that (15), (17), and (18) are equivalent to (9A). This proves (13) in the case $n = 3$.

Now considering equation (9B) applied in the case $Y = e_j$, $X = e_i$, we get $b(e_i) = R_{jjii}$ and so

$$R_{jjii} \quad \text{is independent of} \quad j \quad \text{for} \quad j \neq i\,. \tag{20}$$

Equation (17) now implies that R_{ijji} is independent of $j \neq i$ and then (19) gives

$$R_{ijji} \quad i \neq j \quad \text{is independent of} \quad i \quad \text{and} \quad j\,. \tag{21}$$

Thus (12) is proven in the case $n \geq 4$.

There are no more relations among the components of the curvature for the following reason. If ϕ is a C^∞ real valued function on M, make the conformal change of metric $\widetilde{\langle,\rangle} = e^\phi \langle,\rangle$. The curvature form of $\widetilde{\langle,\rangle}$

is $\tilde{\Omega} = \Omega - \partial\bar{\partial}\phi I$ where I is the $n \times n$ identity matrix. Therefore $\tilde{\Omega}$ satisfies the conditions of Theorem 1 iff Ω does. For an arbitrary $p \in M$ and hermitian symmetric $n \times n$ matrix (C_{jk}), ϕ can be chosen so that $(\partial^2\phi/\partial z_j \partial \bar{z}_k)|p = (C_{jk})$.

Q.E.D.

Theorem 3: *A Kähler manifold* M *of dimension* ≥ 3 *has a Frenet frame for every holomorphic curve iff it has constant holomorphic sectional curvature.*

Proof: The Kähler identities $R_{kji\ell} = R_{ijk\ell} = R_{i\ell kj}$ applied to (15) through (18) yield

$$R_{ijji} = R_{jjii} = R_{jiij} = R_{iijj} \tag{22}$$

$$R_{iik\ell} = R_{kii\ell} = 0 \qquad k \neq \ell \tag{23}$$

$$R_{iiii} = 2R_{iijj} \qquad j \neq i \tag{24}$$

and so the components in (22) and (24) are independent of i and j. Therefore the curvature of M has the form

$$\Omega_{ij} = \rho\,[\omega_j \wedge \omega_i + \delta^i_j \sum_K \omega_k \wedge \bar{\omega}_k].$$

This is precisely the curvature form of a Kähler manifold of constant holomorphic sectional curvature 2ρ.

4. HOLOMORPHIC CURVES IN $\mathbb{P}^n$

The classical case of holomorphic curves in complex projective case will now be examined. Let $Z = (z_0, \ldots z_n) \in \mathbb{C}^{n+1} - \{0\}$ be a set of homogeneous coordinates for a point $[Z]$ in $\mathbb{P}^n$. Introduce the usual scalar product on $\mathbb{C}^{n+1}$, $\langle Z,W\rangle = \sum_{i=0}^{n} z_i \bar{w}_i$, and identify the unitary group $U(n+1)$ with the space of all unitary frames for $\mathbb{C}^{n+1}$. $\mathbb{P}^n$ can be regarded as the base space of the bundle

$$U(n+1) \xrightarrow{p_1} U(n+1)/U(n) \xrightarrow{p_2} U(n+1)/U(1) \times U(n) = \mathbb{P}^n$$

where p_1 takes a frame $Z_0, \ldots Z_n$ to its first vector Z_0 and p_2 is the Hopf map. The Maurer-Cartan forms of the unitary group are defined by

$$dZ_\alpha = \sum_\alpha \theta_{\alpha\beta} Z_\beta \qquad \alpha,\beta = 0 \text{ to } n \tag{25}$$

with

$$\Theta_{\alpha\beta} + \overline{\Theta}_{\beta\alpha} = 0 \tag{26}$$

and they satisfy the Maurer-Cartan equations

$$d\Theta_{\alpha\beta} = \sum_{\gamma} \Theta_{\alpha\gamma} \wedge \Theta_{\gamma\beta} . \tag{27}$$

In this setup, $T\mathbb{P}^n_{[Z]}$ is identified with $\{W \in \mathbb{C}^{n+1} : \langle W,Z \rangle = 0\}$ and so the fiber of the map P_1 over Z becomes the space of unitary frames for $T\mathbb{P}^n_{[Z]}$ where the hermitian metric on $\mathbb{P}^n$ is the Fubini-Study metric

$$ds^2 = \sum_{j=1}^{n} \Theta_{oi} \wedge \overline{\Theta}_{oi}.$$

The connection forms can now be read off from (27):

$$\omega_{ij} = \Theta_{ij} - \delta^i_j \Theta_{oo} \qquad 1 \le i,j \le n . \tag{28}$$

The curvature forms can be read off from the Maurer-Cartan equations:

$$\Omega_{ij} = \Theta_{oj} \wedge \overline{\Theta}_{oi} + \delta^i_j \sum_{k=1}^{n} \Theta_{ok} \wedge \overline{\Theta}_{ok} . \tag{29}$$

Therefore $\mathbb{P}^n$ has constant holomorphic sectional curvature equal to 2.

Let $f: X \to \mathbb{P}^n$ be a holomorphic curve. Using a local coordinate ζ on X and homogeneous coordinates on $\mathbb{P}^n$, f is given by an $n + 1$ dimensional holomorphic vector function $Z(\zeta) = (z_o(\zeta),\ldots,z_n(\zeta))$. We assume that $f(X)$ does not belong to any hyperplane, i.e., $Z(\zeta) \wedge Z'(\zeta) \wedge \ldots \wedge Z^{(n)}(\zeta) \not\equiv 0$. The classical way to define a unitary frame along f is by requiring that $Z_o(\zeta) \equiv \|Z(\zeta)\|^{-1} Z(\zeta)$, $Z_1(\zeta),\ldots Z_k(\zeta)$ form a unitary basis for the vector space spanned by $Z(\zeta)$, $Z'(\zeta),\ldots Z^{(k)}(\zeta)$ (the kth osculating space at ζ) for $k = 1$ to n and for all $\zeta \in X$. We will show that this is the Frenet frame defined in section 3. Equations (25) restricted to f simplify to give

$$dZ_i = \Theta_{ii-1} Z_{i-1} + \Theta_{ii} Z_i + \Theta_{ii+1} Z_{i+1} \tag{30}$$

for Θ_{ii+1} a multiple of $d\zeta$. This along with (28) implies that $\omega_{ij} = 0$ unless $|i - j| \le 1$ and that ω_{ii+1} is a multiple of $d\zeta$. Therefore

$$DZ_i = \omega_{ii-1} Z_{i-1} + \omega_{ii} Z_i + \omega_{ii+1} Z_{i+1}$$

and so $Z_1,\ldots Z_n$ is a Frenet frame.

5. COVARIANT DERIVATIVES OF A HOLOMORPHIC SECTION ALONG A CURVE

Let $L \to M$ be a holomorphic line bundle and s a holomorphic section. Assume that the holomorphic curve $f: X \to M$ has a Frenet frame. In value distribution theory one is often interested in how the image of f intersects the zero set of s in M. Thus it is natural to study the section s restricted to f (or, more precisely, the holomorphic section $s \circ f$ of the bundle $f^*(L) \to X$). We will define successive covariant derivatives of s along f. The definition will not work for all sections because the existence of the covariant derivative will imply conditions on s. We will then show how our definition applies to the classical case of the hyperplane bundle H over $\mathbb{P}^n$ and examine its failure in the case of higher powers (≥ 2) of H.

Let $E_k \to X$ be the holomorphic sub-bundle of f^*TM consisting of the kth osculating spaces of f, i.e., the spaces spanned by $e_1, \ldots e_k$, and let $F_k \to X$ be the holomorphic line bundle defined by $F_k \equiv E_k^*/E_{k-1}^*$ $k = 1$ to n (E_o defined to be trivial). The fact that E_k is a holomorphic bundle is equivalent to the existence of a Frenet frame along X. The hermitian metric and connection on TM induce those on F_k; the connection form is $-\omega_{kk}$. The connection form on $f^*L \otimes F_k$ is therefore $\phi - \omega_{kk}$. The kth covariant derivative s_k of s is a C^∞ section of $f^*L \otimes F_k$ defined recursively by

$$Ds_k = ds_k + s_k(\phi - \omega_{kk}) = s_{k+1}\,\omega_{kk+1} + t_{k+1}\,d\bar{\zeta} \tag{31}$$

$$0 \leq k \leq n-1, \; s_o \equiv s, \; t_1 = 0,$$

$$\omega_{o1} \equiv \omega_1 \text{ and } \omega_{oo} \equiv 0.$$

Since ω_{kk+1} may have isolated zeroes (we consider only curves for which none of the ω_{kk+1} vanish identically) the existence of s_{k+1} implies an assumption of a high order zero of the $d\zeta$ component of Ds_k at all the stationary points of f of order k.

Now consider the hyperplane bundle $H \to \mathbb{P}^n$. The connection form on H is $\phi = -2\,\partial \log \|z\|$ and from [4], p. 77, we have

$$\Theta_{oo} = (\partial - \bar{\partial}) \log \|z\| .$$

A holomorphic section of H restricted to f has the form $s(\zeta) = \langle Z(\zeta), A\rangle$ (the zero set of s is the hyperplane in $\mathbb{P}^n$ given by

$0 = \sum_{\alpha=0}^{n} \overline{A}_\alpha z_\alpha)$. Define the functions $w_\alpha(\zeta), \alpha = 0$ to n, by $A = \sum_{\alpha=0}^{n} w_\alpha(\zeta) Z_\alpha(\zeta)$. Since A is constant we get

$$d\overline{w}_\alpha = \sum_\beta \Theta_{\alpha\beta} \overline{w}_\beta . \tag{32}$$

We will show that the covariant derivative s_k is related to w_k by

$$s_k = ||Z|| \overline{w}_k . \tag{33}$$

Let $\tilde{s}_k \equiv ||Z||\overline{w}_k$ $k = 0$ to n. Then

$$d \log \tilde{s}_k = \partial \log ||Z|| + \overline{\partial} \log ||Z|| + d \log \overline{w}_k$$

$$= \partial \log ||Z|| + \overline{\partial} \log ||Z|| + \frac{||Z||}{\tilde{s}_k} \sum_{\mu=0}^{n} \Theta_{k\mu} \overline{w}_\mu$$

$$d\tilde{s}_k = - \Theta_{oo} \tilde{s}_k - \phi \tilde{s}_k + \sum_{\mu=0}^{n} \Theta_{k\mu} \tilde{s}_\mu$$

$$d\tilde{s}_k + (\phi - \omega_{kk})\tilde{s}_k = \tilde{s}_{k+1} \omega_{kk+1} + t_{k+1} d\overline{\zeta} .$$

This is exactly the equation (31) defining s_k.

The functions w_k play an important role in the value distribution theory of holomorphic curves in $\mathbb{P}^n$. Formula (33) gives them a new geometric interpretation. In the classical theory introduce

$$\phi_k \equiv \sum_{\nu=0}^{k} |w_\nu|^2 \qquad 0 \le k \le n.$$

The fundamental inequalities of Ahlfors are integral inequalities involving the ϕ_k. Their original proof by Ahlfors made use of integral geometry, cf [1]. Recently Cowen and Griffiths [5] gave a proof based on the "method of negative curvature" where the basic analytic step is to find formulas for $\partial\overline{\partial} \log \phi_k$ and $\partial\overline{\partial} \log \log \phi_k$.

We will now examine the higher tensor powers of the hyperplane bundle. The line bundle $H^m \to \mathbb{P}^n$ $(m \ge 2)$ restricted to the holomorphic curve f has connection form $\phi = - m \partial \log ||Z||^2$ and a holomorphic section of H^m along f is given by $s(\zeta) = P(Z(\zeta))$, for P a homogeneous polynomial of degree m. Let $Z_o \equiv ||Z||^{-1} Z$ and $Z_1, \ldots Z_n$ be the Frenet frame along f. Replacing the constant vector $\overline{A}$ in the $m = 1$ case we have the gradient of P, denoted $\frac{\partial P}{\partial z} = \left(\frac{\partial P}{\partial z_o}, \ldots \frac{\partial P}{\partial z_n}\right)$. Define the functions $w_k(\zeta)$ by $\overline{\frac{\partial P}{\partial z}}(Z(\zeta)) = \sum_{k=0}^{n} w_k(\zeta) Z_k(\zeta)$.

Define the hessian of P to be the n + 1 by n + 1 matrix given by

$$\text{Hess } P \equiv \left(\frac{\partial^2 P}{\partial z_i \partial z_j} \right) .$$

Using Euler's formula we get

$$s = \frac{1}{m} < Z, \overline{\frac{\partial P}{\partial z}} > \tag{34}$$

$$\frac{\partial P}{\partial z} = \frac{1}{m-1} Z \cdot \text{Hess } P = \frac{\|Z\|}{m-1} Z_o \cdot \text{Hess } P \tag{35}$$

where $Z \cdot \text{Hess } P$ means the product of the 1 by n + 1 matrix Z with the n + 1 by n + 1 matrix Hess P to produce a 1 by n + 1 matrix. We will also use the formula

$$dZ = d(\|Z\| Z_o) = d(\|Z\|) Z_o + \|Z\| (\theta_{oo} Z_o + \theta_{o1} Z_1). \tag{36}$$

The first two covariant derivatives of s will now be computed.

$$\begin{aligned} Ds &= ds + \phi s \\ &= \langle dZ, \overline{\frac{\partial P}{\partial z}} \rangle - m \frac{<dZ, Z>}{\|Z\|^2} \frac{1}{m} < Z, \overline{\frac{\partial P}{\partial z}} > \\ &= d(\|Z\|) \langle Z_o, \overline{\frac{\partial P}{\partial z}} \rangle \\ &\quad + \|Z\| \theta_{oo} \langle Z_o, \overline{\frac{\partial P}{\partial z}} \rangle + \|Z\| \theta_{o1} \langle Z_1, \overline{\frac{\partial P}{\partial z}} \rangle \\ &\quad - (d(\|Z\|) + \|Z\| \theta_{oo}) \langle Z_o, \overline{\frac{\partial P}{\partial z}} \rangle \\ &= \|Z\| \overline{w}_1 \theta_{o1} \end{aligned}$$

so that $s_1 = \|Z\| \overline{w}_1 = \|Z\| \langle Z_1, \overline{\frac{\partial P}{\partial z}} \rangle$ as in the $m = 1$ case.

$$\begin{aligned} Ds_1 &= ds_1 + (\phi - \omega_{11}) s_1 \\ &= d(\|Z\|) \langle Z_1, \overline{\frac{\partial P}{\partial z}} \rangle \\ &\quad + \|Z\| \langle \theta_{1o} Z_o + \theta_{11} Z_1 + \theta_{12} Z_2, \overline{\frac{\partial P}{\partial z}} \rangle \\ &\quad + \|Z\| \langle Z_1, \overline{dZ} \cdot \overline{\text{Hess } P} \rangle \\ &\quad + \|Z\| \langle Z_1, \overline{\frac{\partial P}{\partial z}} \rangle \left(- m \frac{\langle dZ, Z \rangle}{\|Z\|^2} - \theta_{11} + \theta_{oo} \right) \end{aligned}$$

which, using (34), (35), and (36), reduces to

$$Ds_1 = - mP(Z) \overline{\theta}_{o1} + \|Z\| w_2 \theta_{12} + \langle Z_1 \cdot \text{Hess } P, \overline{Z}_1 \rangle \theta_{o1} .$$

Therefore the $d\zeta$ component of Ds_1 is not a multiple of θ_{12} and so s_2 cannot be defined according to the procedure of Section 4. The coefficient of the extra term, $\langle Z_1 \cdot \text{Hess } P, \bar{Z}_1 \rangle$, has a geometric significance: at points where f is tangent to the projective hypersurface $V(P)$ defined by P, i.e., $P(Z) = 0$ and $\langle Z_1, \overline{\frac{\partial P}{\partial z}} \rangle = 0$, the second fundamental form of $V(P) \subset \mathbb{P}^n$ in the direction Z_1 is given by (cf. [6])

$$\frac{\langle Z_1 \cdot \text{Hess } P, \bar{Z}_1 \rangle}{2\left\|\frac{\partial P}{\partial z}\right\|} .$$

REFERENCES

1. Ahlfors, L. V., *The theory of meromorphic curves*, Finska Vetenskaps-Societeten, Helsingfors: Acta Soc. Sci. Fenn. Ser. A, vol. 3, no. 4 (1941), 3-31.

2. Chern, S.-S., *Complex Manifolds Without Potential Theory*, van Nostrand, 1967.

3. __________, *On the minimal immersions of the two sphere in a space of constant curvature*, Problems in Analysis, in Honor of S. Bochner, Princeton 1970, 27-40.

4. __________, *Holomorphic curves in the plane*, Differential Geometry in Honor of K. Yano, Tokyo 1972, 73-94.

5. Cowen, M., and P. Griffiths, *Holomorphic curves and metrics of negative curvature*, to appear.

6. Vitter, A., *On the curvature of complex hypersurfaces*, to appear in Indiana University Mathematics Journal.

7. Weyl, H., and J. Weyl, *Meromorphic Functions and Analytic Curves*, Princeton, 1943.

8. Wu, H., *The equidistribution theory of holomorphic curves*, Annals of Math. Studies, Princeton, 1970.

THE KOBAYASHI METRIC ON $\mathbb{P}_n - (2^n + 1)$ HYPERPLANES

Michael Cowen

Princeton University

INTRODUCTION

The theory of hyperbolic manifolds [13] contains many beautiful results but, as yet, few methods for determining when a given complex manifold is hyperbolic. This paper demonstrates how the "method of negative curvature" (c.f. [8] for discussion of this method) can be used to prove the hyperbolicity of $\mathbb{P}_n - A_1 \cup \ldots \cup A_m$, where m is $2^n + 1$ and the A_i are hyperplanes in general position in the complex projective space $\mathbb{P}_n$. An estimate of the infinitesimal Kobayashi metric [14] on $\mathbb{P}_n - A_1 \cup \ldots \cup A_m$ is given which, although not sharp, does give some quantitative idea of the behavior of the Kobayashi metric.

The Kobayashi metric on projective space minus hyperplanes has a long history, stretching back to one of the most striking results in complex variables:

Picard's Theorem: *Let* $f\colon \mathbb{C} \to \mathbb{P}_1 - A_1 \cup A_2 \cup A_3$ *be holomorphic, then* f *is a constant.**

In 1896 Borel extended Picard's result to n dimensions:

Borel's Theorem [4]: *Let* $f\colon \mathbb{C} \to \mathbb{P}_n - A_1 \cup \ldots \cup A_{n+2}$ *be holomorphic, then* f *lies in a hyperplane.*

Taking $2n + 1$ hyperplanes $n + 2$ at a time we can use Borel's Theorem to show [10]:

* Note that a hyperplane in $\mathbb{P}_1$ is just a point, and "general position" reduces to "distinct".

Let $f\colon \mathbb{C} \to \mathbb{P}_n - A_1 \cup \ldots \cup A_{2n+1}$ *be holomorphic, then* f *is constant.*

A natural question arises from these Picard-type results: Let $f\colon \Delta_r \to \mathbb{P}_n - A_1 \cup \ldots \cup A_{2n+1}$ be a non-constant holomorphic map, where $\Delta_r = \{|z| < r\}$ is the disc of radius r. How large can r be?

In the case $n = 1$, the theorems of Schottky and Landau give sharp estimates on an upper bound for r in terms of $f(0)$ and $f'(0)$ (see for example [11]). In general, A. Bloch (1926) [3] showed that if $f\colon \Delta_r \to \mathbb{P}_n - A_1 \cup \ldots \cup A_{n+2}$, then r is bounded in terms of $f(0)$ and $f'(0)$ as long as $f(0)$ does not belong to certain "exceptional sets"; there is no estimate on the bound for r however.

The "exceptional sets" arise as follows, for example in $\mathbb{P}_2$. Let $[z_0:z_1:z_2]$ be homogeneous coordinates in $\mathbb{P}_2$, and put $A_i = \{z_i = 0\}$, $i = 0,1,2,$ and $A_3 = \{z_0 + z_1 + z_2 = 0\}$. Putting the hyperplane A_0 at ∞, we have $\mathbb{P}_2 - A_0 \cup \ldots \cup A_3 = \mathbb{C}^2 - \{z_1 = 0, z_2 = 0, z_1 + z_2 = -1\}$ that is:

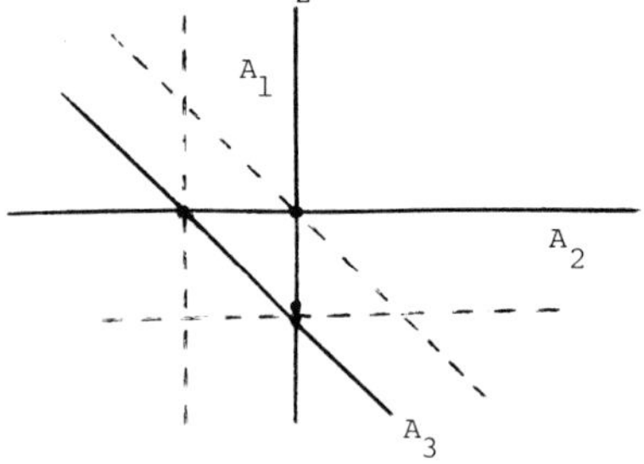

where the dotted lines are the exceptional sets, since we can clearly have a non-constant map, for example, of $\Delta_r \to \{z_1 = -1, z_2 \neq 0\} \subset \mathbb{P}_2 - A_0 \cup \ldots \cup A_3$, with $r = \infty$. Note that a map of $\mathbb{C}$ into any line other than the three exceptional ones would be a map into $\mathbb{C}$ - 2 points = $\mathbb{P}_1$ - 3 points, so the map would be constant. Note also that by Borel's Theorem a map of $\mathbb{C}$ into $\mathbb{P}_2 - A_0 \cup \ldots \cup A_3$ must lie in a line.

It is now clear, when we take the hyperplanes 4 at a time, that r is bounded for $f\colon \Delta_r \to \mathbb{P}_2 - A_1 \cup \ldots \cup A_5$, since the exceptional sets have empty intersection. Indeed, H. Cartan (1928) [6] (see also [9]) showed that r is bounded in terms of $f(0)$ and $f'(0)$ for $f\colon \Delta_r \to \mathbb{P}_n - A_1 \cup \ldots \cup A_{2n+1}$.

We can formulate Cartan's result in terms of the Kobayashi metric. Let M be a complex manifold, $x \in M$, $\xi \neq 0 \in T(M)$. Then the infinitesimal Kobayashi metric $||\xi||_x^2$ is defined as follows [14]:

$$||\xi||_x^2 = \inf 1/r^2, \qquad (0.1)$$

where r ranges over all positive real numbers for which there is $f\colon \Delta_r \to M$ holomorphic with $f(0) = x$, $f'(0) = \xi$. The Kobayashi pseudometric d_M [13] is the integrated form of $|| \; ||^2$, i.e. (see [14]) $d_M(p,q) = \inf \int_a^b \sqrt{||\dot{x}(t)||^2_{x(t)}}\, dt$, where the inf is taken over all piecewise differentiable curves $x(t)$ joining p to q. M is *hyperbolic at* x if there is a neighborhood U of x and a hermitian metric ds^2 on U such that $|| \; ||_y^2 \geq ds_y^2$ for all $y \in U$. M is *hyperbolic* if it is hyperbolic at each point. This is equivalent to d_M being a metric [14].

Cartan's result can thus be formulated as follows: let $M = \mathbb{P}_n - A_1 \cup \ldots \cup A_{2n+1}$, then there exists a constant c_n such that

$$|| \; ||_x^2 \geq c_n ds_x^2 \quad \text{for all} \quad x \in M. \qquad (0.2)$$

where ds^2 is the Fubini-Study metric on $\mathbb{P}_n$, i.e. not only is M hyperbolic, but it is *hyperbolically imbedded* in $\mathbb{P}_n$, which implies M is complete hyperbolic (i.e. complete in the Kobayashi metric) [12].

In Section 1 the Ahlfors Lemma will be used to prove Borel's Theorem by means of constructing metrics of negative curvature. In Section 2 an estimate on $|| \; ||$ for $\mathbb{P}_2 - A_1 \cup \ldots \cup A_5$ is derived, again by using metrics of negative curvature. In Section 3, the method of Section 2 is used for $\mathbb{P}_n - (2^n + 1)$ hyperplanes; a proof of Cartan's result by curvature methods for $\mathbb{P}_n - (2n + 1)$ hyperplanes is not yet known.

The author would like to thank the Department of Mathematics at Tulane University for the opportunity to visit with them during 1972-73, and gratefully acknowledges the many helpful suggestions and discussions with P. Griffiths and A. Vitter.

1. BOREL'S THEOREM

Let $\tau = \frac{i}{2\pi} h(\xi) d\xi d\bar{\xi}$ be a pseudo-hermitian* metric on Δ_r, $d = \partial + \bar{\partial}$ the usual splitting of the exterior derivative, and $d^c = \frac{i}{4\pi}(\bar{\partial} - \partial)$ the "twisted" real differential operator. Define

---

* The "pseudo" refers to the fact that $h(\xi) > 0$ except at isolated points.

$$\text{Ric}\ \tau = dd^c \log h. \tag{1.1}$$

Note that the Gaussian curvature of τ is given by

$$K(\tau) = -\frac{1}{h}\frac{\partial^2 \log h}{\partial\xi\partial\bar{\xi}}$$

so that

$$\text{Ric}\ \tau = -K(\tau)\tau. \tag{1.2}$$

The Poincaré metric on Δ_r is given by

$$ds_r^2 = i/\pi \frac{r^2}{(r^2-|\xi|^2)^2} d\xi d\bar{\xi} \tag{1.3}$$

and satisfies:

$$\text{Ric}\ ds_r^2 = ds_r^2. \tag{1.4}$$

Ahlfors' Lemma asserts that the Poincaré metric is maximal with respect to (1.4), where we use (1.3) to reformulate the result in terms of Ric:

Ahlfors' Lemma [1]: *Let* τ *be a pseudo-metric on* Δ_r *such that* $\text{Ric}\ \tau \geq \tau$. *Then* $\tau \leq ds_r^2$. *In particular, at* 0:

$$h(0) \leq \frac{2}{r^2}. \tag{1.5}$$

The general philosophy of this paper is to construct pseudo-metrics on Δ_r and use (1.5) to give an upper bound on r.

Note that by (1.3) $ds_r^2 \to 0$ as $r \to \infty$, so we have the following corollary to Ahlfors' Lemma:

Corollary 1.6 : *Let* τ *be a pseudo-metric on* $\mathbb{C}$ *such that* $\text{Ric}\ \tau \geq \tau$. *Then* $\tau = 0$.

In order to prove Borel's Theorem by constructing suitable metrics we must first discuss holomorphic curves in $\mathbb{P}_n$. A holomorphic mapping $f: \Delta_r \to \mathbb{P}_n$ is called a *holomorphic curve* and can be represented in homogeneous coordinates by

$$Z(\xi) = [Z_0(\xi) : \ldots : Z_n(\xi)],$$

where Z maps Δ_r into $\mathbb{C}^{n+1} - \{0\}$ holomorphically (see [2], [15], or [16] for details on holomorphic curves). The k-th derived curve $\Lambda_k: \Delta_r \to Gr(k+1, n+1)$, the Grassmann manifold of $(k+1)$-planes in $\mathbb{C}^{n+1}$ is given by:

$$\Lambda_k = Z^{(0)} \wedge \ldots \wedge Z^{(k)}. \tag{1.7}$$

Then $\Lambda_1(\xi)$ is the tangent plane of the curve, $\Lambda_2(\xi)$ the osculating space, etc. The curve f is *non-degenerate* if $\Lambda_n \not\equiv 0$.

The hermitian inner product on $\mathbb{C}^{n+1}$ is given by

$$<z,w> = \sum_{i=0}^{n} z_i \overline{w}_i .$$

If $A \in \mathbb{C}^{n+1} - \{0\}$, then A defines a hyperplane in $\mathbb{C}^{n+1}$:

$$\{z \in \mathbb{C}^{n+1} \mid <z,A> = 0\}$$

which will also be denoted by A. A hyperplane A in $\mathbb{P}_n$ is just the image under the residual map of a hyperplane A in $\mathbb{C}^{n+1}$. The curve f is non-degenerate iff the image of f does not lie in a hyperplane.

The (1,1) form $dd^c \log|z|^2$ on $\mathbb{C}^{n+1} - \{0\}$ is invariant under $z \to \lambda z$, $\lambda \in \mathbb{C} - \{0\}$, and induces a hermitian metric ds^2 on $\mathbb{P}_n$ called the *Fubini-Study metric*, which has been normalized so that the volume of $\mathbb{P}_n$ is 1. If f is non-degenerate then for $e_0, \ldots, e_n$ an orthonormal basis for $\mathbb{C}^{n+1}$,

$$\Lambda_k = \sum_{i_0 < \ldots < i_k} \alpha_{i_0 \ldots i_k} e_{i_0} \wedge \ldots \wedge e_{i_k}$$

and the α's define a curve in $\mathbb{P}_{\binom{n+1}{k+1}-1}$ via the *Plücker imbedding* of $Gr(k+1, n+1)$ in $\mathbb{P}_{\binom{n+1}{k+1}-1}$. We pull back the Fubini-Study metric on on $\mathbb{P}_{\binom{n+1}{k+1}-1}$ and define the non-negative $C^\infty(1,1)$-form

$$\Omega_k = dd^c \log|\Lambda_k|^2. \tag{1.8}$$

The essential ingredient in the Second Main Theorem of Value Distribution Theory is that

$$\Omega_k = \frac{i}{2\pi} \frac{|\Lambda_{k-1}|^2 |\Lambda_{k+1}|^2}{|\Lambda_k|^4} d\xi d\overline{\xi} ; \tag{1.9}$$

c.f. [7] or [8] for the proof. This clearly implies by (1.8) that

$$\text{Ric } \Omega_k = \Omega_{k-1} + \Omega_{k+1} - 2\Omega_k . \tag{1.10}$$

The hermitian inner product $< , >$ on $\mathbb{C}^{n+1}$ extends to $\Lambda^{k+1}\mathbb{C}^{n+1}$ by

$$<a_o \wedge \ldots \wedge a_k , b_o \wedge \ldots \wedge b_k> = \det(<a_i , b_j>).$$

If B is a $(k+1)$-vector in $\mathbb{C}^{n+1}$, $A \in \mathbb{C}^{n+1}$ then the *interior product* $\langle B,A\rangle$ is the unique k-vector such that

$$\langle\langle B,A\rangle,C\rangle = \langle B,A \wedge C\rangle$$

for all k-vectors C. We denote $|\langle B,A\rangle|$ by $|B,A|$.

Let $A \in \mathbb{C}^{n+1}$ be a unit vector and put

$$\phi_k(A) = |\Lambda_k, A|^2 / |\Lambda_k|^2 \tag{1.11}$$

$$\Omega_k(A) = dd^c \log|\Lambda_k, A|^2. \tag{1.12}$$

Note that $\phi_{i+1}(A) \geq \phi_i(A)$, since if $Z_o,\ldots,Z_{i+1}$ are orthonormal such that $Z_o \wedge\ldots\wedge Z_i = \Lambda_i / |\Lambda_i|$ and $Z_o \wedge\ldots\wedge Z_{i+1} = \Lambda_{i+1} / |\Lambda_{i+1}|$ then $\phi_i = \sum_{j=0}^{i} |Z_j, A|^2$ and $\phi_{i+1} = \sum_{j=0}^{i+1} |Z_j, A|^2$. It is clear that the curve has contact of order k at ξ with the hyperplane A iff $\phi_k(A)(\xi) = 0$ and that $\Omega_k(A)$ is a C^∞ (1,1)-form (since $\langle\Lambda_k, A\rangle \not\equiv 0$, we can write $\langle\Lambda_k, A\rangle = (\xi - \xi_o)^k h(\xi)$ where $|h(\xi_o)| \neq 0$ and $dd^c \log|\Lambda_k, A|^2 =$ $= dd^c \log|h(\xi)|^2$). We need the following lemma whose proof is given in [8]:

Lemma 1.13 : $\Omega_k(A) = \dfrac{\phi_{k-1}(A)\phi_{k+1}(A)}{\phi_k^2(A)}\Omega_k$, $k \geq 1$, $\Omega_o(A) = 0$.

Let $\Delta^* = \{0 < |\xi| < 1\}$ be the punctured disc. The universal cover of Δ^* is Δ_1 and the Poincaré metric ds_1^2 induces the metric

$$ds_*^2 = \frac{i}{\pi} \frac{d\xi \wedge d\bar{\xi}}{|\xi|^2\left(\log \dfrac{1}{|\xi|^2}\right)^2} \tag{1.14}$$

on Δ^* which satisfies

$$\text{Ric } ds_*^2 = ds_*^2.$$

Carlson and Griffiths [5] point out that if we wish to construct a metric of negative curvature on $\mathbb{P}_n - \cup A_\nu$, for example, then locally, near the missing hyperplane, it should be modelled after ds_*^2. Thus in our case, the metric should have a term $\dfrac{1}{\phi_o(A_\nu)\left(\log \dfrac{1}{\phi_o(A_\nu)}\right)^2}$ in it. In order to compute Ric of such a metric we use the main inequality of [8]:

Lemma 1.15 : *Let* $\mu > e = 2.7\ldots$ *be a constant. Then*

$$dd^c \log \left[\frac{1}{\log \frac{\mu}{\phi_k(A)}}\right]^2 \geq \frac{2\phi_{k+1}(A)}{\phi_k(A)\left[\log \frac{\mu}{\phi_k(A)}\right]^2}\,\Omega_k - \varepsilon\Omega_k$$

where $\varepsilon = \varepsilon(\mu)$ *is a constant such that* $\varepsilon \to 0$ *as* $\mu \to \infty$.

Let $A_1,\ldots,A_q \in \mathbb{C}^{n+1}$ define hyperplanes in general position in $\mathbb{P}_n$, $q \geq n+2$; where general position means that no $n+1$ of the A_ν are linearly dependent. Let $f\colon \Delta_r \to \mathbb{P}_n - A_1 \cup\ldots\cup A_q$ be a non-degenerate holomorphic curve, and define metrics σ_i on Δ_r by

$$\sigma_i = c_i \prod_{\nu=1}^{q} \left(\frac{\phi_{i+1}(A_\nu)}{\phi_i(A_\nu)\left[\log \frac{\mu}{\phi_i(A_\nu)}\right]^2}\right)^{\alpha_i} \Omega_i \tag{1.16}$$

where the α_i are constants satisfying

$$\frac{1}{n-i} \geq \alpha_i \geq 1/q\,, \quad i = 0,\ldots,n-1, \tag{1.17}$$

and the c_i are also constants.

Remarks: 1) We would like to have $\operatorname{Ric}\sigma_i \geq \sigma_i$, which is not true; the log term was included so that $\operatorname{Ric}\sigma_i$ at least has σ_i appearing as a term by Lemma (1.15). 2) For the proof of the Borel Theorem we may think of α_i as $1/(n-i)$, though for technical reasons the α_i are perturbed slightly. The α_i are needed to prevent σ_i from blowing up too quickly as the curve approaches intersections of the hyperplanes to various degrees of contact, e.g. in $\mathbb{P}_3$ the curve can approach the intersection of 3 hyperplanes (thus the 1/3 exponent in σ_o) or approach being tangent to the intersection of 2 hyperplanes (thus the 1/2 exponent in σ_1).

We make one metric from the σ_i by taking a geometric mean, i.e. we put

$$\omega = c \prod_{i=0}^{n-1} \sigma_i^{\beta/\alpha_i} \tag{1.18}$$

where

$$\beta = 1/\Sigma\,\alpha_i^{-1}, \tag{1.19}$$

and c is a constant.

Since the product telescopes and $\phi_n(A) \equiv 1$ we have

$$\omega = c\left(\prod_{\nu} \frac{1}{\phi_0(A_\nu)}\right)^{\beta} \prod_i \left(\prod_\nu \frac{1}{\left[\log \frac{\mu}{\phi_i(A_\nu)}\right]^{2\beta}}\,\Omega_i^{\beta/\alpha_i}\right) \tag{1.20}$$

Proposition 1.21 : *For* $q \geq n+2$, *the constants* μ, α_i , *and* c *can be chosen (independent of the curve and of* Δ_r) *so that for* ω *defined by* (1.18),

$$\operatorname{Ric} \omega \geq \omega .$$

Proof:

$$\operatorname{Ric} \omega = -\beta \sum_{\nu=1}^{q} dd^c \log \phi_o(A_\nu)$$

$$+\beta \sum_i \sum_\nu dd^c \log \left[\frac{1}{\log \frac{\mu}{\phi_i(A_\nu)}}\right]^2$$

$$+\beta \sum_i 1/\alpha_i \operatorname{Ric} \Omega_i$$

Now $dd^c \log \phi_o(A_\nu) = -\Omega_o$ by (1.11) and (1.8). Using Lemma (1.15) and (1.10) we have

$$\operatorname{Ric} \omega \geq \tag{1.22}$$

$$\beta\left(q\Omega_o + 2\sum_i \sum_\nu \frac{\phi_{i+1}(A_\nu)\Omega_i}{\phi_i(A_\nu)\left[\log \frac{\mu}{\phi_i(A_\nu)}\right]^2} - q\varepsilon \Sigma \Omega_i + \sum_{i=0}^{n-1} 1/\alpha_i \{\Omega_{i+1} - 2\Omega_i + \Omega_{i-1}\}\right),$$

where we have put $\Omega_{-1} = \Omega_n = 0$.

We now need the following easy lemma whose proof can be found in [8]:

Lemma 1.23 (Sum into Product): *If* $A_1,\ldots,A_q$ *are* $q \geq n+2$ *hyperplanes in general position and* $1/q \leq \alpha_i \leq 1/(n-i)$, *then*

$$\sum_\nu \frac{\phi_{i+1}(A_\nu)}{\phi_i(A_\nu)\left[\log \frac{\mu}{\phi_i(A_\nu)}\right]^2} \geq c_i \left[\prod_{\nu=1}^{q} \frac{\phi_{i+1}(A_\nu)}{\phi_i(A_\nu)\left[\log \frac{\mu}{\phi_i(A_\nu)}\right]^2}\right]^{\alpha_i}$$

for c_i *a universal constant (depending, continuously, only on* α_1).

Since

$$2\beta \sum_i \sum_\nu \frac{\phi_{i+1}(A_\nu)}{\phi_i(A_\nu)\left[\log \frac{\mu}{\phi_i(A_\nu)}\right]^2} \Omega_i$$

$$\geq 2\beta \sum_i c_i \sigma_i \tag{1.24}$$

$$\geq \omega ,$$

by (1.23) and the inequality on arithmetic and geometric means, we have that (1.22) yields:

$$\operatorname{Ric}\omega \geq \beta\left\{q\,\Omega_0 + \sum_i \left({}^1\!/\alpha_{i-1} - {}^2\!/\alpha_i + {}^1\!/\alpha_{i+1} - \varepsilon q\right)\Omega_i\right\} \tag{1.25}$$

$$+ \omega, \quad \text{where} \quad {}^1\!/\alpha_{-1} = {}^1\!/\alpha_n = 0.$$

Since the Ω_i are ≥ 0 and $\varepsilon \to 0$ as $\mu \to \infty$, we will be done if we can choose the α_i so that

$$\begin{cases} q - {}^2\!/\alpha_0 + {}^1\!/\alpha_1 > 0 \\ {}^1\!/\alpha_{i-1} - {}^2\!/\alpha_i + {}^1\!/\alpha_{i+1} > 0 \\ {}^1\!/\alpha_{n-2} - {}^2\!/\alpha_{n-1} > 0 \end{cases} \tag{1.26}$$

and such that (1.17) is satisfied.

Note that in order to solve (1.26) we must have $q \geq n + 2$, since if (1.26) holds then:

$$\begin{aligned} 0 &< n\left(q - {}^2\!/\alpha_0 + {}^1\!/\alpha_1\right) + (n - 1)\left({}^1\!/\alpha_0 - {}^2\!/\alpha_1 + {}^1\!/\alpha_2\right) + \cdots \\ &\quad + 2\left({}^1\!/\alpha_{n-3} - {}^2\!/\alpha_{n-2} + {}^1\!/\alpha_{n-1}\right) \\ &\quad + \left({}^1\!/\alpha_{n-2} - {}^2\!/\alpha_{n-1}\right) \\ &= nq - (n + 1)/\alpha_o , \end{aligned}$$

i. e.,

$$\begin{aligned} q &> {}^{n+1}\!/n\alpha_0 \\ &> n + 1 \end{aligned}$$

by (1.17).

Conversely, when $q \geq n + 2$, (1.26) can be solved, for example by putting

$$\alpha_i = 1/\left(n-i + (n-i-1)^2\,\delta\right) \tag{1.27}$$

for $\delta > 0$ small enough, since

$$\begin{aligned} {}^1\!/\alpha_{i-1} - {}^2\!/\alpha_i + {}^1\!/\alpha_{i+1} &= 2\delta \\ {}^1\!/\alpha_{n-2} - {}^2\!/\alpha_{n-1} &= \delta \\ q - {}^2\!/\alpha_0 + {}^1\!/\alpha_1 &= q - (n + 1) + (2 - n^2)\delta. \end{aligned}$$

Q. E. D.

To conclude this section we note that (1.5) of the Ahlfors Lemma applied to the metric ω of Proposition (1.21) implies:

Corollary 1.28 (Borel's Theorem in finite terms) *If* $f\colon \Delta_1 \to \mathbb{P}_n - A_1 \cup \ldots \cup A_q$ *is non-degenerate,* $q \geq n+2$, *the* A_ν *in general position, then* r *is bounded above in terms of* $f(0), \Lambda_1(0), \ldots, \Lambda_{n-1}(0)$.

Borel's Theorem now follows trivially from Corollary (1.6) applied to the ω of Proposition (1.21), since $\omega \equiv 0$ is impossible if $f\colon \mathbb{C} \to \mathbb{P}_n - A_1 \cup \ldots \cup A_q$ is non-degenerate.

2. HEURISTIC REMARKS AND $\mathbb{P}_2$ - 5 HYPERPLANES

We wish to estimate from below the infinitesmal Kobayashi metric on $\mathbb{P}_2 - A_1 \cup \ldots \cup A_5$, the A_ν in general position. That is to say, if $f\colon \Delta_r \to \mathbb{P}_2 - A_1 \cup \ldots \cup A_5$ is a non-constant (possibly degenerate) holomorphic map we want to bound $\frac{1}{r^2}$ away from zero in terms of $f(0)$ and $f'(0)$ where $f'(0)$ is assumed non-zero. In Corollary (1.28) we derived a lower bound even on $\mathbb{P}_2$ - 4 hyperplanes, but the estimate involves $\Lambda_2(0)$, i.e. involves $f''(0)$, and there is no *a priori* reason for $\Lambda_2(0)$ to be non-zero. Indeed, the metric ω is identically zero if f is degenerate. Thus the metric ω, via Corollary (1.28), gives the desired estimate if $\Lambda_2(0)$ is "large" but not if it is "small."

On the other hand, consider the metric σ_0 defined by (1.16), where $1/2 \geq \alpha_0 \geq 1/5$. Since

$$dd^c \log \phi_1(A_\nu) = \Omega_1(A_\nu) - \Omega_1$$

by (1.11), (1.12) and (1.8), then a computation along the lines of the proof of Proposition (1.21) yields

$$\begin{aligned} \operatorname{Ric} \sigma_0 \geq \alpha_0 \Bigg\{ & \sum_{\nu=1}^{5} (\Omega_1(A_\nu) - \Omega_1) \\ & + 5(1-\varepsilon)\Omega_0 \Bigg\} + \Omega_1 - 2\Omega_o \\ & + \sigma_o \, . \end{aligned} \tag{2.1}$$

Now if $\alpha_0 > 2/5$, we can make ε small enough so that

$$\text{Ric } \sigma_0 \geq \sigma_0 + \Omega_1 + \alpha_0 \sum (\Omega_1(A_\nu) - \Omega_1)$$

$$\geq \sigma_0 - (5\alpha_0 - 1)\Omega_1 .$$

If Λ_2 is small, so that Ω_1 is small then Ric $\sigma_0 \geq \gamma\sigma_0$, $0 < \gamma < 1$. But Ric $(\gamma\sigma_0)$ = Ric σ_0, so applying (1.5) to the metric $\gamma\sigma_0$ we have a bound on $1/r$. Thus we can use σ_0 to give the estimate if Λ_2 is small and ω to give the estimate when Λ_2 is large.

To make this all precise, we use a variant of Ahlfors Lemma (see [1]):

Ahlfors' Lemma 2.3 for metrics with support: *Let* $\tau = {}^{i}/_{2\pi}\, h(\zeta)d\zeta\, d\bar{\zeta}$ *be a continuous pseudo-metric on* Δ_r *such that for each* $x \in \Delta_r$ *there exists a neighborhood* U_x *of* x *and a* C^∞ *pseudo-metric* τ_x *on* U_x *with the following properties:*

i) $\tau \geq \tau_x$ *on* U_x,

ii) $\tau = \tau_x$ *at the point* x,

iii) Ric $\tau_x \geq \tau_x$ *on* U_x.

Then $\tau \leq ds_r^2$ and thus (1.5) holds.

We now let

$$\Psi = \max\{\gamma\sigma_0, \omega\} , \; 0 < \gamma < 1 , \tag{2.4}$$

where the α_0, α_1 are chosen so that (1.17) and (1.26) hold and in addition $\alpha_0 \geq 2/5$ (which is no problem since α_0 can be taken slightly less than $1/2$ by (1.27)).

If f is non-degenerate, ω always satisfies Ric $\omega \geq \omega$, and when $\gamma\sigma_o \geq \omega$ then $\gamma\sigma_o \geq c\,\sigma_o^{\beta/\alpha_0}\,\sigma_1^{\beta/\alpha_1}$ by (1.18) and $\gamma\sigma_0^{\beta/\alpha_1} \geq c\,\sigma_1^{\beta/\alpha_1}$ by (1.19). But

$$\sigma_1 \geq M\Omega_1 \tag{2.5}$$

for $M > 0$ a constant, since $0 \le \phi_1(A_\nu) \le 1$ and $x(\log x)^2 \to 0$ as $x \to 0$. Therefore $\gamma\sigma_0 \ge \omega$ implies

$$\Omega_1 \le M'\gamma^{\alpha_1/\beta}\,\sigma_0\,, \quad M' > 0\,. \tag{2.6}$$

Finally (2.2) yields

$$\begin{aligned} \operatorname{Ric}\gamma\sigma_0 &= \operatorname{Ric}\sigma_0 \\ &\ge \sigma_0 - M''\gamma^{\alpha_1/\beta}\,\sigma_0 \quad \text{by (2.6)} \\ &\ge \gamma\sigma_o \quad \text{for } \gamma \text{ sufficiently small.} \end{aligned} \tag{2.7}$$

Thus we can apply Lemma (2.3) to ϕ and we have

$$\begin{aligned} ds_r^2 &\ge \Psi \\ &\ge \gamma\sigma_o\,. \end{aligned} \tag{2.8}$$

But (2.8) is certainly true if f is degenerate by Ahlfors' Lemma applied to $\gamma\sigma_0$, using (2.2), since $\Omega_1 \equiv 0$. Thus (1.5) and (1.9) imply

$$2/r^2 \ge \gamma c_0 \prod_1^5 \left(\frac{\phi_1(A_\nu)}{\phi_0(A_\nu)\left[\log \frac{\mu}{\phi_0(A_\nu)}\right]^2} \right)^{\alpha_0} \left. \frac{|\Lambda_1|^2}{|\Lambda_0|^4} \right|_0 \tag{2.9}$$

Now let $x \in \mathbb{P}_2 - A_1 \cup \ldots \cup A_5$, say $x = [1: x_1 : x_2]$ in inhomogeneous co-ordinates, and let $\xi = (o, \xi_1, \xi_2)$ be co-ordinates for the tangent space at x, i.e. $\xi \longleftrightarrow \xi_1 \frac{\partial}{\partial x_2} + \xi_2 \frac{\partial}{\partial x_2}$. For A a unit vector in $\mathbb{C}^3$ we define:

$$\begin{aligned} \phi_o(A,x) &= |x,A|^2 \big/ |x|^2 \\ \phi_1(A,x,\xi) &= |x\wedge\xi,A|^2 \big/ |x\wedge\xi|^2 \end{aligned} \tag{2.10}$$

By considering all possible holomorphic curves $f: \Delta_r \to \mathbb{P}_2 - A_1 \cup \ldots \cup A_5$ such that $f(0) = x$, $f'(0) = \xi$ we have shown by (2.9):

Proposition (2.11): *Let* $A_1, \ldots, A_5$ *be hyperplanes in general position in* $\mathbb{P}_2$, *then the infinitesmal Kobayashi metric* $\|\xi\|_x^2$ *on* $\mathbb{P}_2 - A_1 \cup \ldots \cup A_5$ *satisfies*

$$\|\xi\|_x^2 \geq c \prod_{\nu=1}^{5} \left[\frac{\phi_1(A_\nu, x, \xi)}{\phi_0(A_\nu, x) \left[\log \frac{\mu}{\phi_0(A_\nu, x)} \right]^2} \right]^{\alpha_0} ds^2(\xi) \tag{2.11}$$

for any $1/2 < \alpha_0 < 2/5$; c, μ *constants depending on* α_0, *and* ds^2 *the Fubini-Study metric on* $\mathbb{P}_2$.

Note that $\phi_1(A_\nu, x, \xi)$ is homogeneous in ξ so that the right hand side of (2.11) is a hermitian metric times a homogeneous function on the tangent bundle of $\mathbb{P}_2 - A_1 \cup \ldots \cup A_5$.

The estimate (2.11) is not sharp since the R. H. S. can go to zero as x approaches a missing hyperplane tangentially (i. e. $\phi_0(A_\nu, x) \to 0$ and $\phi_1(A_\nu, x, \xi) \to 0$ at exactly the same rate) and thus we cannot conclude what is known from Cartan [6], that $\|\xi\|_x^2 \geq c\, ds^2(\xi)$. In addition, as x approaches a missing hyperplane A_ν non-tangentially and x is not near an intersection, the right hand side blows up like

$$\left\{ 1/\phi_0(A_\nu, x) \left[\log \frac{\mu}{\phi_0(A_\nu, x)} \right]^2 \right\}^{\alpha_0}$$

but the Kobayashi metric blows up like

$$1/\phi_0(A_\nu, x) \left[\log \frac{\mu}{\phi_0(A_\nu, x)} \right]^2 .$$

3. $\mathbb{P}_n - A_1 \ldots A_{2^n+1}$

Let $A_1, \ldots, A_q$ be hyperplanes in general position,

$$f: \Delta_r \to \mathbb{P}_n - A_1 \cup \ldots \cup A_q$$

a non-constant holomorphic map, $n \geq 3$. We generalize the construction in §2 by letting, for $0 \leq i \leq n - 1$,

$$\omega_i = \left(\frac{1}{\prod \left[\log \frac{\mu}{\phi_i(A_\nu)} \right]^2} \right)^{\sum_{j=i+1}^{n-2} \beta_j} \cdot \sigma_0^{\beta_i/\alpha_0} \cdots \sigma_i^{\beta_i/\alpha_i} \tag{3.1}$$

where the σ_i are defined by (1.16), the α_i satisfy (1.17) and

$$\beta_i \left(1/\alpha_0 + \ldots + 1/\alpha_i \right) = 1 \tag{3.2}$$

Since f may be degenerate we use the convention that $\omega_i \equiv 0$ if $\Lambda_i \equiv 0$. Let

$$\Psi = \max\{\gamma_0 \ldots \gamma_{n-2}\omega_0, \ldots, \gamma_{n-2}\omega_{n-2}, \omega_{n-1}\} \tag{3.3}$$

We wish to show that if $q \geq 2^n + 1$ we can choose the α_i, and universal constants $\gamma_0, \ldots, \gamma_{n-2}$ such that Ψ satisfies the three conditions of Lemma (2.3). So assume ω_i is not identically zero. If $i = n - 1$, then $\operatorname{Ric} \omega_i \geq \omega_i$ by Proposition (1.21) when $q \geq n + 2$ and the α_i satisfy (1.26). For $i < n - 1$, we have

$$\begin{aligned}
&\operatorname{Ric} \gamma_i \ldots \gamma_{n-2}\omega_i \\
&\geq \sum_{j=i+1}^{n-2} \beta_j \sum_\nu \left\{ \frac{2\phi_{i+1}(A_\nu)}{\phi_i(A_\nu) \left[\log \frac{\mu}{\phi_i(A_\nu)} \right]^2} - \varepsilon\Omega_i \right\} \\
&+ \beta_i \sum_{j=0}^{i} \sum_\nu \left\{ \frac{2\phi_{j+1}(A_\nu)}{\phi_j(A_\nu) \left[\log \frac{\mu}{\phi_j(A_\nu)} \right]^2} - \varepsilon\Omega_j \right\} \\
&+ q\beta_i\Omega_0 + \beta_i \sum_\nu \left(\Omega_{i+1}(A_\nu) - \Omega_{i+1} \right) \\
&+ \beta_i \sum_j 1/\alpha_j \left(\Omega_{j+1} - 2\Omega_j + \Omega_{j-1} \right) \\
&\geq \left(\sum_{j=i+1}^{n-2} \beta_j \right) \{\sigma_i - \varepsilon q \Omega_i\} \\
&+ \beta_i \sum_{j=0}^{i} \{\sigma_j - \varepsilon q \Omega_j\} \\
&+ \beta_i \left(\sum_\nu \Omega_{i+1}(A_\nu) - \Omega_{i+1} \right)
\end{aligned}$$

$$
\begin{aligned}
&+ \beta_i\Big\{(q - 2/\alpha_0 + 1/\alpha_1)\Omega_0 \\
&+ \sum_{j=1}^{i-1} (1/\alpha_{j-1} - 2/\alpha_j + 1/\alpha_{j+1})\,\Omega_j \\
&+ (1/\alpha_{i-1} - 2/\alpha_i)\,\Omega_i + 1/\alpha_i\Omega_{i+1}\Big\} \qquad (3.4)\\
&\geq \sigma_0^{\beta_i/\alpha_0} \cdots \sigma_i^{\beta_i/\alpha_i} - \beta_i(q - 1/\alpha_i)\,\Omega_{i+1} \\
&+ \beta_i\Big\{(q - 2/\alpha_0 + 1/\alpha_1 - \varepsilon q)\Omega_0 \\
&+ \sum_{j=1}^{i-1} (1/\alpha_{j-1} - 2/\alpha_j + 1/\alpha_{j+1} - \varepsilon q)\Omega_j \\
&+ \Big(1/\alpha_{i-1} - 2/\alpha_i - \varepsilon q\big(1 + \sum_{j=i+1}^{n-2} \beta_j/\beta_i\big)\Big)\Omega_i\Big\}.
\end{aligned}
$$

Assume that the following is satisfied:

$$
\begin{aligned}
&q - 2/\alpha_0 + 1/\alpha_1 > 0 \\
&1/\alpha_{j-1} - 2/\alpha_j + 1/\alpha_{j+1} > 0 \qquad 1 \leq j \leq i - 1 \qquad (3.5)\\
&1/\alpha_{i-1} - 2/\alpha_i > 0 .
\end{aligned}
$$

Then since ε can be made arbitrarily small, (3.4) and (3.5) imply

$$
\begin{aligned}
\operatorname{Ric} \gamma_i \cdots \gamma_{n-2}\omega_i &\geq \sigma_0^{\beta_i/\alpha_0} \cdots \sigma_i^{\beta_i/\alpha_i} \qquad (3.6)\\
&\quad - \beta_i(q - 1/\alpha_i)\,\Omega_{i+1}
\end{aligned}
$$

Lemma 3.7 : *Assume* (3.5) *is satisfied and* $\gamma_i \cdots \gamma_{n-2}\omega_i \geq \gamma_{i+1} \cdots \gamma_{n-2}\omega_{i+1}$. *Then for suitable* γ_i,

$$\operatorname{Ric} \gamma_i \cdots \gamma_{n-2}\omega_i \geq \gamma_i \cdots \gamma_{n-2}\omega_i .$$

Proof: If $\omega_{i+1} = 0$ at $\xi \in \Delta_r$ then $\Omega_{i+1} \equiv 0$ so (3.6) gives the desired result, since $\sigma_0^{\beta_i/\alpha_0} \cdots \sigma_i^{\beta_i/\alpha_i} \geq \omega_i$. If $\omega_{i+1} \neq 0$ at ξ, then $\gamma_i\omega_i \geq \omega_{i+1}$ which implies

$$\gamma_i\left(\prod \frac{1}{\left[\log \frac{\mu}{\phi_i(A_\nu)}\right]^2}\right)^{\sum_{i+1}^{n-2}\beta_j} \sigma_0^{\beta_i/\alpha_0 - \beta_{i+1}/\alpha_0} \cdots \sigma_i^{\beta_i/\alpha_i - \beta_{i+1}/\alpha_i} \tag{3.8}$$

$$\geq \left(\prod \frac{1}{\left[\log \frac{\mu}{\phi_{i+1}(A_\nu)}\right]^2}\right)^{\sum_{i+2}^{n-2}\beta_j} \sigma_{i+1}^{\beta_{i+1}/\alpha_{i+1}} .$$

But $\frac{1}{\left[\log \frac{\mu}{\phi_i(A)}\right]^2} \leq \frac{1}{\left[\log \frac{\mu}{\phi_{i+1}(A)}\right]^2}$ since $\phi_{i+1}(A) \geq \phi_i(A)$, and for $0 \leq j \leq i$:

$$\beta_i/\alpha_j - \beta_{i+1}/\alpha_j = (\beta_{i+1}/\alpha_{i+1})(\beta_i/\alpha_j)$$

by (3.2). Thus (3.8) implies:

$$\gamma_i\left(\prod \frac{1}{\left[\log \frac{\mu}{\phi_i(A_\nu)}\right]^2}\right)^{\beta_{i+1}} \left\{\sigma_0^{\beta_i/\alpha_0} \cdots \sigma_i^{\beta_i/\alpha_i}\right\}^{\beta_{i+1}/\alpha_{i+1}} \tag{3.9}$$

$$\geq \sigma_{i+1}^{\beta_{i+1}/\alpha_{i+1}} ,$$

i.e.

$$\gamma_i^{\alpha_{i+1}/\beta_{i+1}} \sigma_0^{\beta_i/\alpha_0} \cdots \sigma_i^{\beta_i/\alpha_i} \geq \prod \left[\log \frac{\mu}{\phi_i(A_\nu)}\right]^{2\alpha_{i+1}} \sigma_{i+1}$$

$$\geq c_{i+1} \prod \frac{\phi_{i+2}(A_\nu)}{\phi_{i+1}(A_\nu)} \Omega_{i+1} \tag{3.10}$$

$$\geq c_{i+1}\Omega_{i+1}$$

Note: If $i = n - 2$ then (3.9) has no log factor on the left hand side and the proof of (3.10) is the same as for (2.5) Thus (3.6) yields

$$\text{Ric } \gamma_i \cdots \gamma_{n-1}\omega_i \geq \sigma_0^{\beta_i/\alpha_0} \cdots \sigma_i^{\beta_i/\alpha_i}$$

$$\cdot \left\{1 - \beta_i(q - 1/\alpha_i)c_{i+1}^{-1}\gamma_i^{\alpha_{i+1}/\beta_{i+1}}\right\} \tag{3.11}$$

$$\geq \omega_i\left(1 - \gamma_i^{\alpha_{i+1}/\beta_{i+1}} M\right),$$

which for γ_i small enough proves (3.7).

Q. E. D.

Now (3.5) holds for each i iff

$$\begin{cases} q - 2/\alpha_0 > 0 \\ 1/\alpha_{i-1} > 2/\alpha_i \qquad i = 1,\dots, n-1, \end{cases} \tag{3.6}$$

(Note (3.5) degenerates to $q - 2/\alpha_0 > 0$ when $i = 0$). Thus Ψ defined by (2.3) satisfies Ahlfors' Lemma for metrics with supports if we can choose the α_i so as to satisfy (3.6) and (1.17). But then

$$q > 2/\alpha_0 > \dots > 2^n/\alpha_{n-1} > 2^n$$

so we must have $q \geq 2^n + 1$, and conversely, if $q \geq 2^n + 1$ we can clearly solve (3.6) and (1.17)

Since $\Psi \geq c\omega_0$ by definition and $ds_r^2 \geq \Psi$ by Lemma 2.3 we have shown via (1.5) that

$$2/r^2 \geq c\left(\prod \frac{\phi_1(A_\nu)}{\phi_0(A_\nu)}^{\alpha_0} \frac{1}{\left[\log \frac{\mu}{\phi_0(A_\nu)}\right]^{2\alpha_0 + 2\Sigma_1^{n-2}\beta_j}}\right)\cdot \left.\frac{|\Lambda_1|^2}{|\Lambda_0|^4}\right|_0 . \tag{3.7}$$

Proceeding as we did for $\mathbb{P}_2$, we can translate (3.7) into an estimate on the Kobayashi metric:

Proposition: *Let* $A_1,\dots,A_{2^n+1}$ *be hyperplanes in* $\mathbb{P}_n$ *in general position,* $x \in \mathbb{P}_n - A_1 \cup \dots \cup A_{2^n+1}$, ξ *a tangent vector at* x *and* α_0 *a constant,* $2^{-n+1} > \alpha_0$. *Then the infinitesmal Kobayashi metric satisfies*

$$\|\xi\|_x^2 \geq c\left(\prod \left[\frac{\phi_1(A_\nu,x,\xi)}{\phi_0(A_\nu,x)}\right]^{\alpha_0} \frac{1}{\log\left[\frac{\mu}{\phi_0(A_\nu,x)}\right]^{2\beta}}\right)\cdot ds^2(\xi) \tag{3.8}$$

where ds^2 *is the Fubini-Study metric on* $\mathbb{P}_n$, c *and* β *are constants.*

Corollary 3.9 : $\mathbb{P}_n - A_1 \cup \dots \cup A_{2^n+1}$ *is hyperbolic.*

Remarks: Corollary 3.9 is far from sharp, since actually $\mathbb{P}_n - (2n+1)$

hyperplanes is hyperbolic. The estimate (3.8) also fails to be sharp, (see the remarks at the end of §2). The problem seems to arise in the rather crude way we have treated what happens at intersections of the hyperplanes. Even in the $\mathbb{P}_2$ case there is as yet no way to replace Bloch's analysis with "curvature" methods.

REFERENCES

1. Ahlfors, L., *An extension of Schwarz's lemma*, Trans. Am. Math. Soc., 43 (1938), 359-364.

2. __________, *The theory of meromorphic curves*, Finska Vetenskaps - Societeten, Helsingfors: Acta Soc. Sci. Fenn., Series A, vol. 3, no. 4 (1941), 3-31.

3. Bloch, A., *Sur les systèmes de fonctions holomorphes à variétés linéaires lacunaires*, Ann. Ecole Normale, 43(1926), 309-362.

4. Borel, E., *Sur les zéros des fonctions entières*, Acta Math., 20 (1896), 357-396.

5. Carlson, J., and P. Griffiths, *A defect relation for equidimensional holomorphic mappings between algebraic varieties*, Ann. of Math. 95 (1972), 557-584.

6. Cartan, H., *Sur les systèmes de fonctions holomorphes à variétés lacunaires et leurs applications*, Ann. Ecole Normale, 45 (1928), 255-346.

7. Chern, S.-S., *Holomorphic curves in the plane*, Differential Geometry, in honor of K. Yano, Kinokuniya, Tokyo, 1972, 73-94.

8. Cowen, M. and P. Griffiths, *Holomorphic curves and metrics of negative curvature* (to appear).

9. Dufresnoy, J., *Théorie nouvelle des familles complexes normales*, Ann. Ecole Normale, 61 (1944), 1-44.

10. Green, M., *Holomorphic maps into complex projective space omitting hyperplanes*, Trans. Amer. Math. Soc. 169 (1972), 89-105.

11. Hayman, W. K., *Meromorphic Functions*, Oxford University Press, London, 1964.

12. Kiernan, P. and Kobayashi, S., *Holomorphic mappings into projective space with lacunary hyperplanes*, Nagoya Math. J., 50 (1973), 199-216.

13. Kobayashi, S., *Hyperbolic Manifolds and Holomorphic Mappings*, Marcel Dekker, New York, 1970.

14. Royden, H. L., *Remarks on the Kobayashi metric*, International Mathematical Conference (Maryland, 1970), Springer-Verlag, Berlin-Heidelberg, New York, 1971.

15. Weyl, H. and Weyl, J., *Meromorphic Functions and Analytic Curves*, Princeton University Press, Princeton, 1943.

16. Wu, H., *The Equidistribution Theory of Holomorphic Curves*, Princeton University Press, Princeton, 1970.

THE ORDER FUNCTIONS FOR ENTIRE HOLOMORPHIC MAPPINGS

James A. Carlson
Phillip A. Griffitns

Brandeis University
Harvard University

INTRODUCTION

In this paper we shall study the notion of *growth* of an entire holomorphic mapping

$$f: \mathbb{C}^n \longrightarrow M$$

into a compact Kähler manifold M. *Order functions* $T_q(r)$ will be introduced for $q = 1, \ldots, n$. These have various geometric interpretations, but perhaps the most illuminating one is that $T_q(r)$ *measures the average amount which the image of* f *meets subvarieties of codimension* q *in* M. Our basic problem is to understand in what way growth of the various $T_q(r)$'s may be related. Our main results, which are not definitive, on this question are stated in section 6 and proofs are given in section 7.

Our underlying motivation is an attempt to gain some insight into the global behavior of a holomorphic mapping in higher codimension, a problem which is thus far most notable for its negative conclusions--the *Fatou-Bieberbach* and *Bezout* counterexamples (cf. [4] for references and further discussion). Although we are unable to settle the basic question concerning the relative growth of the $T_q(r)$'s, we are able to find partial results in which the related notions of negative curvature and plurisubharmonic functions play the essential role.

Following the definition and elementary properties of the order functions in section 1, we have given in sections 2 through 4 three geometric interpretations of them which should help justify their use in measuring the growth of a holomorphic mapping. None of these interpretations is particularly new, and we have only given the formal proofs of the various integral formulae (cf. [6] for a complete discussion). In section 5 we have discussed in more detail our main question and given several illustrative examples and comments.

1. DEFINITION AND FORMAL PROPERTIES OF THE ORDER FUNCTIONS

Let D and M be complex manifolds and

$$f: D \longrightarrow M$$

a holomorphic mapping. To measure the *growth* of f, we assume given an Hermitian metric on M with associated (1, 1) form ω and plurisubharmonic exhaustion function $\tau: D \longrightarrow \mathbb{R} \cup \{-\infty\}$ with Levi form $\psi = dd^c\tau$.* Setting

$$D[r] = \{x \in D: \tau(x) \leqq \log r\} \text{ and}$$

$$\phi_p = \underbrace{\phi \wedge \ldots \wedge \phi}_{p}$$

for a (1, 1) form ϕ on D, the quantities

$$t_q(r) = \int_{D[r]} (f^* \omega)_q \wedge \psi_{n-q} \qquad (n = \dim D)$$
$$T_q(r) = \int_0^r t_q(\rho) \frac{d\rho}{\rho} \tag{1.1}$$

will be called the *unintegrated order function* and *order function* respectively for the holomorphic mapping f. The reason for logarithmically averaging $t_q(r)$ to obtain $T_q(r)$ is because of the First Main Theorem (F.M.T.) reviewed in section 3 below. Since $f^*\omega \geq 0$ and $dd^c\tau \geq 0$, it follows that T_q is increasing and that $T_q(r) \longrightarrow \infty$ as $r \longrightarrow \infty$, unless of course $T_q(r) \equiv 0$. Also, the conditions

$$t_q(r) = O(1)$$

$$T_q(r) = O(\log r)$$

are evidently equivalent.

The most important special case is when M is a compact Kähler manifold, $D = \mathbb{C}^n$, and $\tau = \log \|z\|^2$. The Levi form ψ is then the standard Kähler form on the $\mathbb{P}^{n-1}$ of lines through the origin in $\mathbb{C}^n$. It has a

---

* Recall that $d^c = i(\bar{\partial} - \partial)$.

mild singularity at the origin, but this causes no trouble in our integral formulae, as is easily seen by blowing up the origin so that ψ becomes C^∞. The order functions $T_q(r)$ admit several geometric interpretations which justify their use in measuring the growth of f, and which will be discussed in sections 2 through 4 below. Here we wish to make two elementary observations. The first is that, by the compactness of M, different choices ω and $\tilde{\omega}$ of Kähler metrics on M lead to order functions $T_q(r)$ and $\tilde{T}_q(r)$ which satisfy

$$T_q(r) \sim \tilde{T}_q(r),$$

where the notation $A(r) \sim B(r)$ means that $A(r) \leqq C\,B(r)$ and $B(r) \leqq D\,A(r)$ for positive constants C, D. Consequently, the growth of $\log T_q(r)$ is well-defined.

Our second observation is based on an easy but ubiquitous integral formula. To give it, let Λ be a $C^\infty(q-1, q-1)$ form on $\mathbb{C}^n$. Then

$$\int_0^r \Big(\int_{\mathbb{C}^n[r]} dd^c \Lambda \wedge \psi_{n-q}\Big) \frac{d\rho}{\rho} = \int_{\partial\mathbb{C}^n[r]} \Lambda \wedge d^c \log \|z\|^2 \wedge \psi_{n-q} - \int_{\mathbb{C}^n[r]} \Lambda \wedge \psi_{n-q+1} \tag{1.2}$$

Proof: Using Stokes' theorem, the Fubini theorem, and then Stokes again, we have

$$\begin{aligned}
\int_0^r \Big(\int_{\mathbb{C}^n[r]} dd^c\Lambda \wedge \psi_{n-q}\Big) \frac{d\rho}{\rho} &= \int_0^r \Big(\int_{\partial\mathbb{C}^n[\rho]} d^c\Lambda \wedge \psi_{n-q}\Big) \frac{d\rho}{\rho} \\
&= \int_{\mathbb{C}^n[r]} d^c\Lambda \wedge d \log \|z\|^2 \wedge \psi_{n-q} \\
&= \int_{\mathbb{C}^n[r]} d\Lambda \wedge d^c \log \|z\|^2 \wedge \psi_{n-q} \\
&= \int_{\partial\mathbb{C}^n[r]} \Lambda \wedge d^c \log \|z\|^2 \wedge \psi_{n-q} \\
&\quad - \int_{\mathbb{C}^n[r]} \Lambda \wedge \psi_{n-q+1}
\end{aligned}$$

Q.E.D.

Suppose now that ω and $\tilde{\omega}$ are Kähler metrics in the same cohomology class on M. Then $\omega - \tilde{\omega} = dd^c u$ for some C^∞ function u on M, and consequently

$$(f^*\omega)_q - (f^*\tilde{\omega})_q = dd^c\Lambda$$

where Λ is a C^∞ $(q-1, q-1)$ form on $\mathbb{C}^n$ which is dominated by some constant multiple of $(f^*\omega)_{q-1}$. Thus

$$\text{(i)} \quad \int_{\mathbb{C}^n[r]} \Lambda \wedge \psi_{n-q+1} = 0\left(t_{q-1}(r)\right)$$

$$\text{(ii)} \quad \int_{\partial\mathbb{C}^n[r]} \Lambda \wedge d^c\log\|z\|^2 \wedge \psi_{n-q} = 0\left(\int_{\partial\mathbb{C}^n[r]} (f^*\omega)_{q-1} \wedge d^c\log\|z\|^2 \wedge \psi_{n-q}\right)$$

$$= 0\left(\int_{\mathbb{C}^n[r]} (f^*\omega)_{q-1} \wedge \psi_{n-q+1}\right)$$

$$= 0\left(t_{q-1}(r)\right).$$

Putting this together with (1.2) gives

$$\tilde{T}_q(r) = T_q(r) + 0\left(t_{q-1}(r)\right),$$

so that the growth of $T_q(r)$ is intrinsically defined by the cohomology class $[\omega] \in H^2(M, R)$ in case

$$\lim_{r\to\infty}\left[\frac{t_{q-1}(r)}{T_q(r)}\right] = 0$$

2. A CHARACTERIZATION OF MEROMORPHIC MAPPINGS

Let $f: \mathbb{C}^n \longrightarrow M$ be an entire holomorphic mapping into a compact Kähler manifold M. In order to help justify the word "order function" for the $T_q(r)$, we shall prove the

Proposition 2.1: *The mapping* f *extends to a meromorphic mapping* $f: \mathbb{P}^n \longrightarrow M$ *if, and only if,*

$$T_q(r) = 0(\log r) \qquad (q = 1, \ldots, n) \tag{2.2}$$

Proof: We shall use Bishop's Theorem [11], which in the present context states that the closure $\overline{\Gamma}_f$ in $\mathbb{P}^n \times M$ of the graph $\Gamma_f \subset \mathbb{C}^n \times M$ of f is an analytic set if and only if the volume of Γ_f, computed using any metric on $\mathbb{P}^n \times M$, is finite. As metric on $\mathbb{P}^n \times M$ we use $\eta + \omega$ where $\eta = dd^c\log(1 + \|z\|^2)$ is the standard Kähler form on $\mathbb{P}^n$. We observe that η is also the Levi form of the exhaustion function $\log(1 + \|z\|^2)$ on $\mathbb{C}^n$, which has the same level sets and essentially the same growth as $\log \|z\|^2$.

By Wirtinger's Theorem [11] and the definition of integration over the graph,

$$\mathrm{Vol}(\Gamma_f) = \int_{\Gamma_f} (\eta + \omega)^n = \sum_q \binom{n}{q} \int_{\mathbb{C}^n} (f^*\omega)_q \wedge \eta_{n-q}$$

Setting

$$s_q(r) = \int_{\mathbb{C}^n[r]} (f^*\omega)_q \wedge \eta_{(n-q)}$$

we must prove that

$$t_q(r) = O(1) \longleftrightarrow s_q(r) = O(1) \tag{2.3}$$

This involves relating the unintegrated order functions for the two exhaustion functions $\log \|z\|^2$ and $\log(1 + \|z\|^2)$. Setting

$$I_{p,q}(r) = \int_{\mathbb{C}^n[r]} (f^*\omega)_q \wedge \eta_p \wedge \psi_{n-p-q}$$

then

$$\begin{aligned} I_{0,q}(r) &= t_q(r) \\ I_{n-q,q}(r) &= s_q(r), \end{aligned} \tag{2.4}$$

which suggests that we find an integral formula relating $I_{p,q}$ and $I_{p-1,q}$. The desired formula is

$$\begin{aligned} &\int_0^r I_{p,q}(\rho)\frac{d\rho}{\rho} + \int_{\mathbb{C}^n[r]} \log(1 + \|z\|^2)\ (f^*\omega)_q \wedge \eta_{p-1} \wedge \psi_{n-p-q+1} \\ &= \int_{\partial\mathbb{C}^n[r]} \log(1 + \|z\|^2)\ d^c\log \|z\|^2 \wedge (f^*\omega)_q \wedge \eta_{p-1} \wedge \psi_{n-p-q} \end{aligned} \tag{2.5}$$

This follows from (1.2) when we take $\Lambda = \log(1 + \|z\|^2)\ (f^*\omega)_q \wedge \eta_{p-1}$.

Now we use the obvious inequalities

$$(\log r)\ I_{p,q}(r) \leqq \int_r^{r^2} I_{p,q}(\rho)\frac{d\rho}{\rho} \leqq \int_0^{r^2} I_{p,q}(\rho)\frac{d\rho}{\rho}$$

together with positivity of everything in sight in (2.5) to obtain

$$\begin{aligned} (\log r)\ I_{p,q}(r) &\leqq C\log r \int_{\partial C^n[r^2]} d^c\log \|z\|^2 \wedge \eta_{p-1} \wedge (f^*\omega)_q \wedge \psi_{n-p-q} \\ &= C(\log r)\ I_{p-1,q}(r^2), \end{aligned}$$

where the last step uses Stokes' theorem. This together with a similar argument in the opposite direction gives

$$I_{p,q}(r) \leqq C\, I_{p-1,q}(r^2)$$
$$I_{p-1,q}(r) \leqq C'\, I_{p,q}(r).$$

Using these inequalities recursively and taking (2.4) into account gives (2.3).

Q.E.D.

Remark: Aside from technicalities, the above proof should be read:

$\sum_{q=0}^{n} T_q(r)$ *measures the growth of the volume of the graph of* f.

This provides our first interpretation of the order functions $T_q(r)$.

3. THE F.M.T. AND FIRST CROFTON FORMULA

Let $f: \mathbb{C}^n \longrightarrow \mathbb{P}^m$ be an entire holomorphic mapping and ω the standard Kähler form on $\mathbb{P}^m$. The set of linear subspaces $A \subset \mathbb{P}^m$ of *codimension* q is parametrized by the Grassmanian $G(m-q+1, m+1)$ of $(m-q+1)$ planes through the origin in $\mathbb{C}^{m+1}$. Given $A \in G(m-q+1, m+1)$ we assume that codim $f^{-1}(A) = q$ and set

$$n(A,\rho) = \int_{f^{-1}(A) \cap \mathbb{C}^n[\rho]} \psi^{n-q}$$
$$N(A,r) = \int_0^r \{n(A,\rho) - n(A,0)\} \frac{d\rho}{\rho} + n(A,0) \log r \tag{3.1}$$

where $n(A,0) = \lim n(A,\rho)$ is the multiplicity of $f^{-1}(A)$ at the origin (cf. [6], p. 164). We call $N(A,r)$ the *counting function* of A.

For each A as above, the *Levine form* Λ_A is defined and has the properties (c f. proposition 1.15 in [6]):

(i) Λ_A is a locally $L^1(q-1,q-1)$ form on $\mathbb{P}^m$;

(ii) Λ_A is C^∞ and positive on $\mathbb{P}^m - A$;

(iii) the equation of currents

$$dd^c \Lambda_A = \omega_q - [A]$$

is valid on $\mathbb{P}^m$; and

(iv) $T^*\Lambda_A = \Lambda_{T\cdot A}$ for a unitary linear transformation $T\colon \mathbb{P}^m \longrightarrow \mathbb{P}^m$.

By twice integrating (3.2) just as in the proof of (7.2) one arrives at the First Main Theorem (F.M.T.) (cf. [6], p. 183).

$$N(A, r) + m(A, r) = T_q(r) + S(A, r) + O(1) \tag{3.3}$$

Here the proximity form $m(A, r)$ and remainder term $S(A, r)$ are given respectively by

$$m(A, r) = \int_{\partial\mathbb{C}^n[r]} f^*\Lambda_A \wedge d^c\log\|z\|^2 \wedge \psi_{n-q}$$

$$S(A, r) = \int_{\mathbb{C}^n[r]} f^*\Lambda_A \wedge \psi_{n-q+1}$$

Now assume that codim $f^{-1}(A) = q$ for almost all A, and average (3.3) over the unitary group U_{m+1} acting on $G(m-q+1, m+1)$. Using Fubini's theorem, we obtain

$$\int N(A, r)dA + \int_{\partial\mathbb{C}^n[r]} \left(\int f^*\Lambda_A\, dA\right) d^c\log\|z\|^2 \wedge \psi_{n-q}$$

$$= \int_{\mathbb{C}^n[r]} \left(\int f^*\Lambda_A\, dA\right) \wedge \psi_{n-q+1} + T_q(r)$$

From (iv) we conclude that $\int f^*\Lambda_A\, dA$ is U_{m+1} -invariant, and hence is a constant multiple of $f^*\omega_q$.

Finally, applying Stoke's theorem, we conclude

$$\int m(A, r)dA = \int S(A, r)dA,$$

which gives the *First Crofton Formula:*

$$T_q(r) = \int_{A\in G(m-q+1,\, m+1)} N(A, r)dA \tag{3.5}$$

Thus $T_q(r)$ *measures the average growth of* $f^{-1}(A)$ *as* A *runs over the linear spaces of codimension* q *in* $\mathbb{P}^n$. The First Main Theorem (3.3) and subsequent Crofton formula (3.5) may be viewed as non-compact analogues of the *Wirtinger Theorem* [11], which says that the *volume* (a metric invariant) of an analytic set $Z \subset \mathbb{P}^m$ is equal to its *degree* (an analytic invariant).

In closing this section we should like to derive from (3.3) and (3.5) the following version of the classical *Liouville theorem* due to Chern,

Stoll, and Wu (cf. [12])

If $\displaystyle\lim_{r\to\infty} \frac{t_{q-1}(r)}{t_q(r)} = 0$ *then the image* $f(\mathbb{C}^n)$ *meets almost all* $A \in G(m-q+1,\ m+1)$.

Proof: Let $\Sigma \subset G(m-q+1,\ m+1)$ be the set of A such that $f^{-1}(A)$ is non-empty. We shall compute that the total measure

$$M(\Sigma) = M\left(G(m-q+1,\ m+1)\right) = 1,$$

from which it follows that Σ is dense. Now, writing $G = G(m-q+1,\ m+1)$

$$\begin{aligned} M(\Sigma)\, T_q(r) &= \int_{A\in\Sigma} T_q(r)\,dA \\ &\geqq \int_{A\in\Sigma} N(A,\ r)\,dA - \int_{A\in\Sigma} S(A,\ r)\,dA + O(1) \quad \text{(by (3.3))} \\ &\geqq \int_{A\in G(m-q+1,m+1)} N(A,\ r)\,dA - \int_{A\in G(m-q+1,m+1)} S(A,\ r)\,dA + O(1) \\ &= T_q(r) - t_{q-1}(r) + O(r) \qquad \text{because of (3.5)} \end{aligned}$$

and the fact that

$$\int_{A\in G(m-q+1,m+1)} \Lambda_A\, dA = \omega_{q-1}$$

Combining these, we obtain

$$M(\Sigma) \geqq 1 - \frac{t_{q-1}(r)}{T_q(r)} + O(1),$$

from which the result follows.

Q.E.D.

4. THE SECOND CROFTON FORMULA

In the *equidimensional case* of a non-degenerate holomorphic mapping

$$f\colon \mathbb{C}^m \longrightarrow M_m\ ,$$

the simplest and perhaps most appealing order function is

$$T_m(r) = \int_0^r \Big\{ \int_{\mathbb{C}^m[\rho]} (f^*\omega)_m \Big\} \frac{d\rho}{\rho}\ ,$$

which measures the *volume of the image* and which is the direct generalization of the *Ahlfors-Shimizu characteristic function* [8]. For a point $A \in M$, if we let $n(A,\rho)$ be the number of points in $f^{-1}(A) \cap \mathbb{C}^m[\rho]$ and let

$$N(A, r) = \int_0^r [n(A, \rho) - n(A, 0)] \frac{d\rho}{\rho} + n(A, 0)\log r$$

be the corresponding counting function, then clearly

$$T_m(r) = \int_{A \in M} N(A, r)\omega_m(A)$$

In general, given $f: \mathbb{C}^n \longrightarrow M$ the *order function of highest degree* $T_n(r)$ measures the volume of the image of balls $f(\mathbb{C}^n[r])$. We shall show how the intermediate order functions $T_q(r)$ may be related to such order functions of highest degree. For each $A \in G(q, n)$, the Grassmanian of q-planes through the origin in $\mathbb{C}^n$, we set $A[r] = A \cap \mathbb{C}^n[r]$ and let

$$T_q(A, r) = \int_0^r \{ \int_{A[\rho]} (f^*\omega)_q \} \frac{d\rho}{\rho}$$

be the highest order function of the restriction of f to A. Then the *Second Crofton Formula* (cf. Shiffman [10]) is

$$T_q(r) = \int_{A \in G(q,n)} T_q(A, r)dA \tag{4.1}$$

Here we will give only the formal part of the proof. The integrals which occur will sometimes be improper because certain forms will have a mild singularity along a linear subspace. By blowing up along this subspace, one easily verifies convergence. We now give the proof for $q = 1$.

It will suffice to prove

$$\int_{A \in \mathbb{P}^{n-1}} dA \int_{A[t]} \omega_f = \int_{\mathbb{C}^n[t]} \omega_f \wedge \psi_{n-q} \tag{4.2}$$

Let ω_0 be the standard Kähler form on $\mathbb{P}^{n-1} = G(1, n)$, let $\psi = i\, \partial\bar{\partial} \log ||z||^2$, and observe that $dA = \psi^{n-1}$.

Now $\pi: \mathbb{C}^n - \{0\} \longrightarrow \mathbb{P}^{n-1}$ is a fibration with fiber $A - \{0\}$, and $\pi^* \omega_0 = \psi$. Thus Fubini's theorem gives

$$\int_{A \in \mathbb{P}^{n-1}} dA \int_{A[t]-\{0\}} \omega_f = \int_{\mathbb{C}^n[t]-\{0\}} \omega_f \wedge \psi_{n-1}$$

Writing this identity in terms of improper integrals gives (4.2).

To give the proof for arbitrary q, consider the following diagram:

$$\begin{array}{ccc} \mathbb{C}^n - \{0\} & \longrightarrow & \mathbb{P}^{n-1} \\ \cup & & \cup \\ \mathbb{C}^n - \xi & \longrightarrow & \mathbb{P}^{n-1} - [\xi] \\ \downarrow & & \downarrow \\ \mathbb{C}^n/\xi - \{0\} & \longrightarrow & \mathbb{P}_{\xi}^{n-q} \end{array}$$

where ξ is a subspace of dimension $q-1$, and where $[\xi]$ is the associated $\mathbb{P}^{q-1} \subset \mathbb{P}^{n-1}$. Let dA_{ξ} be the standard invariant measure on $\mathbb{P}^{n-q}$, let ω_{ξ} be the standard Kähler form there, and let ψ_{ξ} be its pullback to $\mathbb{C}^n - \xi$. Now $\mathbb{P}_{\xi}^{n-q}$ parametrizes the set of all q-planes $A_{\xi} \subset \mathbb{C}^n$ which contains ξ. Moreover, $\mathbb{C}^n - \xi \longrightarrow \mathbb{P}_{\xi}^{n-q}$ is a fiber bundle whose fibers are the $A_{\xi} - \xi$. Since $dA_{\xi} = \omega_{\xi}^{n-q}$ and ω_{ξ} pulls back to ψ_{ξ}, the same argument using Fubini's theorem gives the identity of improper integrals

$$\int_{\mathbb{P}_{\xi}^{n-q}} dA_{\xi} \int_{A_{\xi}[t]} \omega_f^q = \int_{\mathbb{C}[t]} \omega_f^q \wedge \psi_{\xi}^{n-q} \tag{4.3}$$

To obtain the desired identity, we integrate (4.3) over all q-1 planes ξ. Thus we let $d\xi$, dA be the standard invariant measures on $G(q-1, n)$, $G(q, n)$, respectively. Then

$$\begin{aligned} \int_{G(q,n)} dA \int_{A[t]} \omega_f^q &= \int_{G(q-1,n)} d\xi \int_{P_{\xi}^{n-q}} d\,A_{\xi} \int_{A_{\xi}[t]} \omega_f^q \\ &= \int_{G(q-1,n)} d\xi \int_{\mathbb{C}^n[t]} \omega_f^q \wedge \psi_{\xi}^{n-q} \\ &= \int_{\mathbb{C}^n[t]} \omega_f^q \wedge \left(\int_{G(q-1,n)} d\xi\, \psi_{\xi}^{n-q} \right) \end{aligned}$$

Referring to the diagram we see that $I = \left(\int_{B(q-1,n)} d\xi\, \psi_{\xi}^{n-q} \right)$ is the pull-back to $\mathbb{C}^n - \{0\}$ of $J = \int_{G(q-1,n)} d\xi\, \omega_{\xi}^{n-q}$, where ω_{ξ} is considered as a form on $\mathbb{P}^{n-1}$ with mild singularities along $[\xi]$. Now J is a form invariant under the highly transitive action of the unitary group on $\mathbb{P}^{n-1}$, and therefore has the form $C\omega^{n-q}$. One easily checks that $C = 1$, from which it follows that $I = \psi^{n-q}$. Thus

$$\int_{G(q,n)} dA \int_{A[t]} \omega_f^q = \int_{\mathbb{C}^n[t]} \omega_f^q \wedge \psi^{n-q}$$

Integrating this with respect to $\frac{dt}{t}$ completes the proof.

A consequence of (4.1) and plurisubharmonicity is the estimate

$$T_1(A, r) \leqq C_\theta \; T_1(\theta r) \quad (\theta > 1) \tag{4.4}$$

Proof: We write

$$f^* \omega = dd^c U$$

for a C^∞ plurisubharmonic potential function U on $\mathbb{C}^n$. This is possible since $\mathbb{C}^n$ has no d or d^c cohomology. Using (1.2),

$$\begin{aligned} T_1(A, r) &= \int_{\partial A[r]} U \sigma_A - U(0) \\ T_1(r) &= \int_{\partial \mathbb{C}^n[r]} U \sigma - U(0) \end{aligned} \tag{4.5}$$

where σ_A and σ are the normalized invariant measures on the spheres in A and $\mathbb{C}^n$ respectively. Since U is plurisubharmonic, it satisfies the sub-mean-value principle, and consequently

$$U(z) \leqq \frac{c}{\theta^{2n} \|z\|^{2n}} \int_{B[z, \theta \|z\|]} U \cdot \Phi$$

where $B[z, \rho]$ is the ball of radius ρ around z and Φ is Euclidean measure. Using (4.6) and the fact that

$$\int_{\partial \mathbb{C}^n[r]} U \sigma$$

is an increasing function of r,* we have

$$\begin{aligned} \int_{\partial A[r]} U \sigma_A &\leqq \frac{c}{\theta^{2n} r^{2n}} \int_{\mathbb{C}^n[(1+\theta)r]} U \cdot \Phi \\ &= \frac{c}{\theta^{2n} r^{2n}} \int_0^{(1+\theta)r} \left(\int_{\partial \mathbb{C}^n[\rho]} U \cdot \sigma \right) \rho^{2n-1} \, d\rho \\ &\leqq \frac{c}{\theta^{2n}} \int_{\partial \mathbb{C}^n[(1+\theta)r]} U \cdot \sigma \\ &= C_\theta T_1(\theta r) \end{aligned}$$

Q.E.D.

---

* $T_1(r)$ is clearly increasing, so by (4.5), $\int_{\partial C^n[r]} U \sigma$ is also

We record the following estimate for later use:

Lemma: *Let* U *be a solution of* $dd^c U = \omega_f$. *Then*

$$M_U(r) = \max_{\|z\| \leqq r} U(z) \leqq C_\theta T_1(\theta r) + \text{const.}, \tag{4.7}$$

where the constant depends on U(0) *and* $\theta > 1$.

Proof: Again the proof rests on the fact that U is plurisubharmonic. Indeed, if A is a line through the origin in $\mathbb{C}^n$, then U restricted to A is subharmonic, hence

$$U(z) \leqq \int_{\partial A[r]} P(z, \xi)\, U(\xi)\, \sigma_A(\xi)$$

where $P(z, \xi)$ is the Poisson kernel and where $z \in A$. Since $P(z, \xi) \leqq \frac{r + \|z\|}{r - \|z\|}$, we have

$$U(z) \leqq \frac{r + \|z\|}{r - \|z\|} \int_{\partial A[r]} U\, \sigma_A$$

Choosing $r = \theta \|z\|$, $\theta > 1$, gives

$$U(z) \leqq C_\theta \int_{\partial A[\theta \|z\|]} U\, \sigma_A$$

Combining this with (4.5) yields

$$\text{for } z \in A,\ U(z) \leqq C_\theta T_1(A, \theta \|z\|) + \text{constant},$$

Using (4.4), we have

$$U(z) \leqq C_\theta T_1(\theta \|z\|) + \text{constant}$$

for any z (and a different C_θ).

Q.E.D.

Remark: The use of integration over the $\mathbb{P}^{n-1}$ of lines through the origin in $\mathbb{C}^n$ originated with H. Kneser (Jahresbericht der deutcher Mathematiker Vereinigung, Vol. 48 (1938), pp 1-28). Inequalities of the kernel (4.2) are given by Kneser (loc. cit, pp 24-28), and have been used by Kujala in his study of the Cousin II problem with growth conditions (Trans. Amer. Math. Soc., vol. 161 (1971), pp. 327-358).

5. COMPARISON OF ORDER FUNCTIONS; COMMENTS ON THE MAIN PROBLEM

The basic question of this paper is:

> *Given an entire holomorphic mapping* $f\colon \mathbb{C}^n \longrightarrow M$ *into a compact Kähler manifold* M, *what, if any, relation is there among the various order functions* $T_q(r)$?

Although we are unable to give a definitive answer to this problem, we are able to give some partial results. Before stating these, a few examples and comments may be illuminating.

Example 1: We construct a map

$$f\colon \mathbb{C}^2 \longrightarrow \mathbb{C}^2 \subset \mathbb{P}^2$$

where $T_2(r) = O(\log r)$ but where $T_1(r)$ may grow arbitrarily fast. Define

$$f(z, w) = (z + h(w), w)$$

where $h(w)$ is an entire function of 1-variable. This map is a biholomorphic automorphism of $\mathbb{C}^2$, thus f is one-to-one, and so $T_2(r) = O(\log r)$ by the first Crofton formula (3.5). On the other hand, by (4.4) $T_1(r)$ will dominate the order function of f restricted to the line $z = 0$, which is then the order function for the holomorphic curve

$$w \longmapsto (h(w), w),$$

and which may be made to grow arbitrarily fast.

Thus an estimate of $T_p(r)$ in terms of $T_q(r)$ for $p < q$ is in general not possible (cf. example 4 below). However, the following comment shows that one might hope to have an estimate on $T_q(r)$ in terms of $T_1(r)$ for $q > 1$.

Example 2: Let $f\colon \mathbb{C}^n \longrightarrow M$ be an entire holomorphic mapping into a projective algebraic variety M. Since M is a compact Kähler manifold, Proposition 2.1 gives that:

> f *is a rational (= meromorphic) mapping if and only if* $T_q(r) = O(\log r)$ $(q = 1, \ldots, n)$

However, since M is projective algebraic, this assertion may be refined to (cf. section 6, [1])

> f *is a rational mapping if and only if* $T_1(r) = O(\log r)$

The reason for this is (i) f is rational if and only if $f^*(\eta)$ is a raitonal function on C^n for every meromorphic function η on M, which

(ii) is the case if and only if $f^{-1}(D)$ is an algebraic divisor on C^n for every divisor D on M, and (iii) the size of $f^{-1}(D)$ is bounded from above by $0(T_1(r))$ using the F.M.T. (3.3) and $S(D, r) \equiv 0$ in codimension one.

Thus, at least in this case, the growth of $T_1(r)$ determines that of the other $T_q(r)$. The following example shows that an estimate of roughly the sort

$$T_q(r) = 0(T_1(r)^q)$$

would be the best possible.

Example 3: Define $f: \mathbb{C}^2 \longrightarrow \mathbb{C}^2 \subset \mathbb{P}^2$ by

$$f(z, w) = (e^z, e^w)$$

we will check that

$$T_1(r) \leqq Cr$$

$$T_2(r) \geqq C'r^2$$

Proof: Using the second Crofton formula (4.1), to prove that $T_2(r) = 0(r)$ it will suffice to show that

$$T_1(A, r) \leq Cr \tag{5.1}$$

for all lines A through the origin. Using the first Crofton formula (3.5), we may compute $T_1(A, r)$ as the average number of solutions to the equation

$$ac^{\alpha t} + bc^{\beta t} + c = 0$$

$$\alpha^2 + \beta^2 = 1, \ ||t|| \leqq r$$

Now (5.1) follows easily from this.

Using (3.5), to compute $T_2(r)$ we must find the average number of solutions to the equations

$$e^z = a, \ e^w = b$$

$$|z|^2 + |w|^2 \leqq r^2$$

Another easy counting argument shows that $T_2(r) \geqq C' r^2$.

The following comments may help illustrate that an estimate of $T_q(r)$ by $T_1(r)$ is not of a general geometric character, but must rely on the Cauchy-Riemann equations.

Example 4: Let $\phi = \frac{\sqrt{-1}}{2}\left(\sum_{i=1}^{n} dz_i \wedge d\bar{z}_i\right)$ be the standard Kähler form on $\mathbb{C}^n$. An alternate formula for $T_q(r)$ is

$$T_q(r) = \int_0^r \left\{ \int_{\mathbb{C}^n[\rho]} (f^*\omega)_q \wedge \phi_{n-q} \right\} \frac{d\rho}{\rho^{2n-2q+1}} \tag{5.2}$$

The proof of this follows immediately from the definition of T_q and the fact that if α is a closed form, then

$$\frac{1}{\rho^{2k}} \int_{\mathbb{C}^n[\rho]} \alpha \wedge \phi_k = \int_{\mathbb{C}^n[\rho]} \alpha \wedge \psi_k$$

where $\psi = dd^c \log \|z\|^2$ (cf. [7] pp 72-73).

If we write $f^*\omega = \frac{\sqrt{-1}}{2}\left(\sum_{i,j} h_{ij}\, dz_i \wedge d\bar{z}_j\right)$ and let λ_i be the eigenvalues of the Hermitian matrix (h_{ij}), then clearly

$$(f^*\omega)_q \wedge \phi_{n-q} = \sigma_q(\lambda_1, \ldots, \lambda_n)\Phi \tag{5.3}$$

where $\Phi = \phi_n$ is the Euclidean volume form and σ_q is the qth elementary symmetric function of the λ_i. The only general *pointwise* relation among the σ_q's is that provided by *Newton's inequality*

$$\sigma_q^{1/q} \leqq C_{pq}\, \sigma_p^{1/p} \qquad (q \leqq p) \tag{5.4}$$

Using (5.4) in standard integral estimates, such as Hölder's inequality, fails to yield any relation among the $T_q(r)$'s. Indeed, it is pretty evident that the analogously defined quantities for a C^∞ mapping $f\colon \mathbb{R}^n \longrightarrow M$ into a compact Riemannian manifold M will in general be unrelated.

Example 5: Let Λ be a lattice in $\mathbb{C}^n$, let M be the complex torus $\mathbb{C}^n/\Lambda$, and let ω be the Kähler form on M induced from ϕ on $\mathbb{C}^n$. We calculate the order functions of the canonical projection $\pi\colon \mathbb{C}^n \longrightarrow M$ using (5.2). But $\pi^*\omega = \phi$, so

$$T_q(r) = \int_0^r \left(\int_{\mathbb{C}^n[\rho]} \phi_n \right) \frac{d\rho}{\rho^{2n-2q+1}} = C_q r^{2q}$$

for suitable constants $C_q > 0$. Therefore

$$T_q(r) = O([T_p(r)]^{q/p})$$

in this special case.

6. STATEMENT OF PRINCIPAL RESULTS AND COROLLARIES

Let $f: \mathbb{C}^n \longrightarrow M$ be a non-degenerate entire holomorphic mapping into an n-dimensional projective algebraic variety M. We shall use the terminology and notations from section 0 of [6] concerning divisors, line bundles, and Chern classes.

Theorem I: *Suppose that the image of* f *omits a divisor* D *which satisfies*

$$\mu c_1(D) + c_1(K_M) \geqq 0$$

for some $\mu \geq 0$. *Then, given* $\theta > 1$, *an estimate*

$$T_n(r) \leqq C_\theta T_1(\theta r)^n e^{C_\mu T_1(\theta r)}$$

is valid.

Corollary 6.1: *If* $c_1(K_M) \geqq 0$, *then*

$$T_n(r) \leqq C_\theta T_1(\theta r)^n$$

In particular this is the case if M *is a complex-torus, a hypersurface of degree* $\geq n+2$ *in* $\mathbb{P}^n$, *etc.*

We are not sure if the exponential factor $e^{C_\mu T_1(\theta r)}$ is really necessary. It seems reasonable in the general case, when f may not omit such a divisor D, to seek an estimate on $T_n(r)$ where the size of $f^{-1}(D)$ appears on the right hand side, but we have been unable to do this.

To state our second result, we assume given an entire holomorphic mapping $f: \mathbb{C}^n \longrightarrow M$ into a not-necessarily compact complex manifold having a Hermitian metric ds_M^2. The order functions $T_q(r)$ can then be defined relative to the given ds_M^2. Moreover, this metric induces an intrinsic connection on M with curvature Θ, and we denote by

$$\Theta(\xi, \eta) = \sum_{i,j,k,\ell} \Theta_{i,j,k,\ell}\, \xi^i \bar{\xi}^j \eta^k \bar{\eta}^\ell$$

the corresponding biquadratic *curvature form* (cf. [3]) written here relative to a unitary frame.

Theorem II: *If* $\Theta(\xi, \eta) \leqq 0$, *then for* $q > p$

$$T_q(r)^{1/q} \leqq C_\theta T_p(\theta r)^{1/p}$$

for any constant $\theta > 1$.

We remark that the theorem is sharp because of example five in the previous section.

Corollary 6.3: *Assume that* M *is a projective algebraic variety with ample divisor* D *and that* $f: \mathbb{C}^n \longrightarrow M-D$ *is a holomorphic mapping. Then there is an estimate*

$$T_q(r)^{1/q} \leqq C_\theta T_p(\theta r)^{1/p} e^{C_\mu T_1(\theta r)}$$

for $\theta > 1$ *and a suitable* $\mu > \theta$.

Again we are unable to determine if the exponential factor is really necessary. A related result without the exponential factor is given by Theorem III below.

To state our final result, we assume that M is a projective variety and D is an ample divisor on M.

Theorem III: *For any holomorphic mapping* $f: \mathbb{C}^n \longrightarrow M-D$ *and constant* $\theta > 1$,

$$T_q(r) \leqq C_\theta T_1(\theta r)^q$$

This result is similar to Corollary 6.3, but is stronger in that the exponential factor $e^{C_\mu T_1(\theta r)}$ does not occur. Example 3 in section 5 shows that this estimate is sharp. As repeatedly suggested, we are unable to decide if the conclusion of Theorems III is valid for an arbitrary holomorphic mapping into a compact Kähler manifold.

7. PROOFS OF MAIN RESULTS

Proof of Theorem I: Let $f: \mathbb{C}^n \longrightarrow M-D$ be a non-degenerate holomorphic mapping as in the statement of the theorem, ω a Kähler form on M, and $\Omega = \omega_m$ the corresponding volume element. By assumption there exists a fibre metric in the line bundle $[D] \longrightarrow M$ such that

$$\mu c_1([D]) + \text{Ric}\,\Omega \geqq 0 \qquad (7.1)$$

Let $\delta \in \theta(M, [D])$ be a holmorphic section which defines D and $|\delta|^2$ the corresponding length function. We may assume that $|\delta|^2 \leqq 1$, and we define the singular form

$$\Psi = \frac{\Omega}{|\delta|^{2\mu}} \geqq \Omega \tag{7.2}$$

Setting

$$\Phi = \prod_{i=1}^{n} \left(\frac{\sqrt{-1}}{2} dz_i \wedge d\bar{z}_i \right)$$

$$f^* \Psi = \xi \cdot \Phi$$

the condition (7.1) is equivalent to saying that $\log \xi$ *is a plurisubharmonic function on* $\mathbb{C}^n$. By the sub-mean-value principle for such functions

$$\log \xi(z) \leq \frac{c}{r^{2n}} \int_{B[z,r]} \log \xi \ \Phi \tag{7.3}$$

where $B[z,r]$ is the ball of radius r centered at z.

Now we let λ_i be the eigenvalues of $f^*\omega$ and σ_q the qth elementary symmetric function of the λ_i as in example 4 of section 5. Then by (5.4)

$$\sigma_n^{1/n} \leqq c \cdot \sigma_1 \tag{7.4}$$

On the other hand, by (7.2)

$$\xi = \frac{\sigma_n}{|\delta|^{2\mu}} \geqq \sigma_n$$

This together with (7.4) yields

$$\log \xi \leqq n \log \sigma_1 + \mu \log \frac{1}{|\delta|^2} + O(1). \tag{7.5}$$

From (7.5), (7.3), and concavity of the logarithm we obtain

$$\log \xi(z) \leqq n \log \left\{ \frac{c}{r^{2n}} \int_{B[z,r]} \sigma_1 \Phi \right\} + \frac{Mc}{r^{2n}} \int_{B[z,r]} \log \frac{1}{|\delta|^2} \Phi + O(1) \tag{7.6}$$

We shall now estimate the first term on the right of (7.6). By (5.2) and (5.3), for $\alpha > 1$

$$\frac{1}{s^{2n-2}} \int_{\mathbb{C}^n[s]} \sigma_1 \Phi \leqq C_\alpha \int_s^{\alpha s} \left(\int_{\mathbb{C}^n[\rho]} \sigma_1 \Phi \right) \frac{d\rho}{\rho^{2n-1}}$$

$$\leqq C_\alpha \int_0^{\alpha s} \left(\int_{\mathbb{C}^n[\rho]} \sigma_1 \Phi \right) \frac{d\rho}{\rho^{2n-1}}$$

$$= C_\alpha T_1(\alpha s)$$

where $C_\alpha = \dfrac{c(n)\alpha^{2n-2}}{\alpha^{2n-2}-1}$ tends to one as $\alpha \longrightarrow \infty$. (Thus we may let α vary provided that it is kept bounded away from one). This gives the inequality

$$\frac{c}{s^{2n}} \int_{\mathbb{C}^n[s]} \sigma_1 \Phi \leqq \frac{C_\alpha}{s^2} T_1(\alpha s) \quad (\alpha > 1) \tag{7.7}$$

Choosing $r = \beta \|z\|$ and setting $\alpha = 1 + \beta$, (7.7) gives

$$\begin{aligned} n\log\left[\frac{c}{r^{2n}} \int_{B[z,r]} \sigma_1 \Phi\right] &\leqq n\log\left[\frac{C\alpha}{\|z\|^{2n}} \int_{C^n[\alpha\|z\|]} \sigma_1 \Phi\right] \\ &\leqq n\log\left[\frac{T_1(\alpha\|z\|)}{(\alpha\|z\|)^2}\right] + O(1) \end{aligned} \tag{7.8}$$

provided that α is kept bounded away from one.

To estimate the second term on the right of (7.6), we observe that

$$\int_{\partial\mathbb{C}^n[r]} \log \frac{1}{|\delta|^2} \sigma = T_1(r) + O(1)$$

where σ is the invariant volume element on the sphere $\|z\| = r$ with $\int \sigma = 1$. This is the F.M.T. in this situation, and results from (1.2) and $dd^c \log \frac{1}{|\delta|^2} = f^* \omega$. Since $|\delta|^2 \leqq 1$, we can integrate the F.M.T. to obtain

$$\frac{c}{s^{2n}} \int_{\mathbb{C}^n[s]} \log \frac{1}{|\delta|^2} \Phi \leqq T_1(s) + O(1). \tag{7.9}$$

Arguing as in the proof of (7.8), we have

$$\frac{\mu c}{r^{2n}} \int_{B[z,r]} \log \frac{1}{|\delta|^2} \Phi \leqq C_\alpha \mu T_1(\alpha\|z\|) + O(1) \tag{7.10}$$

Combining (7.6), (7.8), and (7.10) gives

$$\xi(z) \leqq C \frac{T_1(\alpha\|z\|)^n}{(\alpha\|z\|)^{2n}} e^{C_\mu T_1(\alpha\|z\|)}$$

which, upon choosing $\alpha\|z\| = \theta r$ where $\|z\| \leqq r$, leads to

$$\xi(z) \leqq \frac{C\, T_1(\theta r)^n e^{C_\mu T_1(\theta r)}}{r^{2n}} \quad (\|z\| \leqq r) \tag{7.11}$$

Now

$$\begin{aligned} T_n(r) &= \int_0^r \left(\int_{\mathbb{C}^n[\rho]} \sigma_n \Phi\right) \frac{d\rho}{\rho} \quad \text{(by (5.2))} \\ &\leqq \int_0^r \left(\int_{\mathbb{C}^n[\rho]} \xi \Phi\right) \frac{d\rho}{\rho} \quad \text{(by (7.2))} \end{aligned}$$

$$\leqq C\ T_1(\theta r)^M e^{C_\mu T_1(\theta r)} \qquad \text{(by 7.11))}$$

This completes the proof of Theorem I.

Q.E.D.

Proof of Theorem II: Let $f: \mathbb{C}^n \longrightarrow M$ be as in the statement of Theorem II, and write

$$(f^* \omega)_q \wedge \phi_{n-q} = \sigma_q \Phi$$

as was done in example 4 of section 5. Using the curvature assumption $\Theta(\xi, \eta) \leqq 0$ together with (6.2) and formula (4.13) on page 307 in Lu [8], we find that

$$\Delta \log \sigma_q \geqq 0$$

Thus $\log \sigma_q$ is a sub-harmonic function on $\mathbb{C}^n \cong R^{2n}$, and by the sub-mean-value principle

$$\log \sigma_q(z) \leqq \frac{c}{r^{2n}} \int_{B[z,r]} \log \sigma_q \Phi$$

Combining this with Newton's inequality (5.4) and concavity of the logarithm gives

$$\log \sigma_q(z) \leqq \frac{q}{p} (\log \frac{c}{r^{2n}} \int_{B[z,r]} \log \sigma_p \Phi) \tag{7.12}$$

Arguing as in the proof of Theorem I, (7.12) leads to an estimate

$$\log \sigma_q(z) \leqq \frac{q}{p} \log \left\{ \frac{T_p(\theta r)}{r^{2p}} \right\} + O(1), \tag{7.13}$$

valid for $||z|| \leqq r$ and $\theta > 1$. By (5.2)

$$T_q(r) = \int_0^r (\int_{\mathbb{C}^n[\rho]} \sigma_q \cdot \Phi) \frac{d\rho}{\rho^{2n-2q+1}}$$

$$\leqq C \left(\frac{T_p(\theta r)}{r^{2q}}\right)^{q/p} \int_0^r (\int_{\mathbb{C}^n[\rho]} \Phi) \frac{d\rho}{\rho^{2n-2q+1}}$$

$$= C\ T_p(\theta r)^{q/p},$$

which gives Theorem II.

Q.E.D.

Proof of Corollary 6.3: Given $f: \mathbb{C}^n \longrightarrow M\text{-}D$, choose a metric ds_M^2 and set

$$d\tilde{s}^2 = \frac{1}{|\delta|^{2M}} ds_M^2$$

where $\delta \in \phi(M, [D])$ defines D as in the proof of Theorem I. The curvature form $\tilde{\Theta}$ for $d\tilde{s}^2$ is given by (cf. [3])

$$\tilde{\Theta}(\xi, \eta) = \Theta(\xi, \eta) - \mu|\xi|^2 |\eta|^2,$$

where we have chosen the given Kähler metric on M to be $c_1([D]) = dd^c \log \frac{1}{|\delta|^2}$. For M sufficiently large, $\tilde{\Theta}(\xi, \eta) \leqq 0$, and so we may apply Theorem II to $N = M-D$ and $d\tilde{s}^2$. When we do this, we obtain (6.3) where the exponential factor appears exactly as in the proof of Theorem I.

Q.E.D.

Proof of Theorem III: Using the second Crofton formula (4.1), we find that to prove the result for $T_q(r)$, it suffices to verify it for the restriction of f to A for all q-planes A through the origin.

Relabelling, we will assume given $f: \mathbb{C}^n \longrightarrow M-D$ and prove Theorem III for the highest order function $T_n(r)$. The argument will be by induction, assuming given the result for $T_{n-1}(r)$.

Lemma: *Let* V *be a Kähler manifold with Kähler form* ω. *Let* $f: \mathbb{C}^n \longrightarrow V$ *be holomorphic and let* U *be a solution of* $dd^c U = \omega_f$ *such that* $U(0) = 0$. *Then*

$$T_n(r) \leqq C_\theta T_1(\theta r) T_{n-1}(r) - \int_{\mathbb{C}^n[r]} U(\omega_f)_{n-1} \wedge dd^c \log \|z\|^2 \qquad (7.14)$$

Proof:

$$T_n(r) = \int_0^r \left(\int dd^c U \wedge (f^* \omega)_{n-1} \right) \frac{d\rho}{\rho}$$

$$= \int_{\partial\mathbb{C}^n[r]} U(f^* \omega)_{n-1} \wedge d^c \log \|z\|^2 - \int_{\mathbb{C}^n[r]} U(f^* \omega)_{n-1} \wedge dd^c \log \|z\|^2 \quad \text{(by (1.2))}$$

$$\leqq M_U \int_{\mathbb{C}^n[r]} (f^* \omega)_{n-1} \wedge dd^c \log \|z\|^2 - \int_{\mathbb{C}^n[r]} U(f^* \omega)_{n-1} \wedge dd^c \log \|z\|^2 \quad \text{(using Stokes' theorem)}$$

Now for any increasing function $g(r)$, we have

$$(\log \theta)\, g(r) \leqq \int_r^{\theta r} g(\rho) \frac{d\rho}{\rho} \leqq \int_0^{\theta r} g(\rho) \frac{d\rho}{\rho} \qquad (7.15)$$

Using (4.7) and (7.15) with

$$g(r) = \int_{\mathbb{C}^n[r]} (f^* \omega)_{n-1} \wedge dd^c \log \|z\|^2,$$

we obtain (7.14).

Q.E.D.

Thus, to obtain a bound for T_n by T_{n-1} and T_1, we must bound U from below in some fashion. For $M-D$ we do this as follows. We imbed M into $\mathbb{P}^m$ in such a way that D is the hyperplane section $Z_0 = 0$ relative to suitable homogeneous coordinates $Z = [Z_0, \ldots, Z_n]$. Then we view f as a map $f: \mathbb{C}^n \longrightarrow \mathbb{P}^m - \{Z_0 = 0\}$, and we calculate the T_q relative to the Fubini-Study metric $\omega = dd^c \log \|z\|^2$. This does not affect the relative magnitudes of the order functions in an essential way because of the properties of the T_q under change of metric discussed in section 1. Now the functions $\phi_\alpha = f^*(Z_\alpha/Z_0)$ are holomorphic on $\mathbb{C}^n$, and we have

$$\begin{cases} f(z) = [1, \phi_1(z), \ldots, \phi_m(z)] \\ \omega_f = dd^c \log(1 + \Sigma|\phi_\alpha(z)|^2) \end{cases}$$

Thus we can choose $U = \log(1 + \Sigma|\phi_\alpha|^2)$ in the lemma. Since $U \geqq 0$, the second term on the right of (7.14) is nonpositive, giving

$$T_n(r) \leqq C_\theta \, T_1(\theta r) \, T_{n-1}(r)$$

By the induction hypothesis, we are done.

Remark (1): One other case in which the potential function can be satisfactorily estimated is for V a complex torus. In fact, let $\phi = \frac{\sqrt{-1}}{2} \Sigma \, dz_i \wedge d\bar{z}_i$ be the standard Kähler form on C^n, ω the Kähler form on V induced by the canonical projection in the universal covering $\pi: \mathbb{C}^n \longrightarrow V$. Given $f: \mathbb{C}^m \longrightarrow V$, we can lift to $\tilde{f}: \mathbb{C}^m \longrightarrow \mathbb{C}^n$, and $\omega_f = \phi_{\tilde{f}} = dd^c \|z\|^2$. Thus $U = \|z\|^2 \geqq 0$, and the same argument applies.

(2): For a general map $f: \mathbb{C}^n \longrightarrow \mathbb{P}^m$, we may use the Cousin II problem with growth conditions to write

$$f(z) = [f_0(z), \ldots, f_m(z)]$$

where the $f_\alpha(z)$ are holomorphic, have no common zeroes, and grow essentially like $T_1(\theta r)$. However, we are unable either to bound

$$U = \log \|f\|^2$$

from below, or failing to do this, to estimate the growth of the integral

$$\int_{\mathbb{C}^n[r]} U(f^* \omega)_{n-1} \wedge \psi \tag{7.15}$$

in terms of $M(r) = \max_{z \leqq r} \log ||f||^2$.

Such an estimate will fail when f is too close to zero on a set whose volume grows too fast. Now Cornalba and Shiffman [2] construct a function $f: \mathbb{C}^2 \longrightarrow \mathbb{C}^2$ such that $M(r) = O(r^\varepsilon)$ and $n(D, r) \geqq e^r$ where $D = \{z: f(z) = 0\}$. Therefore, it seems possible that there exist functions f which are (i) of finite order, (ii) nowhere zero, but such that (iii) *the attraction of* f *to zero* is so strong that the integral

$$\int_{\mathbb{C}^n[r]} \log \frac{1}{||f||^2} (f^* \omega_{n-1}) \wedge \psi$$

has infinite order. In conclusion, it seems that the counterexample to the transcendtal Bezout problem works against the possibility of estimating T_q by T_1 in the general case.

REFERENCES

1. Carlson, J., and P. Griffiths, *A defect relation for holomorphic mappings between algebraic varieties,* Ann. of Math., 95(1972).

2. Cornalba, M., and B. Shiffman, *A counterexample to the "Transcental Bezout problem"* , Anals of Math. 96(1972) pp 402-406.

3. Griffiths, P. A., *Hermitian differential geometry, Chern classes, and positive vector bundles,* Global Analysis, a volume in honor of K. Kodaira, Tokyo, University of Tokyo Press, 1969.

4. _________ , *Two theorems about holomorphic mappings,* to appear in Studies in analysis, a volume in honor of Lipman Bers.

5. _________ , *Two theorems on extensions of holomorphic mappings,* Inventiones Math. 14(1971) pp 27-62.

6. _________ , and J. King, *Nevanlinna theory and holomorphic mappings between algebraic varieties,* Acta Math. 130(1973) pp 145-220.

7. Lelong, P., *Plurisubharmonic Functions and Differential Forms,* Gordon and Breach, New York, 1969.

8. Lu, Y. C., *Holomorphic mappings of complex manifolds,* J. Diff. Geom. 2 (1968) pp 299-312.

9. Nevanlinna, R., *Analytic Functions*, Springer-Verlag, New York, 1970.

10. Shiffman, B., *Applications of geometric measure theory to the value distribution theory of meromorphic maps*, in this volume, 63-95.

11. Stoltzenberg, G., *Volume, Limits and Analytic Sets*, Springer-Verlag, New York, 1966.

12. Wu, H., *Remarks on the First Main Theorem in equidistribution theory*, II, Journal of Differential Geometry 2(1968) pp. 369-384.

GENERALIZED BLASCHKE CONDITIONS ON THE UNIT BALL IN $\mathbb{C}^p$

Robert O. Kujala
Tulane University

The purpose of this paper is to provide a brief introduction to two interrelated problems which arise in Value-Distribution Theory but which are not explicitly treated in the other contributions to these Proceedings.

Let us adopt the following terminology and notation: A *divisor* (or *index-function*) on a connected complex manifold M is an integer-valued function on M which is locally the difference in the zero-multiplicity functions of two holomorphic functions. Let $\mathcal{D}(M)$ be the set of all divisors on M. This is a subgroup of the group of all integer-valued functions on M with respect to addition. Let $\mathcal{D}^+(M)$ denote the semigroup of nonnegatively valued elements of $\mathcal{D}(M)$. Let $\text{Mer}(M)$ and $\text{Hol}(M)$ be the field of all meromorphic functions on M and the integral domain of all holomorphic functions on M, respectively; let $\text{Mer}^*(M)$ denote the multiplicative group of $\text{Mer}(M)$, that is, the meromorphic functions which do not vanish identically on M, and let $\text{Hol}^*(M)$ be the semigroup of holomorphic elements of $\text{Mer}^*(M)$. Then we have *the divisor map* ν from $\text{Mer}^*(M)$ into $\mathcal{D}(M)$ defined by letting $\nu(f)$ for f in $\text{Mer}^*(M)$ be ν_f, *the divisor of* f, which is the difference between the zero-multiplicity and pole-multiplicity functions of f. This map is a group homomorphism carrying $\text{Hol}^*(M)$ into $\mathcal{D}^+(M)$. If M is a Cousin II domain, then, by definition, ν maps $\text{Hol}^*(M)$ *onto* $\mathcal{D}^+(M)$ and, therefore, maps $\text{Mer}^*(M)$ *onto* $\mathcal{D}(M)$.

The first problem we will consider is what I choose to call a *Generalized Blaschke Problem*, namely: Given a subset A of $\text{Hol}^*(M)$, characterize $\nu(A)$ as a subset of $\mathcal{D}^+(M)$, that is, give necessary and sufficient conditions that μ in $\mathcal{D}^+(M)$ is $\nu(f)$ for some f in A. This could also be called a *Restricted Second Cousin Problem* but the desired solution is a characterization of divisors *as divisors* and not a characterization of a particular set of Cousin data which belongs to a divisor (as in Stout [13] which may be viewed as a localization of the problem). Generally speaking, the other papers in this volume deal with describing $\nu(f)$ for a given f, that is, they deal with necessary conditions in a solution to a Generalized Blaschke Problem. Thus, the new element here is the "reverse" problem, namely, the "construction" of "canonical" representatives of $\nu^{-1}(\mu)$ for a given μ in $\mathcal{D}^+(M)$.

The second type of problem is a *Quotient Representation Problem* (or *Restricted Poincaré Problem*), namely: Given a subfield F of Mer(M), compare F and the field of quotients of the integral domain of the holomorphic elements of F. This problem is related to the preceeding as follows: Let A be the holomorphic elements of F^*. If we can show that for each f in F^* the pole divisor of f is dominated by some μ in $\nu(A)$ (which involves at least the sufficient condition part of the solution to the Generalized Blaschke Problem for A), then we can choose g in A so that $\nu(g) = \mu$ and then $f = (fg)/g$ will give the desired quotient representation. A stronger solution may be desired, namely, a quotient representation in which the numerator and denominator are locally relatively prime in the ring of germs of holomorphic functions at each point. This is the same as requiring that the pole divisor of each function in F^* actually be an element of $\nu(A)$. The "value" of solving such a problem is the fact that such quotient representations reduce the study of classes of meromorphic functions to the much more highly developed theory of holomorphic functions. For example, the study of characteristic functions can be replaced by maximum-modulus statements.

Of course, the most extensive work on problems of this kind has been done in the case when $M = \mathbb{C}$. The works of Rubel and Taylor [8] and Miles [6] show that (weak) quotient representations are always possible when F is a set of meromorphic functions f such that the characteristic of f, T_f, satisfies an inequality of the form $T_f(r) \leq A\lambda(Br)$ for r sufficiently large and some pair of positive constants A and B when λ is any given nondecreasing continuous function from $(0,\infty)$ to $(0,\infty)$. These are the functions of finite λ-type in the plane. They also treat cases when strong representations are impossible. Extensive work of a similar nature for the case $M = \mathbb{C}^p$ will be found in the papers of Stoll [11] and myself [3].

Some work on the case $M = B_1$ and $M = B_p$ where B_p is the unit ball in $\mathbb{C}^p$ will be found in Tsuji [14] and Mueller [7], respectively.

Actually, the Blaschke Problem has a well-known solution in the case $M = B_1$ and $A = H^p(B_1)^*$ when $0 \le p \le \infty$, namely, $\nu(H^p(B_1)^*) = \nu(H^0(B)^*)$ for $0 \le p \le \infty$, where $H^0(B_1)$ is the Nevanlinna class of functions of bounded characteristic, and $\mu \in \nu(H^0(B_1)^*)$ if and only if

$$N_\mu(r) = \sum_{|z| \le r} \mu(z) \log \frac{r}{|z|} \quad \text{is bounded for all } r \in (0,1)$$

(in the case $\mu(0) = 0$). These results may be found in Rudin [9], Chapters 15 and 17.

If, however, we turn our attention to B_p for $p \ge 2$ then very little of a positive nature is known concerning the Generalized Blaschke Problem or the Quotient Representation Problem for $H^q(B_p)$. In order to examine this situation more carefully and show some of the behavior which can occur let us introduce the following notations: for $f \in \text{Mer}\ (B)$, with $B = B_p$ and $f(0) \in \mathbb{C}$

$$T_f(r) = \int_0^r A_f(t) \frac{dt}{t} \quad \text{for} \quad 0 \le r < 1$$

$$\text{where} \quad A_f(r) = \frac{1}{\pi^p r^{2p-2}} \int_{rB} (1 + |f|^2)^{-2} \frac{i}{2} df \wedge d\bar{f} \wedge \omega$$

$$\text{and} \quad \omega(z_1, \ldots, z_p) = \left[\sum_{j=1}^{p} \frac{i}{2} dz_j \wedge d\bar{z}_j\right]^{p-1}.$$

And for μ in $\mathcal{D}^+(B_p)$ with $\mu(0) = 0$ let

$$N_\mu(r) = \int_0^r n_\mu(t) \frac{dt}{t} \quad \text{for} \quad 0 \le r < 1$$

$$\text{where} \quad n_\mu(r) = \sum_{|z| \le r} \mu(z) \quad \text{if} \quad p = 1 \quad \text{and}$$

$$n_\mu(r) = \frac{1}{\pi^p r^{2p-2}} \int_{\{\mu > 0\} \cap rB} \mu\omega \quad \text{for} \quad p \ge 2.$$

Remark: When A is an analytic set of codimension one in B for $p \ge 2$, then $\int_A \omega$ is $(p - 1)!$ times the real $(2p - 2)$-dimensional Hausdorff measure of (the regular points of) A.

Moreover, if ξ is a unit vector in $\mathbb{C}^p$ and σ is the Euclidean volume element on the unit sphere ∂B in $\mathbb{C}^p$ (oriented with outward normal) normalized so that the volume of the sphere is one, then we define:

$$f|\xi(z) = f(z\xi) \quad \text{for} \quad z \in B_1 \quad \text{and} \quad f \in \mathrm{Mer}(B_p) \quad \text{with} \quad f(0) \in \mathbb{C}.$$

$$\mu|\xi(z) = \nu_{f|\xi}(z) \quad \text{for} \quad z \in B_1 \quad \text{and} \quad \mu = \nu_f.$$

$$A_f(r;\xi) = A_{f|\xi}(r)$$

$$T_f(r;\xi) = T_{f|\xi}(r)$$

$$n_\mu(r;\xi) = n_{\mu|\xi}(r)$$

$$N_\mu(r;\xi) = N_{\mu|\xi}(r).$$

Then $F(r) = \int_{\partial B} F(r;\xi)\sigma(\xi)$ when $F = A_f, T_f, n_\mu,$ or N_μ. (See Stoll [11] or [12] or Shiffman's Crofton Formula in this volume for a discussion of these results.)

Thus, one analogue of the solution of the Blaschke Problem for $H^\infty(B_1)^*$ is the condition

(I) N_μ is bounded on $(0,1)$ ($\mu \in \mathcal{D}^+(B)$ with $\mu(0) = 0$).

This is easily seen to be equivalent to

$$\text{(I')} \qquad \sup_{0<r<1} \int_{rB\cap\{\mu>0\}} \mu(\xi)(1 - \|\xi\|)\,\omega(\xi) < \infty$$

which is the analogue of the more familiar Blaschke condition for $H^\infty(B_1)^*$,

$$\sum_{|z|<1} \mu(z)(1 - |z|) < \infty .$$

That (I) is indeed a necessary condition on μ for μ to be in $\nu(H^\infty(B)^*)$ is a consequence of the Generalized Jensen Formula (Stoll [11] or [12]). In fact, it follows from the First Main Theorem (op. cit.) that (I) is a necessary condition on μ in order that μ be in $\nu(H^0(B)^*)$. However, there is an example of Chee ([2], p. 262) of a divisor μ on B_2 satisfying (I) but such that μ is determining for $H^\infty(B_2)$, that is, the only bounded holomorphic function on B_2 which vanishes on $\{\mu > 0\}$ is identically zero, or, alternately stated, there is no function f in $H^\infty(B_2)^*$ such that $\nu(f) \geq \mu$ on B_2.

This difference in behavior between one and several variables can be "explained" by the following two results (see Rudin [10] for polydisc analogues):

Proposition 1: *Suppose* μ *is a nonnegative divisor on* B_p *with* $\mu(0) = 0$. *Then* N_μ *is bounded on* $(0, 1)$ *if and only if for every* f *in* $\nu^{-1}(\mu)$, $\log|f|$ *has an harmonic majorant on* B_p.

Proof: Suppose $f \in \nu^{-1}(\mu)$ and $\log|f|$ has an harmonic majorant H on B_2. Then by the Generalized Jensen Formula

$$N_\mu(r) = \int_{\partial B} \log|f(r\xi)|\sigma(\xi) - \log|f(0)|$$
$$\leq \int_{\partial B} H(r\xi)\sigma(\xi) - \log|f(0)| = H(0) - \log|f(0)|$$

by the mean-value property for harmonic functions. Thus N_μ is bounded on (0,1). (For the properties of harmonic and subharmonic functions used in this proof see Carathéodory [1].)

Conversely, suppose N_μ is bounded on (0, 1) and $f \in \nu^{-1}(\mu)$. For each $r \in (0, 1)$ let

$$h_r(\zeta) = r^{2p-2} \int_{\partial B} \log|f(r\xi)| \frac{r^2 - \|\zeta\|^2}{\|r\xi - \zeta\|^{2p}} \sigma(\xi).$$

Then h_r is an harmonic majorant of $\log|f|$ on rB since $\log|f|$ is plurisubharmonic and, therefore, subharmonic, and the Poisson kernel for B appears in the integral. It follows that for $0 < r < s < 1$ and $\zeta \in rB$

$$h_r(\zeta) \leq r^{2p-2} \int_{\partial B} h_s(r\xi) \frac{r^2 - \|\zeta\|^2}{\|r\xi - \zeta\|^2} \sigma(\xi) = h_s(\zeta).$$

Moreover, $h_r(0) = N_\mu(r) + \log|f(0)|$ by the Generalized Jensen Formula. Therefore, $\lim_{r\uparrow 1} h_r(0)$ exists, and, by Harnack's Principle, $\lim_{r\uparrow 1} h_r$ is harmonic on B and clearly majorizes $\log|f|$ there.

Q.E.D.

Remark: Clearly, the majorant produced above is the *least* harmonic majorant of $\log|f|$.

Proposition 2: *Suppose* μ *is a nonnegative divisor on* B_p. *Then* μ *is in* $\nu(H^\infty(B_p)^*)$ *if and only if for every* f *in* $\nu^{-1}(\mu)$, $\log|f|$ *has a pluriharmonic majorant on* B_p.

Remark: $H: B_p \longrightarrow \mathbb{R}$ is *pluriharmonic* means H is of class C^2 and $dd^c H = \partial\bar{\partial} H = 0$ on B_p, or, equivalently, H is the real part of a holomorphic function on B_p, or H is both plurisub- and plurisuperharmonic on B_p.

Proof: Suppose $\mu = \nu(f)$ for f in $H^\infty(B)^*$ and $g \in \nu^{-1}(\mu)$. Then g/f is holomorphic without zeros on B so that $\log|g/f|$ is pluriharmonic on B. But $\log|g| = \log|g/f| + \log|f|$ which is majorized by the pluriharmonic function $\log|g/f| + \log\|f\|_B$.

Conversely, suppose there is some f in $\nu^{-1}(\mu)$ such that $\log|f|$ has a pluriharmonic majorant H on B. Let $H = \operatorname{Re} h$ where h is holomorphic on B. Then fe^{-h} is in $\nu^{-1}(\mu)$ and

$$|fe^{-h}| = \exp(\log|f| - H) \leq 1 \quad \text{so that } \mu \text{ is in } \nu(H^{\infty}(B)^{*}).$$

Q.E.D.

Remark: Since the classes of pluriharmonic and harmonic functions are identical on the disc, the combination of the above Propositions gives a "quick" solution to the Blaschke Problem for $H^{\infty}(B_1)$ without the use of the Blaschke product.

Since $$2T_f(r) = \int_{\partial B} \log(1 + |f(r\xi)|^2)\sigma(\xi) - \log(1 + |f(0)|^2)$$

for f in $\operatorname{Hol}(B_p)$ by the First Main Theorem and since $\log(1 + |f|^2)$ is plurisubharmonic in this case, the technique of the proof of Proposition 1 can be used to obtain (compare Markushevich, pp. 240-241):

Proposition 3: f *is in* $H^0(B_p)$ *if and only if* $\log(1 + |f|^2)$ *has an harmonic majorant on* B_p.

We also recall the following proof of:

Nevanlinna's Theorem: *Every meromorphic function having bounded characteristic on* B_1 *is the quotient of two bounded holomorphic functions which are locally relatively prime at every point of* B_1 (i.e., *that have no common zeros*).

Proof: Suppose f is a meromorphic function with bounded characteristic on B_1. We can assume $f(0) = 1$ (if not, either $f \equiv 0$ and we are done or we can multiply f by a constant after performing a biholomorphism of the disc to move the origin to a non-zero of f). By the First Main Theorem N_μ is bounded on $(0, 1)$ when $\mu = \nu_f^{\infty}$ so by Propositions 1 and 2 there is a bounded holomorphic function g on B_1 such that $\nu(g) = \mu$. It follows that fg is a holomorphic function of bounded characteristic on B_1 and $\nu(fg) = \nu_f^0$. By Proposition 3 $\log(1 + |fg|^2)$ has an harmonic majorant H on B_1. Thus $0 \leq \log^+|fg| \leq \log(1 + |fg|^2) \leq H$ on B_1. Let h be holomorphic on B_1 with $H = \operatorname{Re} h$. Then $\nu(ge^{-h}) = \nu_f^{\infty}$ and $|ge^{-h}| = |g|\exp(-H) \leq |g|$ and $\nu(fge^{-h}) = \nu_f^0$ and

$$|fge^{-h}| \leq \exp(\log^+|fg| - H) \leq 1.$$

Therefore, $f = (fge^{-h})/(ge^{-h})$ is the desired representation.

Q.E.D.

Remark: Thus we have a strong solution to the Quotient Representation Problem for functions of bounded characteristic and even more since the representing quotient contains *bounded* holomorphic functions and holomorphic functions of bounded characteristic need not be bounded (see, for example, Markushevich, p. 241).

We will now exhibit an example of a nonnegative divisor on B_2 (the choice $p = 2$ is merely for notational convenience) which exhibits some remarkable behavior and provides some negative answers to many of the questions under discussion.

We begin by choosing z_j in B_1 for each j in $\mathbb{N}$ so that

(i) $|z_j| = 1 - j^{-a}$ where $2/3 < a < 1$.

(ii) $|z_j - 1/3| = 2/3$

(iii) $\operatorname{Im} z_j > 0$.

Let $Z = \{zj: j \in \mathbb{N}\}$

Remark: Condition (iii) is for uniqueness, condition (i) will be used immediately and the role of condition (ii) will become clear later.

By (i) $\sum(1 - |z_j|)^{(3+2b)/2} < \infty$ for every $b > 1/a - 3/2$ which is negative, so that by a theorem of Tsuji [14] there is a function f holomorphic on B_1 such that $\nu(f) = 1$ on Z and zero otherwise and, moreover, $T_f(r) \le A(1 - r)^{-1/2}$ for some positive constant A and all r in $(0, 1)$. Let $F(z, w) = f(z)$. Then F is holomorphic on $B_1 \times \mathbb{C}$ which contains B_2. Moreover, for each $\xi = (u, v)$ in ∂B_2 we have

$$2T_{F|\xi}(r) = \frac{1}{2\pi}\int_0^{2\pi} \log(1 + |f(rue^{it})|^2)\,dt - \log(1 + |f(0)|^2)$$

$$= 2T_f(r|u|) \le 2A(1 - r|u|)^{-1/2}$$

so that

$$T_F(r) \le \int_{\partial B_2} A(1 - r|u|)^{-1/2}\,\sigma(u, v)$$

Using $S: [0, 1] \times [0, 2\pi]^2 \longrightarrow \partial B_2$ given by $S(t, \theta, \phi) = (te^{i\theta}, \sqrt{1 - t^2}\, e^{i\phi})$ as a parameterization of ∂B_2 and the fact that $\sigma(u, v)$ is the pull-back to ∂B_2 of the global form

$$\frac{1}{2\pi^2}\left(\frac{i}{2}\right)^2 [ud\bar{u} - \bar{u}du + vd\bar{v} - \bar{v}dv]$$

we compute

$$\int_{\partial B_2} A(1 - r|u|)^{-1/2}\,\sigma(u, v) = \frac{1}{2\pi^2}\int_0^{2\pi}\int_0^{2\pi}\int_0^1 \frac{At}{(1 - rt)^{1/2}}\,dt d\theta d\phi$$

$$\leq 2A\int_0^1 (1 - rt)^{-1/2}\,dt \leq 2A\int_0^1 (1 - t)^{-1/2}\,dt = 4A$$

for r in $(0, 1)$.

Thus, F is in $H^0(B_2)$. Let us examine the divisor of F more closely. Let $V_j = [\{z_j\} \times \mathbb{C}] \cap B_2$ and $V = \cup V_j$. Then $\nu(F) \equiv 1$ on V and zero otherwise. We wish to show that V is a determining set for $H^\infty(B_2)$, that is, if g is a bounded holomorphic function on B_2 such that $g \equiv 0$ on V, then $g \equiv 0$ on B_2. So suppose g is such a function. For each w in $\mathbb{C}$ with $|w|^2 < 1/2$ consider $\gamma_w : B_1 \longrightarrow B_2$ defined by

$$(1 + |w|^2)\gamma_w(z) = (z + |w|^2, w(1 - z)).$$

It is easily verified that $\|\gamma_w(z)\| < 1$ when $|z| < 1$ and $\|\gamma_w(z)\| = 1$ when $|z| = 1$. Moreover, the first coordinate function of γ_w, namely, $\gamma_{w1}(z) = (z + |w|^2)/(1 + |w|^2)$, is a Möbius transformation of the unit disc onto the open disc centered at $|w|^2/(1 + |w|^2)$ with radius $1/(1 + |w|^2)$. Since the real axis is a diameter of this disc and $\gamma_{w1}(1) = 1$ it follows that $Z \subset \gamma_{w1}(B_1)$ since $|w|^2 < 1/2$ implies $|w|^2/(1 + |w|^2) < 1/3$ and Condition (ii) holds for Z. Now $g_w = g \circ \gamma_w$ is in $H^\infty(B_1)$ and $g_w(z) = 0$ when $\gamma_{w1}(z)$ is in Z, that is when

$$z = w_j = z_j(1 + |w|^2) - |w|^2 \quad \text{for some } j \text{ in } \mathbb{N}.$$

Now Condition (ii) implies that $z_j + \bar{z}_j = 3|z_j|^2 - 1$ so that

$$|w_j|^2 = |z_j(1 + |w|^2) - |w|^2|^2$$

$$= |z_j|^2(1 + |w|^2)^2 - |w|^2(1 + |w|^2)(3|z_j|^2 - 1) + |w|^4$$

$$= 1 + (|z_j|^2 - 1)(1 - 2|w|^2)(1 + |w|^2)$$

Thus, if $g_w \not\equiv 0$,

$$\sum_{|z|<1} \nu(g_w)(z)(1 - |z|) \geq \Sigma_1^\infty (1 - |w_j|)$$

$$= \Sigma(1 - |w_j|^2)/(1 + |w_j|)$$

$$\geq (1/2)\Sigma(1 - |w_j|^2)$$

$$= (1/2)\Sigma(1 - |z_j|^2)(1 - 2|w|^2)(1 + |w|^2)$$

$$\geq (1/2)(1 - 2|w|^2)\Sigma(1 - |z_j|) = \infty$$

by Condition (i) since $a < 1$. Therefore, $g_w(z) = 0$ for all z in B_1. But γ_w is the complex line through $(1, 0)$ and $(0, w)$. Thus, $g(z, w(1 - z)) = 0$ whenever $|w|^2 < 1/2$ and $|z|^2 + |w|^2|1 - z|^2 < 1$. But if $|u|^2 + |v|^2 < 1/3$ then $(u, v) = (u, \frac{v}{1-u}(1 - u))$ and $\left|\frac{v}{1-u}\right|^2 \leq \frac{|v|^2}{1-|u|^2} < 1/2$ so that g vanishes on $\sqrt{1/3}\, B_2$ and, therefore, vanishes identically on B_2.

This shows that Nevanlinna's Theorem is not true for B_2. In fact, since $F = g/h$ for g and h holomorphic implies that $\nu(g) \geq \nu(F)$ we see that F cannot be written as the quotient of two bounded holomorphic functions even if we drop the requirement that the two functions be locally relatively prime everywhere.

We have not yet arrived at the final example. Next let $\varepsilon = \sqrt{1/2}$ and consider the mappings $\Phi(z, w) = \frac{1}{1-\varepsilon w}(\varepsilon z, w-\varepsilon)$ and $\Psi(z, w) = \frac{1}{1+\varepsilon w}(\varepsilon z, w+\varepsilon)$. By direct verification $\Phi = \Psi^{-1}$ and is a biholomorphic mapping of B_2 onto B_2 taking ∂B_2 onto ∂B_2. Now let $G = F\circ\Phi$ and $W = \Psi(\nu)$. Then G is a holomorphic function with bounded characteristic on B_2 and simple zeros on W, which is a determining set for $H^\infty(B_2)$, since these properties are preserved under biholomorphisms. But if we let $\xi = (u, v)$ be a unit vector and let $W_\xi = W \cap \mathbb{C}\xi = W \cap B_1\xi$, we can see that W_ξ is a finite set as follows: Suppose W_ξ is infinite. Then $\Phi(W_\xi) = V \cap \Phi(B_1\xi)$ is infinite so that $\{z \in B_1: z\xi \in V\}$ is infinite. But $\Phi(z\xi)$ is in V if and only if $\phi_\xi(z) = \varepsilon uz/(1 - \varepsilon vz)$ is in Z. Therefore, $u \neq 0$ and ϕ_ξ is a Möbius transformation of B_1 into B_1 since Φ maps B_2 into B_2. Thus, if W_ξ is infinite, then $\phi_\xi(\overline{B_1})$ contains 1 since the closure of any infinite subset of Z contains 1 by definition of Z. But ϕ_ξ is continuous and monic on $\overline{B_1}$ so $1 = \phi_\xi(e^{it})$ for some t in $[0, 2\pi]$. Thus, since $e^{it}\xi$ is in ∂B_2 and Φ maps ∂B_2 onto ∂B_2 we must have that $\Phi(e^{it}\xi) = (1, 0)$, that is, $e^{it}\xi = \Psi(1, 0) = (\varepsilon, \varepsilon)$. But then $W_{(\varepsilon,\varepsilon)} = W_\xi$ is infinite.

Now $\phi_{(\varepsilon,\varepsilon)}(z) = z/(2 - z)$ takes $\mathbb{R}$ to $\mathbb{R}$, -1 to $-1/3$ and fixes 1 so that $\phi_{(\varepsilon,\varepsilon)}(B_1) = \{z \in \mathbb{C}: |z - 1/3| < 2/3\}$. Thus, $\phi_{(\varepsilon,\varepsilon)}(B_1) \cap Z = \emptyset$ so that $W_{(\varepsilon,\varepsilon)} = \emptyset$! (This is the reason for Condition (ii).)

Finally, examining W more closely we see that $W = \Psi(V) = \cup\Psi(V_j)$ and

$$\Psi(z_j, w) - \Psi(z_j, 0) = \frac{\varepsilon^2 w}{1 + \varepsilon w}(-z_j, 1) \quad \text{since } 1 - \varepsilon^2 = \varepsilon^2,$$

so that $W = \cup B_2 \cap [(\varepsilon z_j, \varepsilon) + \mathbb{C}(-z_j, 1)]$, a union of hyperplanes (complex lines) which are disjoint in B_2.

Remark: Choosing ε in $(0, 1)$ arbitrarily and using

$$\Phi(z, w) = \frac{1}{1 - \varepsilon w}(\sqrt{1 - \varepsilon^2}\, z, w - \varepsilon), \quad \text{or}$$

$$\frac{1}{1 - \varepsilon z_p}\left(\sqrt{1 - \varepsilon^2}\, z_1, \sqrt{1 - \varepsilon^2}\, z_2, \ldots, \sqrt{1 - \varepsilon^2}\, z_{p-1}, z_p - \varepsilon\right)$$

in p variables, and altering condition (ii) appropriately will produce a similar example in B_p with or without the condition that W be a union of hyperplanes depending upon whether $\varepsilon^2 = 1/2$ or not.

Summarizing, we have produced a holomorphic function G on B_2 and a nonnegative divisor $\mu = \nu(G)$ on B_2 such that:

(i) G *is in* $H^0(B_2)^*$, that is, G *has bounded characteristic.*

(ii) N_μ *is bounded on* $(0, 1)$ (by (i)).

(iii) G *has only simple zeros on* B_2, *that is,* $0 \le \mu \le 1$ *on* B_2.

(iv) $G^{-1}(0) \cap \mathbb{C}\xi$ *is finite for every* ξ *in* ∂B_2, that is, $n_{\mu|\xi}$ *is bounded on* $(0, 1)$ *for each* ξ *in* ∂B_2.

(v) $G^{-1}(0)$ *is a determining set for* $H^\infty(B_2)$, that is, *there is no function* f *in* $H^\infty(B_2)^*$ *such that* $\nu(f) \ge \mu$ *on* B_2.

(vi) $G^{-1}(0)$ *is a union of hyperplanes (complex lines) disjoint in* B_2.

However, we can produce a sufficient condition that μ be in $\nu(H^\infty(B_p)^*)$ in a special case, namely:

Proposition 4: *Let* $\{L_j\}$ *be a sequence of hyperplanes in* $\mathbb{C}^p$ *none of which contain the origin and let* $\{\mu_j\}$ *be a sequence of natural numbers. Let* $n^*(r) = \Sigma\{\mu_j : L_j \cap r\bar{B}_p \neq \phi\}$ *and*

$$N^*(r) = \int_0^r \frac{n^*(t)}{t}\, dt \quad \text{for} \quad 0 < r < 1.$$

If N^* *is bounded on* $(0, 1)$ *then there is a bounded holomorphic function* f *on* B_p *such that* $\nu_f(\xi) = \mu_j$ *when* ξ *is in* $L_j \cap B_p$ *for some* j *in* $\mathbb{N}$ *and* $\nu_f(\xi) = 0$ *otherwise.*

Proof: Let $r_j = \min\{\|\xi\| : \xi \in L_j\}$ and choose ξ_j in ∂B_p so that $r_j\xi_j$ is in L_j. Then $0 < r_j$ for each j in $\mathbb{N}$ and the condition N^*

is bounded on (0, 1) is equivalent to saying that

$$\Sigma\mu_j(1 - r_j) < \infty.$$

Moreover, ξ is in L_j if and only if

$$<\xi - r_j\xi_j \mid \xi_j> = 0, \quad \text{that is,} \quad <\xi \mid \xi_j> = r_j,$$

where $<\xi \mid \eta>$ is the standard hermitian product on $\mathbb{C}^p$. And if ξ is in $r\overline{B}$, then

$$\left|1 - \frac{r_j - <\xi \mid \xi_j>}{1 - r_j<\xi \mid \xi_j>}\right| = \frac{|1 + <\xi \mid \xi_j>|}{|1 - r_j<\xi \mid \xi_j>|} (1 - r_j)$$

$$\leq \frac{1 + r}{1 - r} (1 - r_j)$$

so that (Rudin [9], 15.6)

$$f(\xi) = \prod \left[\frac{r_j - <\xi \mid \xi_j>}{1 - r_j<\xi \mid \xi_j>}\right]^{\mu_j}$$

converges uniformly on compact subsets of B_p to a holomorphic function with the desired divisor.

Q.E.D.

Remark: It is not clear whether this sufficient condition is also necessary in this special case. If, in fact, it is necessary, then a new question arises, namely, is the failure of this condition equivalent to saying that the divisor is determining for H^∞? It apparently might be possible to find a union of hyperplanes for which the conditon fails but which is contained in the zero set of some bounded holomorphic function. If the above condition *were* necessary then it is clear that the required containing zero set cannot be obtained by adjoining *hyperplanes*.

Current work on Blaschke Problems generally avoids the apparently intractible case of H^∞ and even H^0 and proceeds by looking at known methods of producing "canonical" representatives of a divisor class (such as techniques involving the Kneser integral) and estimating the growth of the canonical function in terms of some growth conditions on the divisor. This works for "large" classes of holomorphic functions with H^0 being "small" and H^∞ extremely "small" in this context. The difficulties arise because the construction techniques are L^2 whereas the characteristic function is (essentially) an L^1 norm and there is the usual loss of control over estimates. Moreover, the modulus of a holomorphic function is not reasonably estimated by its characteristic. These problems are reflected in the work of Mueller [7] and myself [4].

In conclusion let me list some open questions and a reasonable conjecture:

1) The conditon that $N_\mu(r;\ \xi)$ be bounded on $(0, 1) \times \partial B_p$ is necessary for μ to be in $\nu(H^\infty(B_p)^*)$ (by the Jensen Formula). Is this condition also sufficient? It is not clear if $\nu(G)$ constructed above satisfies this condition.

2) The conditon that N_μ be bounded on (0, 1) is necessary for μ to be in $\nu(H^0(B_p)^*)$. Is this condition also sufficient? This appears to be a reasonable conjecture in the light of Propositions 1 and 3. Note that an affirmative answer also provides a strong quotient representation for the meromorphic functions of bounded characteristic.

3) Can determining sets (or divisors) for $H^\infty(B_p)$ or $H^0(B_p)$ be reasonably characterized?

The interested reader is particularly directed to Stoll [12] for its comprehensive treatment of some of these questions and its extensive bibliography.

REFERENCES

1. Caratheordory, C., *On Dirichlet's problem*, Amer. J. Math. 59(1937), 709-731.
2. Chee, P. S., *The Blaschke condition for bounded holomorphic functions*, Trans. Amer. Math. Soc. 148(1970), 249-263.
3. Kujala, R. O., *Functions of finite λ-type in several complex variables*, Trans. Amer. Math. Soc. 161(1971), 327-358.
4. ________, *Functions of finite λ-type on the unit ball in* $\mathbf{C}^p$, to appear.
5. Markushevich, A. I., *Theory of Functions of a Complex Variable*, vol. II, Prentice-Hall, Inc., Englewood Cliffs, N. J.,1965.
6. Miles, J., *Quotient representations of meromorphic functions*, J. d'Analyse Math. 25(1972), 371-388.
7. Mueller, G., *Functions of finite order on the ball*, 1971 disseration, Univ. of Notre Dame.
8. Rubel, L. A., and B. A. Taylor, *A Fourier series method for meromorphic and entire functions*, Bull. Soc. Math. France 96(1968), 53-96.
9. Rudin, W., *Real and Complex Analysis*, McGraw-Hill, inc., New York, 1966.

10. _________, *Function Theory in Polydiscs*, W. A. Benjamin, Inc., New York, 1969.

11. Stoll, W., *About entire and meromorphic functions of exponential type*, Proc. Sympos. Pure Math., vol. 11, Amer. Math. Soc., Providence, R. I., 1968, 392-430.

12. _________, *Holomorphic Functions of Finite Order in Several Complex Variables*, Regional Conf. Series in Math., Amer. Math. Soc., Providence, R. I., 1974.

13. Stout, E. L., *The Second Cousin problem with bounded data*, Pac. J. Math. 26(1968), 379-387.

14. Tsuji, M., *Canonical product for a meromorphic function in a unit circle*, J. Math. Soc. Japan 8(1956), 7-21.

INDEX OF CITED AUTHORS

Work(s) of the following authors are listed as references on the pages indicated:

INDEX